Bioavailability, Toxicity and Risk Relationships in Ecosystems

Editors

R. Naidu[1]
V.V.S.R. Gupta[1]
S. Rogers[1]
R.S. Kookana[1]
N.S. Bolan[2]
D.C. Adriano[3]

[1]*CSIRO Land and Water, Glen Osmond, Adelaide, South Australia*

[2]*Soil & Earth Science Group, Institute of Natural Resources*
Massey University, Palmerston North, New Zealand

[3]*University of Georgia, Savannah River Ecology Lab.*
Aiken, South Carolina, USA

Science Publishers, Inc.

Enfield (NH), USA   Plymouth, UK

SCIENCE PUBLISHERS, Inc.
Post Office Box 699
Enfield, New Hampshire 03748
United States of America

Internet site: *http://www.scipub.net*

sales@scipub.net (marketing department)
editor@scipub.net (editorial department)
info@scipub.net (for all other enquiries)

Library of Congress Cataloging-in-Publication Data

Bioavailability, toxicity, and risk relationship in ecosystems/
    editors R. Naidu ... [et al.]. p.cm.
    Includes bibliographical references (p.).
    ISBN 1-57808-192-0
    1. Metals--Environmental aspects. 2. Metals--Toxicology.
3. Bioavailability. 4. Health risk assessment. I. Naidu, R.

    QH545.M45 B58 2002
    577'.14--dc21
                                                2002030601

Published by Science Publishers Inc., Enfield, NH, USA
Printed in India.

Foreword

This book is a compendium of information that elucidates the role of bioavailability in determining toxicity, and in turn its significance in risk assessment. Bioavailability in this context is loosely defined as the transport and uptake of an element by an organism.

The book is based on a symposium that Drs Ravi Naidu and Domy Adriano organized during the 4th International Symposium on the Biogeochemistry of Trace Elements held on June 23-26, 1997 at the University of California, Berkeley, California. Since this meeting Drs Naidu and Adriano have promoted the concept of the dynamic nature of bioavailability in terrestrial ecosystems at numerous conferences, which have led to the compilation of the present book.

The importance of the source term and chemical speciation, which in turn is heavily influenced by various factors on bioavailability of metal(loid) contaminants in soils is discussed.

Contaminants from varied sources and in various forms, either in aquatic, groundwater or terrestrial systems, and their exposure effects on organisms—from microbial, small invertebrates, plants, animals and to humans are discussed.

The topics were debated by an international group of scientists who, although might have different perspectives in addressing environmental issues, such as soil pollution, have convergent views about the significance of bioavailability in the bioaccumulative effects and its role in evaluating risk.

This special symposium also served as a precursor to the International Workshop on Chemical Bioavailability in the Terrestrial Environment, first held at the CSIRO, Adelaide in 1999.

Professor Albert Page
Albert L. Page, Professor-Emeritus
Department of Environmental Science
University of California
Riverside, CA 92521

Preface

This book describes the bioavailability, toxicity and risk relationships of metal contaminants in ecosystems. It discusses bioavailability within the context of environmental health and ecotoxicological risk assessment and the potential impact that metals may have on soil ecosystem. Specific discussions focus on fundamental principles and scope of bioavailability, soil, plant and microbial processes that influence metal dynamics, indicators of bioavailability, and selected case studies demonstrating the impact of metals on terrestrial ecosystem and how bioavailability relates to regulatory and site assessment requirements. The definition of the term 'bioavailability' and the scientific concept on which it is based are unclear, the methods adopted vary throughout the world and therefore there is no single standard technique for the assessment of either plant and microbial availability of contaminants or their ecotoxicological impacts on soil biota. In this book, while attempting to define bioavailability, we have taken into consideration that bioavailability is a function of soil, nature of contaminant, species/receptor organisms and environmental perturbations including ageing of the contaminant. The book was written in as detailed and comprehensive manner as possible, but with simplicity as its main feature. It is organised in a systematic manner taking the reader from fundamental principles to case studies employing bioavailability as a tool for risk assessment. This allows the reader to grasp the nature and scope of bioavailability in a single and easy-to-read book.

The book covers fundamental principles controlling bioavailability, environmental and human health risk assessment, bioavailability as affected by chemical speciation, potential role of bioavailability in risk assessment, microbial parameters as indicators of bioavailability and case studies demonstrating how manipulation of bioavailability could mitigate ecotoxicological effects of metals.

This book is intended as a text for post graduate students, remediators and risk assessment experts to understand the application of various conventional and innovative tools for assessing bioavailability and risks posed by contaminants at highly contaminated sites. It is also intended for regulatory authorities and environmental planners who wish to learn more about metal bioavailability, risk relations and site remediation.

R Naidu
VVSR Gupta,
S Rogers,
RS Kookana,
NS Bolan
DC Adriano

Contents

Foreword   *v*
Preface   *vii*
List of Contributors   *xi*
Introduction   *xv*

Section A: Fundamental Principles

1. Risk Assemment in Environmental Contamination and   3
   Environmental Health
   Michael R. Moore

2. Bioavailability of Metals in the Soil Plant Environment   21
   and its Potential Role in Risk Assessment
   R. Naidu, S. Rogers, V.V.S.R. Gupta, R.S. Kookana,
   N.S. Bolan and D.C. Adriano

3. The Role of Chemical Speciation in Bioavailability   59
   Sébastien Sauvé

Section B: Indicators of Bioavailability

4. Microbial Parameters as Indicators of Toxic Effects of   85
   Heavy Metals on the Soil Ecosystem
   P.C. Brookes

5. Metal-Algae Interactions: Implications of Bioavailability 109
   M. Megharaj, S.R. Ragusa and R. Naidu

6. Absorption and Translocation of Chromium by Plants: 145
   Plant Physiological and Soil Factors
   R.H. Loeppert, J.A. Howe, H. Shahandeh, L.C. Wei
   and L.R. Hossner

7. Plant Soil Metal Relationships from Micro to Macro Scale   175
   K. Bujtas, A.S. Knox, I. Kadar and D.C. Adriano

Section C: Case Studies

8. Effects of Mine Wastewaters on Freshwater Biota in
   Tropical Northern Australia
   *Scott J. Markich, Ross A. Jeffree, David R. Jones and
   Sven R. Sewell*   207

9. Inplace Inactivation and Natural Ecological Restoration
   Technologies (IINERT)
   W.R. Berti and J.A. Ryan   253

10. An Assessment of the Revegetation Potential of Acidic
    Basemetal Tailings using Metal-Tolerant Grass
    Species and Lime
    W.J. Morrell, N.S. Bolan, P.E.H. Gregg and R.B. Stewart 271

11. Groundwater Arsenic Contamination in West Bengal–
    India and Bangladesh: Case Study on Bioavailability
    of Geogenic Arsenic
    *Uttam Kumar Chowdhury, Mohammad Mahmudur Rahman,
    Gautam Samanta, Bhajan Kumar Biswas, Gautam Kumar Basu,
    Chitta Ranjan Chanda, Kshitish Chandra Saha, Dilip Lodh,
    Shibtosh Roy, Quazi Quamruzzaman and Dipankar Chakraborti* 291

12. Bioavailability, Toxicity and Risk Relationships
    in Ecosystems: The Path Ahead
    R. Naidu, N.S. Bolan and D.C. Adriano   331

Index   339

List of Contributors

A.S. Knox, University of Georgia, Savannah River Ecology Lab, Aiken, SC 29802, USA

Bhajan Kumar Biswas, School of Environmental Studies, Jadavpur University, Kolkata-700 032, India

Chitta Ranjan Chanda, School of Environmental Studies, Jadavpur University, Kolkata-700 032, India

D.C. Adriano, Savannah River Ecology Laboratory, Drawer E Aitken, South Carolina 29802.

David R. Jones[3], Earth Water Life Sciences, P.O. Box 39443, Winnellie, Northern Territory 0821, Australia

D.C. Adriano, University of Georgia, Savannah River Ecology Lab, Aiken, SC 29802, USA

Dilip Lodh, School of Environmental Studies, Jadavpur University, Kolkata-700 032, India

Dipankar Chakraborti, School of Environmental Studies, Jadavpur University, Kolkata-700 032, India

Gautam Samanta, School of Environmental Studies, Jadavpur University, Kolkata-700 032, India

Gautam Kumar Basu, School of Environmental Studies, Jadavpur University, Kolkata-700 032, India

H. Shahandeh, Soil & Crop Sciences Dept., Texas A&M University, College Station, TX 77843

I. Kadar, Research Institute for Soil Science and Agricultural Chemistry of the Hungarian Academy of Sciences, H-1022 Budapest, Herman Otto ut 15, Hungary

J.A. Howe, Soil & Crop Sciences Dept., Texas A&M University, College Station, TX 77843

J.A. Ryan, U.S. Environmental Protection Agency National Risk Management Laboratory, Cincinnati, OH USA

K. Bujtas, Research Institute for Soil Science and Agricultural Chemistry of the Hungarian Academy of Sciences, H-1022 Budapest, Herman Otto ut 15, Hungary

Kshitish Chandra Saha, School of Environmental Studies, Jadavpur University, Kolkata-700 032, India

L.C. Wei, Soil & Crop Sciences Dept., Texas A&M University, College Station, TX 77843

L.R. Hossner, Soil & Crop Sciences Dept., Texas A&M University, College Station, TX 77843

M. Megharaj, CSIRO Land and Water, Private Mail Bag No. 2, Glen Osmond, Adelaide SA 5064

Michael R. Moore, Director, NH&MRC, National Research Centre for Environmental Toxicology (NRCET), The University of Queensland, PO Box 594, Archerfield, Queensland 4108, Australia.

Mohammad Mahmudur Rahman, School of Environmental Studies, Jadavpur University, Calcutta-700 032, India

N.S. Bolan, Department of Soil Science, Massey University, Palmerston North, New Zealand.

P.C. Brookes, Agriculture and Environment Division, IACR-Rothamsted, Harpenden, Herts., AL5 2JQ, UK.

P.E.H. Gregg, Institute of Natural Resources, Massey University, Palmerston North, New Zealand

Quazi Quamruzzaman, Dhaka Community Hospital, Bara Magh Bazar, Dhaka-1217, Bangladesh.

R.B. Stewart, Institute of Natural Resources, Massey University, Palmerston North, New Zealand

R.H. Loeppert, Soil & Crop Sciences Dept., Texas A&M University, College Station, TX 77843

R. Naidu, CSIRO Land and Water, Private Mail Bag. No. 2, Glen Osmond, Adelaide, South Australia 5064.

R.S. Kookana, CSIRO Land and Water, Private Mail Bag. No. 2, Glen Osmond, Adelaide, South Australia 5064.

Ross A. Jeffree, Environment Division, Australian Nuclear Science and Technology Organization, Private Mail Bag 1, Menai, New South Wales 2234, Australia

S.R. Ragusa, CSIRO Land and Water, Private Mail Bag No. 2, Glen Osmond, Adelaide SA 5064

S. Rogers, CSIRO Land and Water, Private Mail Bag. No. 2, Glen Osmond, Adelaide, South Australia 5064.

Scott J. Markich, Environment Division, Australian Nuclear Science and Technology Organization, Private Mail Bag 1, Menai, New South Wales 2234, Australia

Sébastien Sauvé, Department of Chemistry, Université de Montréal, 2900 Édouard-Montpetit, Montréal, QC, Canada, H3C 3J7.

Shibtosh Roy, Dhaka Community Hospital, Bara Magh Bazar, Dhaka-1217, Bangladesh.

Sven R. Sewell, ERA, Ranger Mine, Locked Bag 1, Jabiru, Northern Territory 0886, Australia

Uttam Kumar Chowdhury, School of Environmental Studies, Jadavpur University, Calcutta-700 032, India

V.V.S.R Gupta, CSIRO Land and Water, Private Mail Bag. No. 2, Glen Osmond, Adelaide, South Australia 5064.

W.R. Berti, DuPont Central Research & Development, Newark, DE USA

W.J. Morrell, Centre for Mined Land Rehabilitation, University of Queensland, Brisbane, 4072, Australia

Introduction

Although transformation and uptake of metal contaminants by plants and soil organisms have been the focus of much research during the past 30 years, methods used to assess bioavailable fraction are just as variable as the definition of the term 'bioavailability' itself. Bioavailability of metals has been linked to their total concentration and free ion activity in soil solution, exchangeable -pool, fraction taken up by organisms and more recently to the concentration that causes ecotoxicity. The methods adopted for measuring the bioavailability of contaminants vary throughout the world and therefore there is no single standard technique for the assessment of either the bioavailability of contaminants or their ecotoxicological impacts on soil biota.

Difficulties with bioavailability assessment probably arise because different groups have different understanding of bioavailability concept. A microbiologist considers microbial component of biosphere, and a plant scientist often focuses on plant biota only, whereas a soil and environmental scientist considers the ecosystem as a whole. The concept of bioavailability is then very much driven by the organisms or pure solution phase used for measurement of the contaminant availability. For example, bioavailability is often related to the fraction of contaminant extractable in a chemical reagent that correlates with the total contaminant uptake by plants. Although this has been quite successful with certain plants and soil types, its applicability has often been found to be fraught with limitations. Firstly, as bioavailability is often assessed by chemical extractions it is likely that extractants may alter the chemical equilbria between different fractions during the extraction procedure, resulting in the release of those pools of contaminants that are not necessarily available to plant roots and micro-organisms in the soil solution. This is supported by the definition that suggests that the concept of bioavailability should consider the overall toxicology to the ecosystem and not just bioavailable fraction that is largely a measure of solubility (Moore, 2001). Secondly and more importantly, chemical extraction techniques do not consider the physiology of plant and microbe including the transport of metals across the membranes. The parameters that control the membrane transport of contaminants are receptor species-specific and possibly dictate that the same bioavailable pool of contaminants in soils may not necessary result in the same rates of either plant uptake or impact on micro-organisms.

However, it is to be pointed out that the bioavailability of metals in soils has recently been examined using the physiologically-based in-vitro chemical fractionation schemes that include Physiologically Based Extraction Test (PBET), Potentially Bioavailable Sequential Extraction (PBASE) and Gastrointestinal (GI) Test. These innovative tests predict the bioavailability of metals in soil and sediments when ingested by animals and humans. Again, as in the case of traditional sequential extraction scheme, the ability of the PBET extractant to solubilize metals increases with each successive extraction step. Metals extracted earlier in the PBET sequence are more soluble and, therefore more potentially bioavailable than metals extracted later by the more aggressive extractants used (Basta and Gradwohl, 2000).

It is also possible that plants and microorganisms can adapt to environmental extremes and with time are able to develop capabilities to make a chemical more or less bioavailable depending on its physiological requirement. Presumably because of this, large differences also exist between contaminant uptake and toxic response between different organisms, such as plants and microorganisms, and also between species and strains of the same species. This may be further complicated by speciation of contaminants in terrestrial and aquatic environments, which affects metal bioavailability and toxicity. For example, trivalent chromium [Cr(III)] is generally less bioavailable and less toxic than hexavalent chromium [Cr(VI)].

In this book an attempt is made to summarise bioavailability in relation to ecotoxicology, soil pore water chemistry and chemical and biological indices. Following detailed discussions of the definitions and techniques used for bioavailability assessment the book focuses on case studies illustrating how bioavailability could be manipulated to reclaim contaminated land. Clearly this issue remains to be very complex and deserves concerted efforts by diverse range of scientific disciplines.

References

Moore, MR (2001) Risk assessment in environmental contamination and environmental health. In: R. Naidu et al (eds) Bioavailability, Toxicity and Risk Relationships in Ecosystems. Science Publishers, Inc., Enfield, NH, USA, Chapter 2, this book).

Basta, N. T. and Gradwohl, R. 2000. Estimation of Cd, Pb, and Zn bioavailability in smelter-contaminated soils by a sequential extraction procedure. J. Soil. Cont. 9, 149-164.

FUNDAMENTAL PRINCIPLES

1

Risk Assessment in Environmental Contamination and Environmental Health

Michael R. Moore

INTRODUCTION

All substances are poisons: There is none which is not a poison.
The right dose differentiates a poison from a remedy
Paracelscus 16 C

Toxicology is concerned with defining the potential of a substance to produce adverse effects. Environmental toxicology is the study of health effects of low level, often long-term, exposure of biological systems and, in particular, of human populations, to environmental toxins or poisons. It is a multidisciplinary subject that draws together a number of medical and scientific disciplines. Environmental toxicology is a major component of public health research. Hazard evaluation determines whether any of the known potential adverse effects will develop under specific conditions of use. Toxicology is thus only one aspect taken into account in hazard evaluation. There are multiple circumstances where one has to assess the risk of otherwise safe chemicals or of the processes leading to their production.

Contaminated sites contain a diversity of materials of both natural and anthropogenic origin. There are two obstacles to sound management of chemical safety. First, there is an inadequate database on which decisions are to be based. Some attempt is being made to address this through information contained in monographs published by the South Australian Health Commission (El Saadi and Langley, 1991; Langley and van Alphen, 1993; Edwards et al., 1994; Langley et al., 1996). Because we cannot predict the dose of a toxin that reaches a site of action in humans, we have to turn to animal surrogates. This is the most reasonable way to predict

Director, NH&MRC, National Research Centre for Environmental Toxicology (NRCET), The University of Queensland, PO Box 594, Archerfield, Queensland 4108, Australia.

bioavailability since prior to biological effects being observed we are properly only looking at presence or availability of a substance. In most cases, we lack the evidence to study toxicokinetics in human populations and must extrapolate from in vivo or in vitro studies by which we can gain valuable qualitative data that is not reconcilable quantitatively in humans because of physiological differences. Second, there is a reluctance to pursue appropriate risk management procedures, particularly in developing countries, because of the economic burden. This is often a sequel of stringent standards established through conservative risk assessment. Ecotoxicity tests are valuable in guiding us to likely toxic events, but these do not provide qualitative evidence of human effect. The bridge between such tests and bioavailability has still to be established. Beyond that we pass to the sphere of metabolism, toxicokinetics, and redistribution for which there is abundant evidence of heterogeneity and polymorphism in human and animal populations and thus diversity of response.

Risk Assessment

The objective of risk assessment is to measure the probability that adverse health effects will develop from known or suspected exposures to toxic substances, including nutrients. They are usually based on extrapolation of data from dose-response relationships in animal experiments and occasionally from human epidemiological studies. From these, the risk of an effect being found is calculated for known or anticipated ranges of occupational or environmental exposure. The data can also be used to establish a risk-free dose of the substance. Risk assessment is fundamentally a probabilistic quantitation of Hazard evaluation, which is the probability that adverse events will occur (Paustenbach, 1989). We live in a world where chemical safety presents a global challenge. Compounds now essential to everyday living, present us with the dilemma of balancing the benefits of their use with the risks inherent in their nature. Toxicology and hazard evaluation are the tools by which we can estimate the potential risks associated with contaminants. It is important to properly understand and evaluate the likely consequences of not only contamination and risk but also the potential additive or synergistic effects of the concurrent presence of materials of natural origin. Beyond this, there is also a need to recognize differences between essential elements and conventional toxins including heavy metals, metalloids and organic contaminants. Some essential elements, such as Se and Zn may become toxic at higher levels of exposure.

One has also to decide whom or what one is trying to protect in risk assessment. What is the value of ecological risk assessment where health-based effects are the principal area of concern? Environmental investigation levels (EILs) need to be formulated which are site-specific, unlike health investigation levels (HILs) which are universal. Ecological measures for protection of vulnerable "in fauna" are appropriately more stringent than those necessary for protection of human health. The importance of this is borne out by the work of advisory groups such as that set up by the Society for Environmental Toxicology and Chemistry (Menzie, 1997).

Language

One of the greatest impediments to reconciliation of health and environmental objectives in the evaluation of contamination is one of communication. Health scientists and environmental scientists speak languages that have terms in common but with quite different meanings. For example, "bioavailable fraction" is not the same as "bioavailability" in higher organisms. Risk assessments are similar but employ differing perspectives on safety factors, particularly those linked to the precautionary principle. This is important to properly evaluate the likely consequences of not only contamination but also the potential additive or synergistic effects of the concurrent presence of materials of natural origin. The health sector needs to balance "prudent public health" with pragmatism with regard to the environmental "precautionary principle". Health cannot endorse a precautionary approach without applying it across all sectors of medicine. The cost implications are substantial. In rehabilitation of contaminated land, for example, costs of implementation of (HILs) and (EILs) frequently differ by an order of magnitude.

Data Quality

The data for risk assessment must be of acceptable quality to ensure avoidance of default information. Generally human-derived information is preferable in HILs to in vitro or in vivo data obtained experimentally. However, there are circumstances where such information is all that is available or is likely to be available. Biomarkers provide a valuable adjunct to the assessment process because they permit evaluation of dose-response relationships and thus provide a bridge between measures of external exposure and internal effect. This is a critical link in the sequence-chemical exposure: internal dose: health effect. The use of biomarkers is particularly important in assessment of susceptible populations that are especially sensitive to chemicals and their metabolites because of either genetic linkage or acquired susceptibility. Such biomarkers of exposure and effect can be used further to evaluate compliance with risk management measures. In health terms, it is the fraction of the dose of contaminant retained that is of concern. Exposure is a poor measure of dose, and the environmental "bioavailable fraction" is not the same as "bioavailability" in higher organisms (Moore, 1996). The fundamental problem for EILs is that the numbers of species involved is large. Consequentially, there is every likelihood that there will be one species that is hypersensitive. HILs relate to a single species, humans, where variation is much lower than the interspecies variation associated with EILs (Gibson et al., 1996). There are also site-specific adaptations to existing natural and anthropogenic pollutants. Frequently, this results in discredited EILs, which lie below background levels. In Australia, decisions will have to be made on whether ecologically based risk assessment is adaptable to human exposure scenarios. For those human-based scenarios, the decision will be whether it is appropriate to ignore ecologically based criteria and rely on human health-related criteria. Ultimately, it may be more appropriate to run EILs and HILs in parallel rather than to attempt any form of integration because of these tensions and incompatibilities.

Bioavailability

The concept of bioavailability of elements and compounds is fundamental to toxicology. It is, however, not the same as the bioavailable fraction, which is largely a measure of solubility. Chapters 3, 4, and 5 consider the measures of bioavailability as they relate to soil biota and plants. Bioavailability is an index of systemic absorption and is generally applicable to humans, but may also be applied to animals. By virtue of the way it is defined the difference in the area under the plasma-concentration curve (AUC) after intravenous and oral dosing (Figure 1.1) it is inapplicable to lower life forms or to plants.

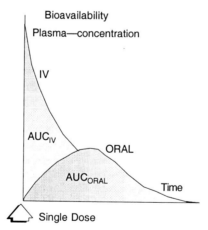

Fig. 1.1: An example of plasma concentration of a substance administered orally or intravenously. The area under the curve (AUC) over time is indicative of bioavailability fraction given in equation 1.

Bioavailability $(F) = (\text{Dose}_{IV})(\text{AUC}_{ORAL})/(\text{Dose}_{ORAL})(\text{AUC}_{IV})$ $\qquad$ (1)

where F is bioavailability, Dose_{IV} is the intravenous dose of toxin, AUC_{IV} is the area under the plasma-concentration curve after IV dosage. Those subscripted ORAL are equivalent values after oral dosage.

Bioavailability is only linear (i.e. dose independent) in a limited number of circumstances, but is often taken to be so for easy manipulation of data. As toxicants are usually delivered to organs by the systemic circulation, the fraction of xenobiotic in the circulation is critical. This can be modified in many ways such as limited oral absorption, rapid first-pass metabolism (both hepatic and intestinal), competitive binding, and the matrix in which the xenobiotic is situated. A further consideration is the rate of release into circulation. Rapid release is more likely to produce a high peak concentration of a chemical (C_{max}), which can exceed the critical concentration for organ-specific toxicity.

The concept of clearance is important to toxicokinetics. Clearance is generally a dose-independent process because toxin elimination processes are not usually saturated. This means that clearance is a linear function of circulating concentrations of toxin, i.e. it is first order. In cases where bioavailability is complete (100%) the dose rate establishes the steady-state concentration of the toxin (equation 2).

$$\text{Dose rate} = [C_{ss}]\, Cl \qquad (2)$$

where C_{ss} is the steady-state concentration of the drug and Cl is clearance.

There is the possibility of saturation of clearance mechanisms when zero-order kinetics prevail and constant amounts are cleared in unit time. In such circumstances, clearance is variable. Equation 2 makes the fundamental assumption that bioavailability is total, i.e. 100%. This makes legislation overly conservative. Therefore the equation is better presented as:

$$F.\, \text{Dose rate} = [C_{ss}]\, Cl \qquad (3)$$

where F is the bioavailability.

Organ-specific clearance of a toxin is contingent on blood flow through the organ. This means that the systemic blood flow in which the toxin is borne must be taken into account together with the circulating concentration of the toxin before and after passing through an organ. Thus,

$$\text{Organ elimination} = Q\,(C_A - C_V)$$

where Q is the blood flow, and C_A the concentration before passing through the organ (arterial) is and after passing through the organ (venous) is C_V. Organ specific clearance

$$Cl_{Organ} = Q\,\frac{(C_A - C_v)}{C_A}$$

The total clearance from the body (Cl) is the sum of individual organ clearances from the liver, kidney, spleen, etc.

$$Cl = \Sigma\, Cl \text{ All Organs}$$

The simple systems and principles utilized in pharmacokinetics and pharmacodynamics can be applied to toxicokinetics. The major deficiency is that the level of uncertainty in the exposure factors associated with contaminated sites renders the simple fixed exposure-time relationships invalid. Classically, the steady kinetics associated with repeated administration of drugs can be used to calculate steady-state concentrations of toxins. Inevitably, toxins are not administered at finite well-defined intervals as is the case with therapeutic drugs. Exposure is more likely to be either erratic or continuous. In the former circumstance, the type of relationship is akin to that shown in Figure 1.2; in the latter case, the time interval can be conceived as infinitely small and the fluctuations in concentrations over a given time equally small.

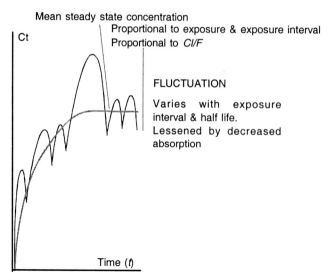

Fig. 1.2: Toxicokinetic relationships for repeated exposures to toxins where C_t (mg/l) is concentrations at the specific time.

The average concentration of a toxicant C_{ss} (av) at steady state is,

$$C_{ss}(\text{av}) = \frac{F.\,\text{exposure}}{Cl.\,t(\text{av})}$$

where F is the bioavailability, "exposure"—is the dose of toxicant, Cl is clearance and $t(\text{av})$ is the mean time interval between exposures.

An indepth study of the process of toxicokinetics is not intended here. The foregoing does, however, give some guidance on the complexity of the process which is made even more complex when factors such as saturable binding, metabolism, etc. result in nonlinear toxicokinetics.

Even when a substance is bioavailable, there is no guarantee of toxicity. One is seeking to establish that once absorbed, substances reach the biological site of action. Presence alone does not mean that a compound is toxic. Consider a simple toxic chemical such as ethanol. In most individuals, only part of the imbibed alcohol reaches the systemic blood. With doses relevant to social drinking, this is due mainly to gastric first-pass metabolism of alcohol, which acts as a barrier against toxic levels of alcohol in the blood. The activity of gastric alcohol dehydrogenase can account for a substantial fraction of this metabolism. Fasting, gender, old age, dilution of alcoholic beverages, and chronic alcohol consumption are some factors that decrease the gastric metabolism of alcohol. Aspirin and some H_2-receptor antagonists also inhibit this gastric activity. In subjects with documented first-pass metabolism, these drugs increase blood alcohol levels, especially after repeated small drinks, and may result in unexpected impairment of ability to perform complex tasks such as driving (Baraona et al., 1994).

Biomagnification

Where materials are accumulated in biota such that they can pass up a food chain, the potential for accumulation or biomagnification can occur. "Biomagnification" is a term used exclusively in ecotoxicological studies. Materials passing through these trophic levels place organisms (such as humans) at the top of the food chain at most risk. A special aspect of biomagnification is the concept of "secondary poisoning", which deals with toxic effects on higher members of a food chain. Secondary poisoning results from ingestion of organisms that contain accumulated substances (indirect exposure). A strategy for the assessment of the potential for secondary poisoning has been developed by Romijn et al. (1993). In this concept, the predicted chemical concentration in food of higher organisms is compared with the mammalian toxicity of the chemical as an indication of possible effects on birds and mammals.

A requirement for biomagnification is the bioaccumulation and/or bioabsorption of chemicals either by direct uptake from the environment or by the uptake of particle-bound chemicals and concentration in the organisms (e.g. microorganisms, algae, invertebrates, vertebrates). There is convincing evidence that nonmetabolized residues that are not completely excreted may be readily transferred to the next trophic stage. It is useful to examine the potential for biomagnification by using schemes such as that of Franke et al. (1994) to evaluate the process. The principal criterion for biomagnification is bioaccumulation. This is more realistic as it is likely that first pass metabolism, biliary excretion, competitive binding, and other metabolisms will inactivate the toxin before it can enter the circulation and reach the sites of action in specific organs.

A good example of barriers and transfer is that of metal contamination of soils. Lead is the most studied metal in biosystems, and can be used to study the likely consequences of the presence of a metallic environmental toxin. Geological sources of lead contain substantial concentrations of the metal, certainly higher than that considered compatible with plant or animal life. Yet, even in heavily contaminated acid soils, plant life can be sustained and plants do not accumulate more than modest amounts of the metal within their cells. Plants often become contaminated typically with dusts, and it is these that probably provide the greater impact of the metal to herbivores. Omnivorous animals such as humans have a further limitation on transfer because much of the retained metal by foodstuff animals is bound up in the inaccessible bony matrix of foodstuff herbivores and not consumed. The net result is decreasing concentrations of the metals through the food chain.

Bioaccumulation

Bioaccumulation is often used as an alternative concept to bioavailability. Assessment of the bioaccumulation of substances in living organisms calls for measurement of concentration and elimination of half-life. It has been suggested that this measurement is a prerequisite for ecotoxicology and the presence of bioaccumulation is an ecological risk (Franke et al. 1994). Bioaccumulation occurs when in the absorption/excretion cycle of toxin exposure, there is some retention of the previous dose at the time of subsequent exposure above baseline values. The possibility of biomagnification in trophic chains also exists.

Bioaccumulation potential (BAP) is a qualitative (not quantitative) indicator of risk of bioaccumulation in living organisms because of physicochemical and structural properties of a substance. Until now, bioaccumulation potential has been estimated on the basis of the n-octanol/water partition coefficient P_{ow} in its logarithmic form (log P_{ow}). If not possible to measure directly, low values of P_{ow} can be calculated from the chemical structure of a substance. This approach assumes that accumulating organic substances are hydrophobic, can freely diffuse through cell membranes, and are enriched in the lipid fraction of organisms. On the other hand, the correlation between the n-octanol/water partition coefficient (calculated as log P_{ow}) and the bioconcentration factor (calculated as log$_{BCF}$), which expresses the enrichment of a substance in an organism relative to concentration of the substance in the medium, has been proved to be very poor for many types of chemicals. The n-octanol/water partition coefficient is generally not a good model of

bioaccumulation behavior of organic chemicals because it does not take into consideration many other factors such as
- active transport,
- diffusion through cell membranes,
- metabolism and accumulation of metabolites,
- organs specific accumulation, and
- uptake and distribution kinetics.

n-Octanol simulates the lipidfractions in organisms, but it is doubtful whether it is an adequate surrogate of lipids.

A special problem is the measurement of log P_{ow} of ionizable substances, which may have multiple partition equilibria. The test guide for log P_{ow} measurement (OECD 1981, 1993) states that log P_{ow} measurements should be made on ionizable substances only in their nonionized form (free acid or free base), thus allowing the determination of maximum lipophilicity of a tested substance. Despite these limitations, it is accepted that log P_{ow} values ≥ 3 indicate that the substance has the potential to bioaccumulate.

Superlipophilic substances are those having a log $P_{ow} > 6$ and with a molecular weight greater than 600 Da or a molecular cross-section larger than 0.95 nm, and cannot be expected to bioaccumulate. However, Geyer et al. (1992) have shown that the bioconcentration for such substances has been underestimated and may be considerably higher if bioaccumulation is tested within the limits of water solubility, i.e. without the use of solubilizers. Additionally, there is a potential for accumulation owing to biosorption of highly lipophilic compounds onto biological surfaces.

Therefore, the following have to be considered while evaluating bioaccumulation:
- metabolism/transformation,
- intra- and interspecies variance,
- organ-specific accumulation (reversible/irreversible),
- incomplete elimination (bound residues), and
- bioaccessibility of the chemical (binding to particulate and dissolved fractions) as well as other unquantifiable criteria.

Availability

There are a number of tests for availability of toxic materials from single components and from mixtures. These tests are generally based on the assumption that greater hydrophilicity enhances availability. Thus, each uses solution into aqueous media at various pH values as a measure of availability. However, such measures seldom give any other information besides qualitative guidance on the likely uptake by organisms. Quantitative data would depend upon actual measures of uptake of toxins, for example in humans. Physiological differences between rodent and human GI tract function guarantees that studies of availability carried out in animal models will only be qualitative. However, such tests are valuable since they provide some comparative measure of availability, but they cannot provide a quantitative assessment of uptake in humans.

The Reality Check

A toxin needs to reach a site of action before it can be properly described as bioactive. This capacity for bioactivity is an essential adjunct to the principles of bioavailability of materials in higher organisms. No bridge is established between these ecotoxicity tests and the necessary toxicokinetics or human susceptibility to toxicity.

There is a need for a reality check. It is little point to proceed with such testing if the reality test shows that the possibility for bioaccess is negligible outside the test tube. The same reality test needs to be applied to higher organisms bereft of gills and transdermal adsorption. The aim should be to provide a means to estimate exposure to various materials. The essential task is therefore to model speciation, transport, and real-life environmental exposure. This calls for actual measurement of concentrations of material under investigation. There is a need to acknowledge that many toxins are insoluble. In normal conditions, there is no reason to suppose that they will speciate to a soluble toxic compound in normal field conditions.

Nutrient Toxin Interactions

While considering food chain transfer of nutrients, one has to bear in mind that not only are there transfers of beneficial material but there may also be transfers of toxins of both natural and anthropogenic origin. What is actually consumed is a composite of the diverse influences of nutrient input from marine aquatic and terrestrial sources on plant and animal foodstuffs. It would be an oversight to assume that toxin transfers are exclusively of manmade origin. There are cases of transfer of digitalis to snails and then to the foodstuff consumed. Central to this process are dose-response curves associated with nutrients and toxic materials. For toxins, the dose-response curve is a characteristic S-shaped curve proceeding from an area of no effect through a threshold of safety into a region of toxicity and thence to death (Figure 1.3). Even essential nutrients can follow this later part of the curve in the region of its toxicity, but essential nutrients in excess can also be toxic.

For such nutrients, however, there is a region below the area of toxicity in which homeostasis is maintained. The essential nutrient can then, however, fall into an area of deficiency, which is effectively just as toxic as when in excess, and which when taken to its ultimate extreme can lead to death. One can thus see that in risk assessment of nutrients, there is a

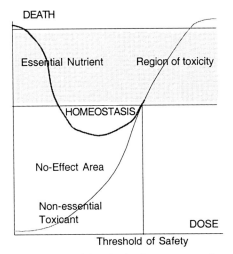

Fig. 1.3:   Dose-response relationships- typical responses of an individual exposed to an essential and a non-essential substance.

need to take into account both portions of the curve within areas of excess and areas of deficiency.

Health

In health terms, the fraction of the dose of contaminant retained is of concern. Exposure is a poor measure of dose, and the environmental "bioavailable fraction" is not the same as "bioavailability" in higher organisms. The fundamental problem for EILs is that the numbers of species involved is large. Consequentially, there is every likelihood that there will be one species that is hypersensitive. As discussed above, HILs, unlike EILs, relate to a single species, humans, where variation is much lower than the interspecies variation associated with EILs. Moreover, there are also site-specific adaptations to existing natural and anthropogenic pollutants. Often this results in discredited EILs which lie below background levels. Decisions will have to be made on whether ecologically based risk assessment is adaptable to human exposure scenarios. For those human-based scenarios, the decision will be whether it is appropriate to ignore ecologically based criteria and rely on human health-related criteria. Ultimately, it may be more appropriate to run EILs and HILs in parallel rather than to attempt any form of integration because of these incompatibilities.

Management of the Process

Major obstacles to sound management of safety of nutrients and other chemicals include (a) the inadequate database on which decisions are to be based and (b) there is a reluctance to pursue appropriate risk management procedures, particularly in developing countries, because of the economic burden leading to a sequel of stringent standards established through orthodox risk assessment. The inadequate data base results in frequent use of conservative default values in risk assessments, with consequent excess of conservatism in the assessments. Thus, one has to decide who or what one is trying to protect in risk assessment. Where health-based effects are the principal area of concern the value of ecological risk assessment needs to be ascertained. Site specific EILs need to be formulated, unlike HILs which are universal. Ecological measures for protection of vulnerable "in fauna", especially where there is a paucity of data are appropriately more stringent than those necessary for protection of human health.

Modelling and Data Quality

The data for risk assessment must be of acceptable quality to ensure avoidance of default information. Generally, human-derived information is preferable in HILs to in vitro or in vivo data obtained experimentally. However, there are circumstances where such information is all that is available or is likely to be available. Biomarkers provide a valuable adjunct to the assessment process because they permit evaluation of dose response relationships and thus provide a bridge between measures of external exposure and internal effect. This is a critical link in the sequence chemical exposure:internal dose:health effect. The use of biomarkers is particularly important in the assessment of susceptible populations that are specially sensitive to chemicals and their metabolites for reasons of either genetic linkage or acquired susceptibility. Such biomarkers of exposure and effect can be used further to evaluate compliance with risk management measures.

A good example of the potential of the miscalculations associated with modeling or predicted data is the uptake of pesticides such as chlorothalonil in celery being grown as a food crop. There are three potential ways by which the likely load given to humans can be calculated on the basis of the available information. First, one can calculate the theoretical maximum residue contribution (TMRC), which will involve modeling based on the amount of pesticide applied to plants, which would give a theoretical figure. Secondly, check the concentrations of chlorothalonil on the vegetables at the farm gate and predict the load on the basis of concentrations at that site, mean intakes of vegetables, again predicted, and calculate concentrations in terms of mean predicted body weight, etc. Finally, one could go the dinner plate and carry out a duplicate diet residue monitoring study by measuring the actual concentrations of chlorothalonil in the cooked food, and thus the actual intake of the pesticide. This last measure is a reality check on what might be otherwise predicted or calculated in the other studies. The figures obtained using these three ways are shown in Table 1. If one has theoretically 100% of the permissible daily intake based on theoretical calculations, but only 0.2% when one does the reality check, that the models used for the theoretical calculations are flawed. Such is often the case in modeling. Most models depend on default values where actual measurements are lacking. Such default values need to be conservative since in no case would one want to underestimate the likely exposure to the chemical. Yet in many cases, the overconservatism of one part of the model is loaded on the overconservatism of another part of the model. This results in estimates that are wildly different from the concentrations actually found, with consequent overassessment of the degree of risk associated with the use of these compounds.

Table 1.1: Concentration of chlorothalonil on celery

Estimate	Concentration (ppm)	% of tolerable intake
TMRC	15.0	100
Farm gate	4.1	27
Residue monitoring	0.12	0.8

OUTCOMES OF MODELLING—THE REALITY CHECK

Environment

A confounding problem that tends to maintain and reinforce the separation between health and environmental approaches to risk assessment is the great disparity between the understanding of "environment" not only between health and environmental sectors but also between agencies within the same sector. Since the Ottawa Charter, and elaborated in innumerable subsequent declarations and documents, the discrepancy between "a healthy environment" and an "environment conducive to health" continues to confuse and confound.

Within Australia, for example, New South Wales defines environment "as the place and surroundings in which we live and work". Victoria has a similar but expanded definition, which incorporates "the social factors of aesthetics as well as the physical environment". Finally, Queensland has an approach quite different in scope which encompasses the New South Wales and Victorian definitions but also includes "social, economic and cultural conditions that affect or are affected by". The National Environment Protection Council legislation that enacts the detail of the Inter Governmental Agreement on the Environment finesses the definition. This has become problematic as individual measures are developed, such as the national environmental protection measure (NEPM) for air. There are different interpretations among environment jurisdictions, environment, and other governmental agencies, particularly those in the development sector. More importantly, no agreement has yet been reached between these environment agencies and the nongovernmental sectors.

The "environmental perspective" is even more complex. Background levels of "hazardous substances" in ecosystems vary widely. Within a particular ecosystem, elevated levels may be critical to maintaining species diversity and ecosystem balance. Geothermal areas and monsoonal environments may be important for mercury, lead, arsenic, and sulfate acidification. Wide seasonal variations are critical determinants in ambient concentrations of toxic metal species essential to flood plain diversity.

The "historical" contamination of agricultural land, often from nutrients or from contaminants carried with nutrients, has heavily impacted on historical species biodiversity both through the introduction of favoured exotic plant species, domesticated mammals, particularly for food, and the feral species, visibly foxes, rabbits, and cats, but potentially of more consequence and more insidious through insects, plant viruses, and microorganisms. Should the "new" ecosystems established by human activities be treated as "environmental contamination"? It depends on whether the beneficiaries—often competing species or those more adaptable are in vogue with human perceptions. The same is not true for contamination introduced with fertilizers such as cadmium in superphosphate which is wholly deleterious.

The inevitable conclusion is that while health guidelines can adopt a single-species approach; albeit with a range of assumptions, environmental guidelines essentially require an ecosystems approach. No two ecosystems however apparently related can be treated identically. In the foreseeable future, pragmatic, implementable environmental guidance for contaminated land will remain largely generic and the adoption of other than local "baseline values" fraught with complexities.

Linkage between Health and Environment

The linkage between environmental and health data confers considerable benefits, but also poses many dangers if not carefully carried out. In linking the data, it is too easy to overlook the statistical problems and inconsistencies of different datasets or to misinterpret their apparent relationships. Valid linkage thus relies on the use of both valid data and appropriate linkage methods.

Since the clearly adversarial approaches of the pre-1992 ANZECC/NHMRC Contaminated Land Guidelines where health and industry had closer agreement than health and environment, there has been great progress in bringing the two sectors together primarily through the jointly funded national workshops (ANZECC 1992). Many methods for data linkage of health and environment have been developed. Their suitability for linking environmental and health data, however, is often limited and always needs to be carefully assessed. Two important criteria must be considered in this context. On the one hand, the methods must be simple, inexpensive to implement, and operable with available data, thus allowing rapid assessment. If the methods are overly complex, requiring extensive resources and large amounts of additional data collection, their use may be costly and result in delays in action. On the other hand, if their results are to be accepted as a basis for action, the methods must be scientifically credible and statistically valid. This means that they should be accurate, sensitive to the variations in the data of interest, and unbiased. They should also produce results that agree with those obtained from more detailed studies for which the statistical precision can be quantified.

In practice, these requirements cannot always be met; indeed, if they could, there would hardly be a need for individual-level studies. Even if they do not meet all these criteria, the methods may be of considerable value. Results from ecological studies, for example, are useful if the potential biases can be identified and evaluated. At the very least, the results can show areas or issues requiring further, more detailed investigations. Where detailed individual-level studies have not been performed, there is an urgent need to access methods that can help to shed light on the extent and health effects of specific forms of environmental pollution.

Although human health may be the major concern, serious consequences can flow from adopting a human-centered approach to ecological thinking. One cannot assume that the goals of environmental protection, occupational health and safety, and public health are always the same. There are many examples, particularly in rapidly developing economies, where improvements in occupational and public health resulting from social and economic improvements have resulted in increased degradation of the environment.

The development of health-based acceptance criteria is still in its infancy, with consideration being given only to protection of physical health (and occasionally discomfort, nuisance, and amenity) and no consideration for societal well being and social valuations. For most substances and circumstances, criteria have not been established to date.

Acceptance criteria for individual substances for contaminated land continue to be developed by health and environmental agencies, but the need to assess the health and environmental effects of policies and programs related to contaminated land remains almost totally neglected. In particular, there has been little attempt to accommodate the problem of mixtures of contaminants (Shaw et al., 1996). As a result in broad health/environmental issues, areas of agreement and conflict surface during the implementation stage rather than in establishing the principal acceptance criteria at the planning stage. Indeed, the effect of mixtures on microbes, especially algae, have been extensively studied. As discussed in Chapter 5, mixtures of metals can have markedly different effects on algae.

Development of international chemical safety programs in which risk assessment is an integral part is essential. Pollution by nutrients is independent of national boundaries. This is particularly true of air-borne and water-borne pollutants. It is also true of nutrient-contaminated soils, which inevitably contribute to air, water, and marine pollution through dust and runoff transfers. International programs are the only solution to globally sound management of chemicals. To respond to the challenge of the evolving chemosphere in which we live, we need to develop more adequate databases from which we can derive more credible environmental and health risk assessments.

REFERENCES

ANZECC (1992) Australian and New Zealand guidelines for the assessment and management of contaminated sites. ANZECC and NH&MRC, Canberra, 159p.

Baraona, E., Gentry, R.T., and Lieber, C.S. (1994) Bioavailability of alcohol - Role of gastric metabolism and its interaction with other drugs. *Digestive Diseases.* **12**: 351–367.

Edwards, J.W., van Alphen, M., and Langley, A. (1994) Identification and assessment of contaminated land. Improving site history appraisal. (*Contaminated Sites Monograph Series* No.2) Adelaide, South Australian Health Commission, Hyde Park Press; 196p.

El Saadi, O., and Langley, A. (1991) (Eds.) The health risk assessment and management of contaminated sites. (Contaminated Sites Monograph Series No. 1). Adelaide, South Australian Health Commission, Hyde Park Press; 240p.

Franke, C., Studinger, G., Berger, Böhling, G., Bruckmann, U., Cohors-Fresenborg, D., and Jöhncke, U. (1994) The assessment of bioaccumulation. *Chemosphere* **29**: 1501–1514.

Geyer, H.J., Steinberg, C.E.W., and Kettrup, A. (1992) Biokonzentration von persistenten superlipophilen Chemikalien in aquatischen Organismen UWSF—Z. *Umweltchemie. Ökotoxologie* **4**: 74–77.

Gibson, E., Strudwick, D., and Walker, P. (1996). Introduction to the draft national framework for ecological risk assessment. In *4th National Workshop Health Risk Assessment and Management of Contaminated Sites*, Langley, A., and Imray, P. (Eds.) *Contaminated Sites Monograph Series* No. 6, South Australian Health Commission. pp 51–64.

Langley, A., and van Alphen, M. (1993) (Eds.) The health risk assessment and management of contaminated sites. (Contaminated Sites Monograph Series No., 2) Adelaide, South Australian Health Commission, Hyde Park Press; 347p.

Langley, A., Markey, B., and Hill, H. (1996) (Eds.) *The health risk assessment and management of contaminated sites, Contaminated Sites Monograph Series* No. 5, South Australian Health Commission, Open Book Publishers. 486p.

Menzie, C. (1997) Environmental assessment of contaminated soils. *SETAC News* 17: 29.

Moore, M.R. (1996) Bioavailability of materials from contaminated soils. In *The Health Risk Assessment and Management of Contaminated Sites.* Langley, A., Markey, B., and Hill, H. (Eds.) Commonwealth Department of Human Services and Health and the Environmental Protection Agency. *Contaminated Sites Monograph Series* No. 5, South Australian Health Commission, pp. 339–353.

National Research Council (1994) *Science and Judgement in Risk Assessment*. NRC National Academy Press, Washington, DC. pp.650.

OECD (Organization for Economic Cooperation and Development) (1981) OECD Guidelines for testing of Chemicals No. 305 A—E. Bioaccumulation, 12 May 1981, Paris.

OECD (1993) OECD Guidelines for Testing of Chemicals No. 107, Draft Partition Coefficient (n-octanol/water): Shake Flask Method, November 1993, Paris.

Paustenbach, D.J. (1989) The Risk Assessment of Environmental and Human Health Hazards. Wiley, New York. pp1121.

Romijn, C.F.A., Luttik, R. Y. D, Meet, D., Slooff, W. and Canton, J.H. (1993) Presentation of a general algorithm to include effect assessment on secondary poisoning in the derivation of environmental quality criteria. Part 1. Aquatic food chains. *Ecotoxicological and Environment Safety*. **26**: 61–85.

Shaw, G.R., Moore, M., and Barron, W. (1996) Chemical mixtures and soil contamination. *The Health Risk Assessment and Management of Contaminated Sites*, Langley, A., Markey, B., and Hill, H. (Eds.) *Contaminated Sites Monograph* Series No. 5. South Australian Health Commission, pp.303–322.

Bioavailability of Metals in the Soil Plant Environment and its Potential Role in Risk Assessment

R. Naidu[1], S. Rogers[1], V.V.S.R. Gupta[1], R.S. Kookana[1], N.S. Bolan[2] and D.C. Adriano[3]

INTRODUCTION

Metal contamination of soil environments and the assessment of its potential risk to terrestrial and aquatic biota and public health is one of the most challenging tasks confronting scientists today. This is partly due to the plethora of soil biochemical and physical processes controlling the fate of metals and partly because of the difficulty associated with the assessment of their availability. A credible method for the assessment of "bioavailability" of metals and other contaminants is lacking.

The definition of "bioavailability" and the concept on which it is based are unclear, the methods adopted vary throughout the world, and therefore there is no single standard technique for the assessment of either plant availability of metals or their ecotoxicological impacts on soil biota. Some of the difficulties probably arise because of confusion with the understanding of the bioavailability concept (see Chapter 1) and relating this to the measurement of metal availabilities. For example, a number of investigators relate bioavailability to that fraction of extractable metals that correlates with the total metal uptake by plants. Although this has been quite successful with certain plants and soil types, its applicability has often been found to be fraught with limitations. Firstly, as bioavailability is often assessed by chemical extractions, it is likely that extractants may release those pools of metals which the plant roots and microorganisms do not exploit in the soil solution. This is supported by the definition proposed in Chapter 1 which suggests that the concept of

---

[1] CSIRO Land and Water, Private Mail Bag. No. 2, Glen Osmond, Adelaide, South Australia 5064.
[2] Department of Soil Science, Massey University, Palmerston North, New Zealand.
[3] Savannah River Ecology Laboratory, Drawer E Aitken, South Carolina 29802.

bioavailability is fundamental to toxicology and it is not the same as bioavailable fraction, which is largely a measure of solubility. Secondly and more importantly, chemical extraction methods do not consider physiological and biochemical factors associated with plants and microbes, e.g. transport of metals across the cell membranes. The parameters that control the membrane transport of metals are species-specific and possibly dictate the rate of uptake by different plant and microbes from the same bioavailable pool of metals in soils.

Bioavailability has also been considered in terms of the soil solution concentration of analyte species (Chapter 3) in contact with plant roots at any given time. Some laboratories use plants and microbes as indicators of bioavailable fraction whereas many tend to place increasing reliance on chemical extraction techniques. The nature and the mode of action of these extractants vary from simple desorption (e.g. $Ca(NO_3)_2$, 0.5M NH_4NO_3) to complexation (e.g. EDTA, DTPA) of metal ions. Increasing attempts have been made to use anion and cation exchange resins which mimic the plant roots to measure the bioavailable fraction of metals in soils. Tiller (1996) concludes that some of the extraction procedures do not have universal acceptance in all countries, "not necessarily because of lack of any usefulness but because of a certain inertia we all share in terms of our mental commitment as well as availability of equipment and the need to support existing data bases."

Markedly different types of processes control the bioavailability of metals in terrestrial and aquatic environments. In a terrestrial environment, adsorption and desorption processes are the predominant factors controlling bioavailability. The bioavailability of metals in the aquatic environment is determined by processes such as adsorption to and desorption from solid phases as well as chemical complexation with inorganics and dissolved organic carbon. As the fate of many contaminants is determined by the surface properties of the particulate matter with which they are associated, their chemistry, bioavailability, and transport depend on the degree of partitioning between solid and solution phases. Metal complexation to naturally occurring humic substances, however, is perhaps of greater significance biologically and environmentally than the adsorption processes, especially in aquatic systems. Thus, understanding the mechanisms involved in the retention, release, and mobility is a prerequisite to determining the bioavailability of metals ions in both terrestrial and aquatic environments. Such information is necessary to predict the environmental impact of metals from anthropogenic sources as well as to develop regulatory measures related to the application and disposal of heavymetal containing materials on agricultural soils.

Of primary concern, however, is the definition of bioavailability and the analytical techniques to accurately predict the potential availability of metals to terrestrial and aquatic biota. The principal goal of this chapter is to present an overview of (a) factors controlling bioavailability and (b) limitations of the existing methods of assessment of their bioavailability.

SOURCES OF HEAVY METALS

Natural Pedogenic Inputs

Metals enter the soil environment as a result of both natural and anthropogenic activities. The natural processes of pedogenesis lead to mineral breakdown and translocation of products as well as accessions from dust storms, volcanic eruptions, and forest fires. The concentrations of metals released into the soil system by pedogenic processes are generally related to the nature of the parent materials and their rate of weathering dissolution. Some soils are enriched with heavymetal containing minerals, the dissolution of which contributes to the heavy metal loading in the soil solution. Examples of pedogenic contamination include the seleniferous soils in the San Joaquin Valley in California (Tanji et al., 1986) and arsenic in groundwater reported in many countries in Asia (see Chapter 11). Both the distribution and mobility of pedogenic metals are influenced by the specific interaction of metals with various soil constituents, in particular at the iron and manganese hydrous oxide and soil solution interface. These oxides play significant role in governing the distribution of metals in the solid phase of the soil. The sorption of metals by the oxidic minerals, particularly in the form of nodules, concretions or layers is most often responsible for their deposition in the soil profile and at the root-soil interface. Such deposition of oxyhydroxides has been reported to retard heavymetal uptake through the immobilization of metals at the soil and root interface (Otte et al., 1989).

Anthropogenic Inputs

Anthropogenic processes comprise all contributions made by human activities. Anthropogenic contributions include atmospheric emissions, land depositions from industries, mining and metallurgy, land disposal of urban wastes and municipal and farm sewage and fertilizer applications. As anthropogenic activities generally lead to metal accumulation at the surface horizons, they are considered a greater threat to the environment because of their accessibility to plants, animal, and human systems. Metals from anthropogenic sources are generally considered to be present in environmentally unstable forms, and are thus more soluble and bioavailable than those derived from natural processes. Surface inputs also have implications for offsite impacts of metals due to contamination of ecosystems by surface runoff and wind erosion.

The most common sources of metal input to cropping and pasture soils are, however, municipal sludge (biosolid) and phosphatic fertilizers. Land disposal of sludge is becoming increasingly common due to the stringent guidelines prohibiting the ocean disposal of sludges. Because of its high nutritional value, municipal sludge has become more attractive for agricultural purposes. This is now being widely practiced in developing countries, although some developed countries also allow disposal onto land within certain regulatory guidelines based on the biosolid metal concentrations (McLaren and Smith, 1996). These guidelines have been developed with the aim of preventing metal toxicity to plants and to human and animal consumers of the plants (Ross et. al. 1991). However, sludge may also contain high concentrations of Cd, Zn, Cu, and other metals (Table 2.1) that may pose potential risks to human and animal health following land application. A major concern relating to sludge application to soils, particularly with repeated applications, is that metals accumulate in the surface layer of soils. Although limited movement of metals have been reported in soils treated with sludge (Chang et al., 1984), Dowdy and Volk (1984) conclude that heavymetal movement was most likely to occur where heavy applications of sludge are made to sandy and acid soils with low organic matter content receiving high rainfall or irrigation. Besides the potential risk of metal mobility, the surface retention of metals also pose risk to the potential toxic effects of metals on plant growth and their introduction into the food chain. There is also increasing concern on the effects of high soil metal concentrations on soil biological activity.

Table 2.1: Heavy metal content of sewage sludge in New Zealand and the maximum amount of sludge that could be applied without exceeding the threshold concentration in soil

Heavy metal	Content (mg kg^{-1})	Threshold level[a] (mg kg^{-1} soil)	Maximum amount of sludge (Mg ha^{-1})[b]
Cadmium	8	3	187
Chromium	1200	600	250
Lead	500	500	500
Mercury	2	1	250
Selenium	5	3	300
Zinc	1600	300	93
Copper	500	130	130
Nickel	150	70	233
Boron	33	3	45

[a] Based on EEC regulations; [b] based on 5cm depth of incorporation and a bulk density of 1 Mg m^{-3}.

In the developing countries such as those in the South Pacific region where traditional fertilizer application practices still persist (Haynes and

Naidu, 1989), metals may not be a major concern. In contrast, there are many areas in Australia where heavy application of phosphatic fertilizers has led to considerable accumulation in the soil genetic horizon. Inputs of heavy metal Cd in Australian soils from long-term fertilizer use have been estimated to be approximately 1.6 g ha^{-1} yr^{-1} compared to 4.3 g ha^{-1} yr^{-1} in UK (Hutton and Symon, 1986) and 8.9 g ha^{-1} yr^{-1} in New Zealand (Bramley, 1990).

The release of heavy metals from biosolids in the environment is dependent on the chemistry of biosolid mineralization in soils. Given the complex nature of biosolids, metals are likely to be present (a) as free metal ions; (b) bound to silicate clays or organic matter by predominantly electrostatic forces (sometimes referred to as the "exchangeable fraction"); complexed or chelated by organic materials; held on specific adsorption sites such as amorphous precipitates of alluminosillicates, Si, Fe, Al, or Mn oxides, or on the zone of alteration on the surfaces of crystalline or microcrystalline oxides, carbonates or phosphates; and held on or occluded in hydrous Al, Fe and Mn oxides (Beckett, 1989).

The differences in metal biosolid chemistry and their impact on bioavailability are often not considered by researchers while comparing different biosolids. Reviewing heavy-metal accumulation in plants resulting from the land application of biosolids, McBride (1995) concluded that relatively little has been learnt from field studies about fundamental processes that determine solubility and chemical forms of heavy metals in biosolids and soil.

Fate of Heavy Metals in Soil Environment

Following entry into the soil system, metal ions rapidly interact with the soil mineral and organic phases and form a quasi-equilibrium with metal ions in soil solution. The concentration and nature of the ions present in the soil solution is a result of these interactions. Thus, the chemical composition of soil solution is dynamic and determined by the multiphase equilibria involving (1) the solid phase comprising clay minerals, poorly ordered inorganic, and organic materials; (2) the liquid phase comprising water; (3) the gaseous phase comprising mainly oxygen and carbon dioxide; and (4) soil : solution interface or the exchange phase. Whatever the mode of entry of metal contaminants into the soil environment, their interaction with the soil involves both sorption and desorption and precipitation and dissolution processes. These reactions also impact on processes that control release, mobility, and plant uptake of metal contaminants. For a detailed review of the fate of metals in soils readers are directed to reviews by McBride (1989); Naidu et al., (1996, 1998).

BIOAVAILABILITY OF HEAVY METALS

Definition of Bioavailability

Considerable controversy exists in the literature on the precise definition of bioavailability. It has been linked to metal ion activity in soil solution, the exchangeable metal fraction, and more recently to the concentration of metals that cause ecotoxicity. However, the scientific community universally accepts none of these relationships. This may be due to the many inconsistencies between the so-called "bioavailable fraction" and the uptake of metals by plants and their impact on soil organisms. This difficulty may also be related, in part, to the large differences in metal uptake and toxic response between different organisms such as plants and microorganisms, and also between species and strains of the same species (Giller et al., 1997). Discussing the toxicological impacts of heavy metals on soil microorganisms, Baath (1989) noted that the range of values for Cu, Pb, and Zn concentrations at the "No Observed Effect Level" (NOEL) was 1 and 10 orders of magnitude in different soils. Witter (1992) noted that the differences in toxic response to a given concentration of metals in soils are entirely due to variation in metal sensitivity between plant species. Although chemical extractions involving dilute salt solutions have been used to measure the bioavailable pool of metals in soils (DIN, 1995), ultimately the bioavailability of metals can be determined only by in situ techniques. Such techniques involve growing the organisms of interest in the contaminated material and quantifying the uptake of metal into the organism or assessing the toxicological response (Giller et al., 1997; Wolt, 1994). A major limitation of the current definition of bioavailability is the assumption that it is a static process. As discussed later, bioavailability is a dynamic phenomenon that besides showing seasonal variations, also varies significantly with changes in soil environment including the chemistry and biology at the root soil interface that is dependent on soil properties (Eq. 2.1). Unless these factors are taken into consideration while devising soil tests, it may not be feasible to have a test that accurately predicts bioavailability of all metal ions in different soils. Bioavailability can be defined as the concentration of a particular substance at a specific time that can cause an effect on the morphology or physiology of a specific organism.

Bioavailability $= f$(soil properties). (plant characteristics). (microbes). dT

$F_{b,t} = F_{b,0} \, K_{t1}$

F_{bt} = fraction bioavailable at time t

$F_{b,0}$ = fraction bioavailable at time 0

$K_{t1} = f(a$ (clay) $+ b$ (organic matter) $+ c$ (yield), etc.)

Measurement of Bioavailability

Many different chemical and biological techniques have been used to assess the bioavailable fraction of metals in soils. However, these techniques are either not universally accepted due to favoured chemical methods or have proved ineffective in defining the bioavailable fraction. The problem also appears to be related to the difficulty in identifying an analytical technique that can provide a good estimate of the pool of toxic metals ions that are accessible to plants and microbes. This is partly constrained by the wide range of forms in which metals are present and partly due to the lack of specificity of the chemical and biological techniques used for assessing the different pools of the metal ions. Moreover, there is also much uncertainty on the precise nature of the ionic species that are taken up by plants. Although plant scientists often for plant uptake define this as the "amounts of chemical species available"; microbiologists often discuss this in relation to the concentration that causes toxicity to soil microorganisms. Whatever the definition of bioavailability, there are considerably large numbers of chemical and biochemical methods that have been used to assess the biologically available metal content of soils. Some of these techniques together with their limitations are discussed next.

Chemical techniques

Metal ions exist in a variety of mineral and organic phases in the soil environment. Soil metals may be present (1) in soil solution as ionic or organically complexed species; (2) on exchange sites of reactive soil components; (3) complexed with organic matter; (4) occluded in oxides and hydroxides of Al, Fe and Mn and (5) entrapped in primary and secondary minerals (Hodgson, 1963; Viets, 1962). These fractions are in dynamic equilibrium and the concentration of metal ions in soil solution is strongly dependent on the equilibrium between these fractions. A number of studies (McLaren et al., 1986; Clark and McBride, 1987) suggests that considerable amount of trace metal ions that enter the soil environment may not be exchanged by alkaline and alkaline earth metals unless there are other factors conducive to their exchange.

It is generally accepted that the trace metals present in water-soluble, exchangeable, and adsorbed fractions are readily available to plants whereas those associated with primary and secondary minerals are relatively unavailable to plants (Viets, 1962). As plant roots absorb metals directly from soil solution, soil fractions that contribute to this pool also control plant availability of trace metals. McBride (1989) concluded that there are six direct measures for the assessment of metal availability.

1. Exchangeability by cations that do not specifically sorb,
2. exchangeability by specifically adsorbing cations,
3. pH reversibility,
4. isotope exchangeability,
5. desorbability by chelating agents, and
6. dissolution by strongly acidic solutions.

Based on these measures, a number of different extractants have been used to investigate the bioavailable metal fractions in soils. Some of these analytical techniques are listed in Table 2.2. There are four main groups (including water) of extractants based on their mode of action:

Table 2.2: Chemicals used for the extraction of metals

Methods	Extractant	Nature of metal extracted
Water	H_2O	Soluble
Weak salt solution	$CaCl_2$, NH_4NO_3, CH_3COONH_4, CH_3COOH, $AlCl_3^-$	Soluble + exchangeable
Weak acids	HCl	Soluble + exchangeable + dissolved
Exchange resins	Ca^{2+}, chelating agents	Soluble + exchangeable + adsorbed
Weak chelating agents	EDTA	Soluble + exchangeable + adsorbed

The mode of action of the extractants varies significantly. While water essentially provides an estimate of soluble metal ions, both dilute salt solutions and weak acid solutions act by displacing metals from exchange sites. Because the majority of trace metals are held in specific sites with high affinity for metals, a large excess of cations is necessary to exchange with them (Shuman, 1992). The major advantage with the neutral salts is that they do not alter the pH at the exchange sites. The measured pH of the equilibrium solution, however, is altered (Beckett, 1989). Depending on the concentrations, salt solutions generally extract both nonspecifically and specifically bound metals. One of the major limitations of the electrolytes is that they alter the native soil solution equilibrium considerably (Helmke and Naidu, 1996). Thus, these extractants do not provide a true reflection of the chemical status of metal ions in soil solution.

The chelating agents act both by dissolution (depending on the pH) and through complexation reactions. Generally, metal ions removed by these extracts are essentially from dissolved minerals, exchange sites, and soluble organic chelates (Adams and Kissel, 1989). Given the widely different mode of actions, it is not surprising that none of the above methods have been found to predict metal bioavailability across all soil types. The limitations of the chemical extraction techniques are discussed later.

Murthy (1982) demonstrated that water-soluble, exchangeable, and complexed forms of Zn are readily available for plant uptake. Khaswaneh

(1971) related plant metal uptake to the ion activity in soil solution. He suggested that (1) relative intensity (2) rate of replenishment, and (3) ionic composition were the major factors controlling ion uptake by plants. Although each of these factors has been studied in isolation using nutrient culture solution, plants grown in soils reflect the effect of a combination of all of these factors. Minnich et al. (1987) studied soil solution Cu^{2+} activity in soil saturation extracts and compared this with Cu accumulation in young snapbeans (*Phaseolus vulgaris* L). They found a non-linear relationship between soil solution Cu in sludge-treated soils and Cu content of snapbean. Comparative study using soils amended with Cu salt gave a linear relationship. A similar relationship was observed between DTPA-extractable Cu and total plant Cu contents. In contrast, there was no significant relationship between total soil Cu and plant Cu content. This suggests that total Cu is not a good predictor of plant available Cu.

Some examples of strong relationship between dilute salt solution extractable metals and plant metal contents include, (1) the use of $MgCl_2$-extractable Zn in relation to Zn uptake in orchards (Neilson et al., 1987), (2) $CaCl_2$ extractable Cd and Cd uptake in certain agricultural crops (Jackson and Alloway, 1991) and (3) NH_4NO_3 extractable Cd and metal uptake by radish (Symeonides and McRae, 1977). Iyengar et al. (1981) found that plant Zn was most strongly correlated with exchangeable Zn and that multiple regression analysis showed that inclusion of Zn in each of the different soil fractions accounted for over 90% of variability in plant Zn uptake. Lagerwerff (1971) reported significant correlations between Cd extracted from a soil treated with various levels of Cd and Cd concentrations in radish plants. A significant correlation was also found between CH_3COONH_4-extractable Cd and Cd concentration in radish and lettuce grown on 30 soils. However, these studies were conducted using soils amended with freshly added Cd salts. Haq et al. (1980) compared extractants for plant-available Zn, Cd, Ni, and Cu in contaminated soils in a glasshouse study. They sampled 46 surface mineral soil materials which, over a period of two years were contaminated to varying degrees with heavy metals by the application of sewage sludge or from air pollution around metal smelters. Subsamples of the contaminated soils were then extracted with M CH_3COONH_4 (pH this 7.0), 0.5 M CH_3COOH, 0.01 M EDTA (in M $(NH_4)_2CO_3$, pH 8.6), 0.005 M DTPA, 0.02 M NTA, 0.6 M HCl + 0.15 M $AlCl_3$, $(COOH)_{2+}$ + $(COONH_4)_2$, and water. Zinc, Cd, Ni, and Cu in Swiss chard was correlated with the amounts of these metals removed by each of the extractants. Of the nine extractants, CH_3COONH_4 was the best predictor of plant-available Zn if only extractable Zn and soil pH were included as independent variables in a regression equation. Acetic acid-extractable metal was the best predictor of both plant-available Cd and Ni when soil pH was included in the equations. Their attempt to find a suitable soil extractant for plant-available Cu was unsuccessful.

In contrast to the weak salt solutions there has been limited success with chelating agents such as DTPA and EDTA. The utility of DTPA (0.005 M pH 7.3) and EDTA (0.01 M, pH 7.0) for measuring the availability of Cu, Fe, Mn, and Zn from uncultivated A- and C-soil horizons to western wheatgrass (*Agropyron smithis* Rybd.), silver sagebrush (*Artemisia cana* Pursh), and above ground biomass was assessed at 21 geochemically diverse sites in the northern Great Plains by Gough et al. (1980). They found that positive relationships between plant metal contents and either DTPA- or EDTA-extractable metals were limited; and based on these data, the authors indicate that caution should be used in the formulation of regulations for plant-available levels of these metals in Great Plain soils based solely on extractable levels. Contrasting relationship has been reported between DTPA- and EDTA- extractable Cd and the plant Cd content. For instance, positive, negative, and no relations have been recorded between metal extracted by chelating agents and plant metal uptake data (Haq and Miller, 1972; Rappaport et al., 1988). In addition, neither of these extractants proved suitable for a mixture of heavy metals. Latterell et al. (1978) found that metal concentration in DTPA extracts was a good predictor of foliar Zn in snapbeans, but not of Cu or Cd. Singh and Narwal (1984) observed that while plant uptake of metals decreased with increasing pH from 5.6 to 7, there was no difference in DTPA-extractable metals with pH. Moreover, the method does not give a reliable measure of plant-available Zn and Cu in sludge-amended soils (Barbarick and Workman, 1987; Bidwell and Dowdy, 1987; Adams and Kissell, 1989). Adams and Kissel (1989) attributed this to (1) the ineffectiveness of DTPA at low pH, (2) the inability of DTPA to buffer soil extract near pH 7.3, and (3) increased amounts of soluble chelated metals at higher sludge rates and higher soil pHs. Chelating agents such as DTPA predict only the probability of whether soils can supply adequate Zn to meet plant demand. Davies (1992) reported that the objective of using DTPA is to determine the required quantity of trace-metal fertilizer applications. He predicted that DTPA may not necessarily predict metal plant uptake.

Despite this lack of correlation between DTPA-extractable metals and metal plant uptake, several investigators have used DTPA extractions for determining plant available Cu, Zn, Fe, and Mn in soils (Lindsay and Norvell, 1978; Wear and Evans, 1968; Brown et al., 1971). Lindsay (1972) suggested that a mixture of DTPA and triethanolamine (TEA) would give a better assessment of plant available Cu and Zn than would EDTA. However, Haq and Miller (1972) found that neither DTPA nor EDTA gave a significant correlation between extractable Cu and crop Cu concentrations. Lack of correlation between DTPA and plant metal content may not be surprising given that this extractant was originally designed for near neutral calcareous soils with insufficient available Zn, Fe, Mn, or Cu (Lindsay and Norvell, 1978) for maximum yield of crops. O'Connor (1988) published an excellent overview on the "Use and Misuse of DTPA Soil Test" and reported that some of the failures are related to the improper chemical makeup of the extracting solution, sample preparation, and extraction procedures. At low pH values, certain metals may not chelate with DTPA; and in such cases no significant relationship has been reported.

Limitations of the existing chemical techniques

There is increasing evidence suggesting that the failure of chemical extraction techniques is related to their improper application rather than the efficacy of the test *per se*. This is clear from both the nature of metal interactions in soils and the mode of action of the chemical extractants. Clearly, the efficacy of the soil tests is dependent on (1) the chemistry of metals and soil interactions, (2) the mode of action of the extractants, (3) chemistry at the soil and root interface, (4) environmental factors, (5) physiology of plants, (6) nature of bioindicators, and (7) nature of soils. None of the existing soil tests consider all of these factors, and, therefore it is not surprising that many investigators have failed to get a strong prediction of plant metal content when employing chemical techniques. The following sections summarize some of these limitations.

Chelating agents. Let us first consider the most frequently used DTPA test. This test was originally designed for the assessment of trace metal availability in calcareous soils. DTPA acts by chelating metal ions. Studies by Lindsay and co-workers showed excellent agreement between extractable Cu and Zn and plant metal content in soils with pH exceeding 8. This is not surprising given that at high pHs, both metal ions exist in solution as soluble metal and organic complexes. However, following the development of the DTPA test, many scientists opted to use this test in acidic soils as well. For instance, Neilson et. al. (1987), assessed the efficacy of DTPA along with $CaCl_2$ in 20 soils ranging in pH from 4.02 to 8.34. They found a poor relationship between DTPA-extractable metal ions and plant metal content. In contrast, a strongly significant relationship was found between the dilute salt solution extractant and the plant metal uptake. The lack of relation between DTPA and plant metal content is predictable given that the test acts by chelating metal ions in high pH soils. Therefore, one may question the validity of using this test for soils with pH less than 7.

There are numerous reports that show that in acid soils, chelating agents may enhance adsorption of metal ions through soil and ligand and metal bridging (e.g. Harter and Naidu, 1995; Naidu and Harter, 1998). Similar to DTPA, EDTA also acts by a chelation mechanism and while this test may be effective in high pH soils, its applicability to low pH soils is questioned. Besides, both these tests probably extract that portion of metal ions which is not available in the short term.

Dilute salt solutions. Dilute salt solutions have been found to be successful with a wide range of soils, and particularly with Cd and Zn. However, such extractants often fail to predict plant available Cu. These differences may be attributed to the widely different chemistry of Cu ions relative to Zn and Cd in soils. Fotovat and Naidu (1997) studied the influence of soil solution composition on heavy-metal chemistry in alkaline sodic and acidic soils. They found contrasting effects of ionic strength and saturating cations on the concentration of Cd, Zn and Cu in soil solution. Although soil solution Zn and Cd increased with increasing saturating cation charge in acidic soils, the concentration of Zn was significantly lower in alkaline sodic soils. Increasing cationic charge and ionic strength led to a marked decrease in the concentration of Cu in soil solution irrespective of the soil pH. These results are consistent with those of Cavallaro and McBride (1978), who found that competing ions such as Ca^{2+} are much less effective in releasing Cu^{2+} into solution than Cd^{2+}.

The results of Fotovat and Naidu (1997) suggest that Zn in acidic soils and Cd in all soils are controlled by ion exchange reactions whereas Cu is controlled by mechanisms other than ion exchange. These investigators attributed the decrease in soil solution Zn in alkaline sodic soils and Cu in all soils to the decrease in the dissolved organic carbon (DOC) concentrations. They concluded that DOC plays a significant role in the behavior of Zn in alkaline soils and that of Cu in all soils. Given that the chemistry of Cu, Zn, and Cd may vary with pH and soil solution composition, the recommendation that single salt solution may be effective for the prediction of uptake of some heavy metals by plants (Novozamsky et al., 1993) or as analytical criteria for guidelines on maximum tolerable load of trace metals in soil (Winistörfer, 1995; Wenzel and Blum, 1995) is questionable. This is supported by the published literature, which shows that chelating agents and dilute salt solutions are selective for certain metals. However, comparison of bioavailability indices as predicted by chemical extractants is fraught with difficulty given the limited information provided by the researchers.

Plants modify the chemistry of root-soil interface. The immediate vicinity of plant roots (rhizosphere) has long been recognized as an area of particular importance for nutrient turnover and bioavailability.

Consequently, extensive work has been done to elucidate root and soil interface chemistry in relation to plant nutrient availability. However, recent studies indicate that the changes in chemistry may also lead to significant variations in the availability of heavy-metal ions (Uren and Reisenauer, 1988). The impact of roots on the adjacent soil medium is mainly through the release of organic and inorganic exudates into the adjacent soil environment and indirectly through the stimulation of microbial activity in the rhizosphere (a result of increased levels of C in the rhizosphere from root exudates). Organic material inputs arise from the sloughage of root material and also from direct root exudation (Hale and Moore, 1979). Plant roots exude a number of organic compounds including amino acids, sugars, and organic acids. Heterotrophic microbial populations use these organic compounds as a carbon source and in turn produce similar compounds such as microbial polysaccharides, etc. (Francis et al., 1992). Although plant root exudates are rapidly turned over in the rhizosphere, numerous investigators have demonstrated that such organic compounds can in themselves contribute to heavy-metal solubility in soils. Figure 2.1 shows that root exudates may enhance solubility of metal ions by two major processes. First, exudates may act as chelating agents and enhance solubility of metal ions through the formation of soluble metal and organic chelates; second, any cations present in the soil solution may displace heavy-metal ions. Chelation processes will shift this equilibria toward the formation of more metal and organic chelates. Although DTPA and EDTA extractants act through chelation of metal ions, neither of the two reagents provide the same binding affinity as the root exudates do. In contrast to chelators, neither soil water extracts nor dilute salt extractants mimic the chemistry at the root and soil interface. Consequently, our expectation for the chemical tests to provide a strong basis for predicting plant available metal ions may be overambitious.

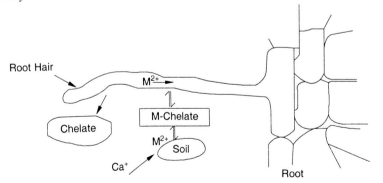

Fig. 2.1: Schematic diagram showing the mechanism of metal release in the presence of a chelating root exudate.

Soil environmental effects. One of the major soil environmental factors influencing heavy metal chemistry is the redox potential. Changes in redox potential have been shown to either increase or decrease the solubility of certain nutrient ions that bind to oxyhydroxide surfaces. Studies by Binghm (1976) demonstrate that such fluctuations in redox potential can have marked effects on the plant availability of metal ions. Bingham et al. (1976) studied Cd availability to rice in $CdSO_4$-enriched sludge-amended soil under "flood" and "nonflood" culture. Grain production under flooded condition was not affected by Cd; however, a 25% reduction in yield was associated with a treatment of 320 μg Cd g^{-1}. Under nonflood management, however, a comparable decrement in grain production was observed even at Cd concentrations of 17 μg Cd g^{-1}. Also, the Cd content of grain under nonflood conditions was 55% greater than that of grain under flood management. The reduced Cd was related to reduced availability of Cd attributed to the precipitation of CdS. More work, however, needs to be done to elucidate the mechanisms that control heavy-metal chemistry under reducing conditions and to assess the seasonal fluctuations in redox potential and its impact on plant Cd availability.

Plant measurements of bioavailability

The amount of metals taken up by plants has often been used as a measure of the bioavailable fraction. However, bioavailability as determined by plant studies should be more correctly termed "phytoavailability" as these pools are not necessarily available to the same degree to other soil organisms. Many studies have demonstrated a positive relationship between the concentration of metals in soils and plant materials. For instance, Heckman et al. (1987) showed that increasing applications of metal-contaminated biosolids to soils linearly increased shoot concentrations of Zn, Cd, Cu, and Ni in soybean (*Glycine max*). Earlier studies by Dowdy et al. (1978), however, reported that increasing soil metal concentrations increased Zn and Cu in snapbean tissues until a point was reached beyond which further increases in metal concentrations did not increase tissue metal content. Ryan and Chaney (1994) also found that the plant metal concentrations reached an optimum level with increasing sludge loading. Beyond this point, further applications of sludge did not increase the metal content of the plant tissue. This observation has been termed as the "plateau" effect. Thus, in terms of plant measurements of bioavailability, uptake does not appear to be always related to total soil metal concentrations.

The nature of plant species has also been shown to influence the capacity of plants to take up metals. For instance, Oliver et al. (1995) showed marked differences in the ability of different wheat cultivars to concentrate Cd metals in soils subjected to many years of P fertilizer applications. This suggests that the bioavailability of metals as measured by plant may vary from species to species. Such a variation would limit the application of plants as indicators of the absolute bioavailability of metals. Brown et al., (1984) implied that plant species-specific mechanisms influencing Cd uptake were not significantly altered as a result of soil type. However, they used $CdCl_2$ salt as the metal source that would be highly bioavailable in relation to Cd from industrial waste or biosolid contamination. The cultivation of plants immediately after spiking does not reflect the effects of ageing commonly observed in contaminated soils. However, recent efforts have been made to develop indices of metal uptake by food crops from more realistic contaminated soil situations.

Brown et al., (1996) studied the uptake of Cd by cabbage, carrot, corn, potato, and tomato from soils receiving a range of biosolid applications. A statistically significant log linear relationship between metal uptakes into vegetable crops was demonstrated, and the authors suggest that lettuce may be used as a suitable indicator plant for Cd uptake. Although the study of Brown et al., (1996) provides some promising evidence towards the development of a metal uptake indicator plant species, chemical techniques (see section on chemical techniques) do give some indication of the plant-available metal pool. Practically, we are still in the situation of having to grow individual species in contaminated soils to assess what portion of the metal loading is phytoavailable.

Microbial measurements of bioavailability

Measurements of metal bioavailability and toxicity in soils using soil microorganisms are receiving increasing attention as a result of evidence that microorganisms are more sensitive to heavy-metal stress than plants or soil macrofauna (Giller et al., 1997). For this reason, responses of different groups of microorganisms have been suggested as surrogates for measurement of bioavailability (see chapters 4, 5 and 6). Metal toxicity alters the composition and activity of different groups of soil biota both (1) directly through their toxic effects on the survival and growth of individual micro- and macroorganisms and (2) by disturbing the linkages that exist among different components of the soil biota foodweb. Microbial techniques for monitoring soil pollution have been reviewed by Pankhurst et al., (1998). Merits and limitations of various techniques involving different groups of micro-, meso-, and macro-fauna, detritus food-web and various biological processes were discussed in detail by van Straalen and Krivolutsky (1996), van Straalen and Lokke (1997), and Peijneenburg et al. (1997). Determination of metal bioavailability and toxicity to soil microorganisms can be estimated by the impact of metals on

1. the size of the soil microbial population (Brookes et al., 1986; Chandler and Brookes, 1991; Martensson and Witter, 1990): indicator

species, functional group of microorganism, or the structure of foodweb;

2. community and physiological functions: (respiration, element cycling, C and N mineralization, metabolic quotient) and

3. cellular biochemical function (Giller et al., 1997): expression of enzymes, development of *lux* based biosensors indicating cellular metabolic stress, microtox assays.

The various groups of soil organisms used as bioindicators for heavy metal toxicity in soils are discussed next.

Soil microflora. The majority of studies investigating heavy-metal toxicity and bioavailability to soil microbial populations focus on the impact of metals at either the population level (Brookes et al., 1986; Chandler and Brookes, 1991; Martensson and Witter, 1990; McGrath et al., 1995) or at the functional level (e.g. Giller et al., 1997). Will and Suter (1994) noted that although metal contaminants may change microbial community structure and species diversity, the functional ability of the community may remain unchanged. Although the population and community function techniques have been extensively used to study the impact of metals on soil microorganisms, these methods only give a crude indication of metal bioavailability to microorganisms, and are more suited to the measurement of toxic response, i.e. at what concentration of total metal in soil does a physiological response occur. Besides, it is difficult to extrapolate results between sites given that both the composition of the microbial community and soil factors controlling metal bioavailablility may vary with sites.

The abundance and species composition of microbial groups reflect the total ecotoxicological effects of heavy metals and not just the results of bioavailability at any particular time. However, a measurement of the metabolic status of microorganisms, their growth rate, and behavior (movement and formation of resting structures) at any particular time or in short-term assays will provide a more meaningful measure of bioavailability of heavy metals.

The measurement of the growth response of bacterial or fungal isolates using agar plate or liquid culture assays generally indicates the ecotoxicological effect of test compound present in the sample. Various modifications of these assays have been tested to determine the bioavailable concentrations of test compound in natural samples based on the growth response of different species of bacteria and fungi. In addition to the negative effects of heavy metals on population dynamics of bacteria and fungi, the presence of higher concentrations of heavy metals (bioavailable forms) also affects the metabolic status of microorganisms (Fliessbach et al., 1994). A number of different techniques have been tested to measure the metabolic status of microorganisms in order to determine the bioavailability of heavy metals in soil. In the substrate-induced respiration method, the ability of microorganisms to assimilate/utilize an easily available carbon substrate such as glucose is determined as a measure of bioavailability of heavy metals (Bardgett and Saggar, 1994).

Soil microfauna (protozoa). The use of protozoan growth and behavior as a measure of bioavailability of heavy metals and pesticides (insecticides, nematicides, and herbicides) in soils has potential as a novel bioassay (Pratt et al., 1988; Forge et al., 1993; Ekelund et al., 1994; Gupta and Naidu, 1997). Some of the features of protozoa that make them potentially suitable for toxicity assays include (1) the delicate external membranes of active-form protozoa being extremely sensitive to a wide range of toxic compounds, (2) relatively easy to observe and enumerate in pure culture, and (3) short generation times (hours) compared to other soil fauna, thus responding rapidly to perturbations. A short-term bioassay is essential in order to avoid interference due to the development of tolerance to the heavy metal.

Pratt et al. (1988) evaluated the potential toxicity to protozoans of soil leachates containing organics and/or heavy metals and found that soil samples from an uncontaminated agricultural field showed no toxicity in the tests whereas soils contaminated with toxic wastes (e.g. heavy metals, coal, organics, etc.) showed different degrees of toxicity. They observed a significant negative relationship between survival of protozoan species and concentration of leachate from the contaminated soils.

Laboratory bioassays using protozoans such as *Colpoda steinii, Acanthamoeba* spp., *Saccamoeba* spp., *Bodo* spp., *Oikomonas* spp. and *Cercomonas* spp. as test organisms have proved to be useful in providing rapid assessment of the bioavailability of heavy metals in soils (Forge et al., 1993; Ekelund et al., 1994; Gupta and Naidu, 1997). Heavy metals such as Ni, Zn, Pb, and Cd cause growth abnormalities to both the active and cystic forms of soil ciliate (*Colpoda* spp.) and amoebae species (*Saccamoeba* spp) and reduce their rate of growth. The magnitude of the inhibitory effect on protozoan growth is strongly related to the concentration of heavy metal in the "active" (bioavailable) form rather than the total concentration of the metal in the sample. For example, the relative growth rate of *Colpoda* spp. at different concentrations of Ni was significantly improved in the presence of phosphate buffer, which reduces the level of Ni in the «active» form, thereby reducing its bioavailability (Fig. 2.2).

Soil meso and macro fauna. The above-mentioned methods using microflora and protozoa have the potential to provide a measure of bioavailability of heavy metals in the short-term and even facilitate the measurement of temporal changes. In contrast, responses by mesofauna

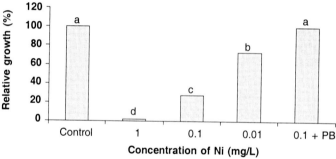

Fig. 2.2: The relative growth of the soil ciliate *Colpoda* spp. at different concentrations of Ni and the effect of phosphate buffer (PB) in laboratory bioassays. Bars with the same letters are not significantly different at P < 0.01 (Gupta and Naidu, 1997).

(microarthropods) and macrofauna (enchytridae, invertebrates, and earthworms) are cumulative effects. These fauna accumulate heavy metals in excess of their growth requirements, and a number of techniques using the data on the tissue concentrations of heavy metals have been suggested as indicators of bioavailability of heavy metals (Lee, 1985; van Wersem et al., 1994). Accumulation of heavy metals by meso- and macrofauna in their bodies can result in the disruption of body function, performance, and death. Standard procedures have been developed using earthworms as test organisms (EEC 1982, 1983) where LD_{50} values are calculated over a set period of time (2 days for contact filter paper bioassay or 2 weeks for artificial soil bioassay). Gupta et al. (2000) reported that LD_{50} values for the earthworm species *Eisenia fetida* after a 48, 72 and 144 h exposure to Cr were 79, 50, and 5.6 μg ml^{-1}, respectively, indicating that LD_{50} values are time dependent. Therefore, LD_{50} values from short-term bioassays do not accurately predict the impact of heavy metals on these larger fauna in soil, mainly because some metals are acutely toxic and cause rapid death whereas effects of others are chronic and death results after a prolonged period of time.

To overcome the problems associated with the above-mentioned assays there has been efforts to search for biomarkers of larger fauna, e.g. the metabolic performance indicators such as enzyme activities and accumulated metal concentrations as affected by the metal concentration, especially at "hidden lethal" and "sublethal" levels (van Straalen and Lokke, 1997; Rundgren and Nilsson, 1997; Gupta et al., 2000). Based on the total concentrations of heavy metals accumulated by macroorganisms, critical levels of toxic chemicals have been estimated using several statistical methods. Van Wensem et al. (1994) suggested that the methods for hazard assessment may be improved by using internal and external threshold concentrations (ITC and ETC, respectively) and concentration factors (CFs). Using data for invertebrate species from laboratory experiments and field studies, they were able to calculate critical levels for cadmium. However, these methods are time consuming and only provide an over-all effect of heavy-metal bioavailability to the species tested.

Measurement of the membrane stability of lysosomes within the coelomocytes of earthworms through the neutral-red retention (NRR) as-say has shown promise as a biomarker for the toxic effects of heavy metals such as Cu and Ni (Weeks and Svendsen, 1996; Scott-Fordsmand et al., 1998). Scott-Fordsmand et al. (2000) reported that even though NRR time reflected the bioavailable Cu fraction, the method needs to be investigated for the species differences and effects of NRR response to soil factors such as salinity.

Lux Biosensors. The measurement of metal impacts at the cellular level may achieve some degree of standardisation in assessing the bioavailability of metals in soils. One promising technology is the use of microbial biolu-minescence to determine metal bioavailability to microorganisms (Isenberg, 1993). The commercially available Microtox assay is currently considered the international standard in microbial ecotoxicity and bioavailability testing. This assay relies on the naturally luminescent ma-rine *Photobacterium phosphoreum*; the decline in light production when the organism is exposed to the contaminant indicates that the toxic metal is interfering with cellular function.

The applicability of a marine organism to study metal bioavailability in the soil environment has been questioned. An alternative is the modifi-cation of soil microorganisms such as *Pseudomonas fluorescens* through the construction of lux-marked bacteria (Amin-Hanjani et al., 1993) by inser-tion of *lux* genes from *Vibrio fischeri* (Paton et al., 1995 a,b). This approach has the advantage of using environmentally relevant organisms to assess metal bioavailability in the soil environment. This technique essentially involves the removal of solution phase from contaminated soils and incu-bation with the bioluminescent test organism. Although this measures bioavailability in the solution phase, it is not an in-situ measurement of bioavailability in the soil environment. Some efforts are being made to measure the in-situ lux response of these organisms by using soil slurry systems. A major drawback is the difficulty in measuring the bioluminies-cence response of the microorganism in the soil system due to light quench-ing (Heitzer et al., 1992). Although lux-based soil metal bioavailability techniques are a promising advancement in the measurement of metal bioavailability to the microbial cell, a major drawback is their reliance on a single test organism (Paton, 1995a). How indicative the bioavailability of a metal to a lux marked bacterium such as *Pseudomonas fluorescens* is to soil fungi, algae, other bacteria, or community response is unclear. As many soil processes are regulated by more than one organism (species, physiological type, trophic group), data from single-species bioassays may have limited potential for field use. What is required is the development of a standard suite of lux-marked organisms that can measure biochemical response to metal exposure, encompassing a range of common physiological and biochemical attributes. Reviewing microbial indicators of soil pollution Brookes (1993) concluded that although there is no single microbial property that is ideal for monitoring soil pollution, molecular techniques such as the use of modified organisms may have considerable future potential for soil contaminant studies. However, we are still a long way from defining the bioavailable pool of metals in a soil using microbial techniques.

Molecular Techniques. Although current soil biological techniques allow us to see changes in the measured parameters in response to metal bioavailability, gross measurement of process does not identify changes at a biochemical or molecular level. In addition, ecotoxicity tests based on lux, microtox, or individual test species in-vitro demonstrate metal impact on a single species. However, it is questionable how metal bioavailability to one species relates to bioavailability to a complex soil microbial community.

The development of new approaches based on molecular biology techniques to measure metal bioavailability to soil microflora offers some potential to determine the bioavailability of metals to whole soil communities (based on community responses) or to assess the bioavailability of metals based on the response of specific microbial genes, such as those involved in biogeochemical transformations (Rogers and Colloff, 1999). A significant advantage of current molecular approaches to soil microbiology is that genetic material (DNA, mRNA) is extracted directly from environmental samples, thus removing the need to isolate and culture individual organisms, and allowing studies of metal bioavailability and microbial response to be carried out in-situ (Innis et al., 1999). Briefly, these techniques involve isolation and purification of DNA from environmental samples, amplification of target gene sequences by the Polymerase Chain Reaction (PCR), and visualisation of PCR products by agarose or polyacrylamide gel electrophoresis. The volume edited by Akkermans et al. (1996) provides an excellent introduction to soil molecular ecology techniques.

Although there are examples of the application of molecular techniques to the study of metal bioavailability to soil microbial communities, such as that by Sandaa et al. (1999) where the impact of metals on community genetic diversity was determined using the sequence diversity of the 16S gene (16S rDNA), due to their recent development molecular techniques are not as widely applied currently to studies of metal bioavailability as are the more traditional approaches described in previous sections.

Factors Affecting Bioavailability

There are a number of soil and plant factors that affect the bioavailability of metals (Table 2.3). Although soil chemical factors have been studied extensively, limited information is available on the plant properties and how these influence metal bioavailability.

Soil Factors

Bioavailability of heavy metals in soil is strongly influenced by the nature and chemical and physical characteristics of soils (Table 2.3). The nature and proportion of soil constituents determine not only the extent but also the mechanism of metal adsorption and therefore the binding strength. Soils have both internal and external binding sites, and bioavailability depends on where the adsorbed metal resides. Metals adsorbed onto the interior of crystal of Mn and Fe oxides (occluded) and silicate clays lead to a fraction that is very resistant to desorption unless very strong extractants are used and therefore can be treated as essentially biologically unavailable (Bruemmer et al., 1986). Bioavailabilty is a function of desorption; in other words, the binding strength with which the metal is held on the soil-solids. Factors that influence the metal binding affinity of soils control their bioavailability.

Table 2.3: Factors affecting metal behavior and bioavailability in soils

Soil properties	Plant properties	Environmental and miscellaneous factors
pH	Nature of plant species	Climatic conditions
Surface charge density	Nature of plant cultivars	Management practices
Organic matter	Plant parts and age	Irrigation practices
Nature and clay content	Ion interactions	Irrigation water/salinity
Oxidic minerals		Topography
Redox potentials		
Soil solution composition		

Numerous studies have demonstrated that the uptake of heavy metals such as Cd is influenced by a wide range of soil variables including temperature (Siriratpuriya et al., 1985), chloride salinity (Bingham et al., 1983), pH (Andersson and Nilsson, 1974; John, 1972; Tyler and McBride, 1982; Jackson and Alloway, 1991), organic matter (McLean, 1976), calcium concentrations (Tyler and McBride, 1982), plant characteristics, and environmental factors.

pH

Plant uptake of Cd has been related to pH in numerous studies. Liming the soil has often been shown to decrease the plant uptake of Cd applied in sewage sludge (Andersson and Nilsson, 1974) or as inorganic salts

(Williams and David, 1976; Hortenstine and Webber, 1981). In an excellent review on the effect of heavy-metal pollution on plants, Page et al. (1981) concluded that liming was the most effective means to minimize absorption of Cd by plants, grown in acidic soils. However, contrasting effects of liming on Cd uptake by plants have also been reported. For instance, McLean (1976) reported no significant effect of liming on Cd uptake by lettuce when Zn and P were also added along with Cd. Similarly, Pepper et al. (1983) did not find any effect of liming on the uptake of Cd added in sludge by corn in a field experiment. Andersson and Persson (1982) and Erriksson (1989) studied the effect of pH induced by liming and fly ash on the amount of Cd taken up by perennial ryegrass (*Lolium perenne*) and winer rape (*Brassica napus*). They found that increasing the pH did not always reduce plant uptake of Cd.

Binding strength

The desorbability and, therefore, the bioavailability of metals is controlled by the strength of metal binding by soils. Although it is difficult to differentiate between nonspecifically and specifically sorbed metals, Tiller et al., (1984) separated the adsorbed metal in two components of differing affinity by a controlled desorption procedure by using 0.01 M $Ca(NO_3)_2$. They found that for Cd and Zn, the fraction with strong affinity (assumed to be specifically adsorbed) varied with the nature of mineral materials present in the soil, the pH of the background solution, and the loading of soil surface with metals. The desorption of Cd adsorbed on clay separates was strongly influenced by soil pH and while the proportion of easily desorbed metal varied from 0.3 to 0.8 at pH 5, it ranged from 0.1 to 0.5 at pH 7 on same clay separates. Using the same procedure, Kookana et al., (1997) have recently found an exponential relationship between the "easily desorbable" or nonspecifically adsorbed fraction and the adsorption coefficient (K_d) of the Cd in eight different soils (Fig. 2.3).

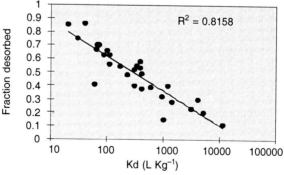

Fig. 2.3:   Desorption of cadmium is determined by its adsorption affinity to soils (Kookana et al.,1997).

This means that the extent of desorption and availability can be empirically estimated from the adsorption coefficient. Because the K_d integrates the influence of the nature of soil solid phase, the soil solution composition, and other factors, it has the advantage over any other single indicator of bioavailability.

The bioavailability of heavy metal is influenced by its time of contact with soil. Many studies show that the desorbability of metals declines with time (aging effect), and this has been attributed to diffusion into the matrix of soil solid phase (Bruemmer et al., 1988). Some researchers claim that with time, the metals may reorganize themselves on soil surfaces and find "comfortable seats" from which they are reluctant to move back into soil solution and be bioavailable. Spatial peculiarities of some minerals may provide specific "key-holes" for the metal "keys" (Bruemmer et al., 1986). Nature of the metals determines the inherent affinity they have for a soil under a given set of soil conditions. Metals can exist in various forms, and different species of metals have different affinity for soil solution and therefore the bioavailability.

Rhizosphere interaction(s)

As discussed above, plant roots modify the chemistry at the soil and root interface through the formation of organic substances and also through changes in the hydrogen ion concentration. Both of these changes can enhance the availability of nutrients and other ionic species. Formation of metalchelate complexes may lead to a reduction in the activity of metal ions in the immediate vicinity of the roots. According to Lindsay (1972), removal of chelated metal establishes a diffusion gradient to transport more chelated metal to the root. A similar gradient in the ligand ion is established at the root, and this causes diffusion of the free ligand to move away from the root where it can form more M-chelates. The role of rhizosphere interactions in micronutrient acquisition has been established in laboratory and glasshouse studies (Curl and Truelove, 1986; Marschner et al., 1986; Rovira et al., 1983). Concentrations of Cu, Zn, Mn, and Co in the soil solution increase when plants are present (Linehan, 1989). Root exudates may influence metal availability through its direct effect by acidification, chelation, precipitation, and oxidation and reduction reactions (Uren and Reisenauer, 1988). There is increasing evidence that soluble root exudates increase the solubility of metals in the rhizosphere depending on plant species and cultivars (Jolley and Brown, 1989; Kawai et al., 1988). Mench and Martin (1991) investigated the mobilization of Cd and other metals from two soils by root exudates of *Zea mays* L., *Nicotiana tabacum* L., and *Nicotiana rustica* L. They found that in the presence of root exudates, the solubility of Fe and Mn was much higher than Cd, Cu,

Ni, and Zn and the extent of Cd extraction by root exudates was similar to the order of Cd bioavailability to the three plants grown on soils. These studies indicate that enhanced uptake of Cd in certain plants may be attributed to the increased mobilization of Cd by root exudates.

Sarkar and Wyn Jones (1982) investigated the effect of rhizosphere pH, the extractable levels of Fe, Mn, and Zn in the soil, and their uptake into the roots and shoots of dwarf french beans. They varied the rhizosphere pH by applying three different sources of nitrogen-choline phosphate, ammonium phosphate, and calcium nitrate to initially homogenized soils. They found that Fe and Mn contents of both shoot and root were inversely proportional to rhizosphere pH. The Mn content showed a typical log relationship with decreasing pH. They found that shoot Fe, Zn, and Mn content were significantly correlated with the extractable levels determined in the rhizosphere and nonrhizosphere soils.

Microbial Processes Affecting Bioavailability

Microorganisms influence the bioavailability of heavy metals either by directly regulating processes that alter the chemical status of the metal or through modification of the soil environment (e.g. by affecting the rate of redox reactions), which alter the availability of heavy metals to other microorganisms or plants. Carbon dioxide produced as a result of microbial activity may influence the chemical status of heavy metals through changes in oxidation and reduction reactions in soil solution. In general, the distribution of micro and macroorganisms is highly heterogenous, both the abundance and activity of organisms is concentrated to micro sites that are rich in C and inorganic nutrients, accessible, and harbor favorable physicochemical conditions. For example, biological activity near plant roots (rhizosphere) and decomposing plant residues (detritusphere) is generally more than double compared to that in bulk soil. Hypothetically, such large differences in the level of microbial activity potentially lead to hotspots of activity with increased/reduced bioavailability of heavy metals; however, this has not been demonstrated experimentally. Management practices that affect the distribution or intensity of these hotspots can also affect the overall bioavailability of heavy metals. However, little information is available on this aspect.

Many bacteria produce extracellular polysaccharides as secondary carbon metabolites for protection against desiccation and to assist in the transport of compounds across the lipopolysaccharide cell membrane. Microbial exocellular polysaccharides have been shown to affect the bioavailability of heavy metals by influencing their precipitation, solubilization, complexation and adsorption reactions (Smith, 1994). Other microbial processes affect metal bioavailability by changing pH of microenvironment or through adsorption or uptake, temporarily removing metals from the bioavailable pool. Microbial cells can concentrate (bioaccumulate) metals through a number of mechanisms including ion exchange at the cell wall, complexation reactions at the cell wall, intra- and extra-cellular complexation, resulting in the removal of metals from solution. These mechanisms of complexing Cd have evolved to allow the cell to survive high extra- and intra-cellular Cd concentrations. Extracellular accumulation of metals on microbial cells generally involves either adsorption of metals in the polysaccharide coating or adsorption to binding sites such as carboxyl residues, phosphate residues, S-H groups, or hydroxyl groups on the cell surface.

Another avenue through which micro-organisms affect the bioavailability of heavy metals to plants is by increasing the rate of up-take of heavy metals and the volume of soil exploration. For example, mycorrhizal fungi are known to increase the uptake of essential nutrients (phosphorus) and other metals (e.g. Zn) by plants. In this case, mycorrhizal fungi may not alter the chemical status of heavy metal per se, but by improving the ability of plants to access a greater portion of the soil volume, they improve the level of accumulation.

A number of microbial genera are able to oxidize or reduce metals. Table 2.4 details some of the more common microbially-mediated metal redox reactions. The main consequence of microbial redox reactions is to change the mobility or toxicity and potential bioavailability of metals. An example of this is the reduction of Cr(VI) to Cr(III) by a chromate-resistant *Pseudomonas fluorescens* strain (Hughes and Poole, 1989). Chromium (VI) is an anion and is therefore less tightly sorbed to surfaces than the positively charged Cr (III) cations. Chromium (VI) is, therefore, the most toxic of the two species due to its higher bioavailability in soils (McGrath, 1995). Microbial reduction of Cr(VI) to Cr(III) will therefore reduce Cr bioavailability in soils.

Table 2.4: Common microbially-mediated metal redox reactions

Metal	Transformation	Reaction
Antimony	$Sb^{3+} \rightarrow Sb^{5+} + 2e^-$	Oxidation
Arsenate	$As^{5+} + 2e^- \rightarrow As^{3+}$	Reduction
Arsenite	$As^{3+} \rightarrow As^{5+} + 2e^-$	Oxidation
Cadmium	$Cd \rightarrow Cd^{2+} + 2e^-$	Oxidation
Manganese	$Mn^{2+} \rightarrow Mn^{7+} + 2e^-$	Oxidation
Mercury	$Hg^0 \rightarrow Hg^{2+} + 2e^-$	Oxidation
Mercury (I)	$Hg^{1+} + 1e^- \rightarrow Hg^0$	Reduction
Mercury (II)	$Hg^{2+} + 1e^- \rightarrow Hg^+$	Reduction

Microbial methylation of As, Cd, Hg, Pb, and Se by species of common soil bacteria such as *Bacillus* and *Pseudomonas* has been recorded by a number of authors and is extensively reviewed by Hughes and Poole

(1989). Although methylation will in part result in some volatilization loss of these metals from soils, methylated forms remaining in the soil matrix are more mobile and known to be more toxic than non-methylated forms (Smith, 1994).

Although there is significant evidence of microbial transformations of metals, the effect of these processes on metal bioavailability, especially the temporal dynamics of bioavailable pool in soils, is unclear. The majority of studies investigating metal transformations have been performed in vitro with pure microbial cultures. Although many of these organisms have been originally isolated from soils, it is almost impossible to extrapolate their biochemical activity in pure cultures to that in the soil environment.

Plant physiological processes affecting metal bioavailability

The impact of plants on the bioavailability of heavy metals in soils is essentially limited to the soil zone surrounding the plant roots and the rhizosphere. Campbell et al. (1990) define the rhizosphere in its simplest form as the volume of soil influenced by the root. The volume of soil influenced by the plant root varies with soil type and species, but generally extends a few millimetres to a few centimetres from the root surface (rhizoplane) (Campbell et al., 1990).

Plants influence heavy-metal bioavailability within the rhizosphere mainly through exudation of compounds from roots. Plant roots exude significant amounts of organic carbon into the rhizosphere system, including sloughed cells, polymeric carbohydrates, and water-soluble exudates such as amino acids and sugars (Whipps, 1990). The excretion of H^+ or HCO_3^- by plant roots can change rhizosphere pH by up to one unit compared to the bulk soil (Brazin et al., 1990). Although changes in rhizosphere pH may influence contaminant sorption, solution phase concentration, and plant uptake, the consequence of local pH changes around the root on contaminant availability in the rhizosphere remains speculative. Jarvis et al. (1976) demonstrated increased Cd uptake into plants when solution pH increased, but those of Chaney and Hornik (1978) demonstrated the opposite. Change in pH and its influence on metal bioavailability remains unclear. It is also important to note that plant species, the form of N supply (NO_3^- or NH_4^+), and soil type will influence the degree of pH change in the rhizosphere. Exudation of organic ligands from plant roots may impact on the activity of free ions in the solution phase. The formation of metal and ligand complexes may complex metals in solution in a non-sorbing form. Newman (1985) estimated that loss of cells, exudates, and polysaccharide gel from roots of maize can account for 60 mg of material lost g^{-1} root d^{-1}. Such carbon inputs may impact on metal bioavailability through blocking sorption sites on mineral colloids or increasing the number of surfaces available for sorption. The stimulation of microbial activity in the rhizosphere due to the carbon inputs from plant has been shown to impact on metal bioavailability.

BIOAVAILABILITY: SOIL INGESTION

Ingestion of soils, whether inadvertently or otherwise, is an important route of exposure for contaminants that are not geochemically or biologically mobile. Although soil ingestion has long been linked to medical problems, including parasitic or trace metal nutrition (Wong et al., 1991; Sheppard et al., 1995), its implication as an important pathway for contaminants was realized only recently. Since then, substantial progress has been made to quantify the soil ingestion pathway, including the nature of contaminant release mechanisms by numerous investigators (Binder et al., 1986; Calabrese et al., 1989, 1990, 1991; Davies et al., 1990; van Wijnen et al., 1990).

IMPLICATIONS OF BIOAVAILABILITY TO TOXICOLOGICAL STUDIES AND CURRENT REGULATORY GUIDELINES

Considerable controversy exists amongst toxicologists and regulators on the basis for the existing guidelines for environmental and health risk assessment. The relationship between the total metal content in the soil and its positive or negative effect on biota is not straightforward. It is now well recognized that the speciation of metal ions plays important role with respect to bioavailability (see Chapter 3). Primarily because of this, many toxicologists and environmental scientists question the significance of total metal concentrations as indicators of the environment and health risks. Despite these claims, the Dutch, the US Environmental Protection Agency (USEPA) and the existing Australian National Health and Medical Research Council (NHMRC) Guidelines are all based on total metal concentrations. There is now extensive debate between toxicologists and regulators on the merits of basing guidelines on total metal concentrations. The schematic diagram in Fig. 2.4 illustrates the distribution relationship between metal concentrations and a range of soil samples from sites A to Z plotted in order of increasing metal concentration. As shown in the figure, soil metal concentrations in all samples exceeding total metal concentrations (Metal P) violate the USEPA and NHMRC maximum permissible concentrations. If the permissible concentrations are based on bioavailability, then the maximum permissible concentration can be either equal to or less than the total concentration as illustrated by line B. However, bioavailability may vary depending on the test species, method used for assessment, and environmental conditions.

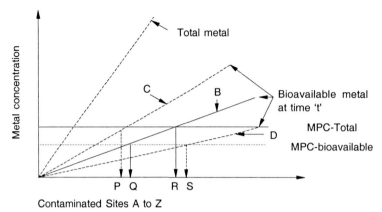

Fig. 2.4: Schematic diagram illustrating relationship between total and bioavailable metal fraction in contaminated sites A to Z in increasing order as related to total and bioavailable MPC

Therefore, metals violating MPC will vary (lines C and D) depending on how bioavailability was assessed (as shown in Fig. 2.4). Thus, the slope of line B may either increase or decrease, leading to widely different predictions on soil samples exceeding the permissible concentration. This uncertainty in bioavailability increases the risk of exposure assessment. This suggests that unless better methods for bioavailability assessments are devised, total metal concentrations may be considered more conservative and perhaps a sound basis for soil risk assessment.

CONCLUSIONS

It is clear from the discussion here that bioavailability means different things to different people. While bioavailability has been linked to various pools, i.e. to metal ion activity in soil solution, the exchangeable metal fraction and more recently to the concentration of metals that cause ecotoxicity, none is universally accepted. A major limitation of the current definition of bioavailability is the assumption that it is a static process. Far from it, bioavailability is a dynamic phenomenon that varies seasonally and with changes in soil chemical and biological environment. Clearly, there is no consensus yet on a single definition among the scientific community, and there is an urgent need to develop an acceptable definition of the concept.

   Numerous chemical and biological techniques have been used to assess the bioavailable fraction of metals in soils, but these have serious limitations in providing a good estimate of the pool of toxic metal ions that are accessible to plants and microbes. This is because of the wide range of forms in which metals are present, uncertainty on the precise nature of the ionic species that is taken up by plants, and the lack of specificity of the chemical and biological techniques. A number of different extractants have been used to investigate the bioavailable metal fractions in soils, but they often do not provide a true reflection of the chemical status of the metal ion in soil solution. The complexities of root and soil interface chemistry and environmental effects, among other factors, makes it difficult for a chemical test to predict plant-available metal ion. More work needs to be done to elucidate the mechanisms that control heavy-metal chemistry under reducing conditions and to assess the seasonal fluctuations in redox potential and its impact on plant Cd availability. Often, the failure of chemical extraction techniques is related to their improper application rather than the efficacy of the test per se.

The bioavailability of metals is best determined by in situ techniques involving growing of the organisms of interest in the contaminated material and quantifying the uptake of metal into the organism or assessing the toxicological response. Measurements of metal bioavailability and toxicity in soils using soil microorganisms are receiving increasing attention, as microorganisms are more sensitive to heavy-metal stress than plants or soil macrofauna. The methods using microflora and protozoa have the potential to provide a measure of bioavailability of heavy metals in the short-term and even facilitate the measurement of temporal changes. In contrast responses by mesofauna (microarthropods) and macrofauna (enchytridae, invertebrates and earthworms) are cumulative effects. These methods, however, are time consuming and can only provide an overall effect of heavy-metal bioavailability to the species tested. Bioavailability, as determined by plant studies should be more correctly termed "phytoavailability" as these pools are not necessarily available to the same degree to other soil organisms.

In conclusion, there is an urgent need to provide a better definition of bioavailability, especially keeping these questions in mind, bioavailability to what, for what, and when. Development of sound tests for bioavailability poses a real challenge to the scientific community. Once again, the above questions are pertinent in this regard. There is now extensive debate between toxicologists and regulators on the merits of basing guidelines on total metal concentrations. Obviously, from the standpoint of risk assessment at contaminated sites, proper assessment of bioavailability is critical to estimate potential exposure. As bioavailability may vary depending on the test species, method used for assessment, and environmental conditions, the uncertainty in bioavailability can lead to erroneous estimate of the risk of exposure. Unless better methods for bioavailability assessments are devised, total metal concentrations remain to be a more conservative and perhaps a sound basis for soil risk assessment.

REFERENCES

Adams, J.F., Kissel, D.E. (1989) Zinc, copper and nickel availabilities as determined by soil solution and DPTA extraction of a sludge-amended soil. *Comm in Soil Sci and Pl. Anal* **20**,(12): 139–258.

Akkermans, A.D., van Elsas, J.D., and de Bruijn, F.J. (1996) *Mol Micro Eco Manual*. Kluwer Academic, Dordrecht.

Amin-Hanjani, S., Meikle, A., Glover, L.A., Prosser, J.I., and Killham, K. (1993) Plasmid and chromosomally encoded luminescence marker systems for detection of Pseudomonas fluorescens in soil. *Mol Eco* **2**: 47–54.

Andersson, A, and Nilsson, K.O. (1974) Influence of lime and soil pH on Cd availability to plants. *Ambio* **3**: 198–200.

Andersson, A, and Persson, J. (1982) Uptake in agricultural crops of trace elements from flyash: In Coal Health-Environment Project (Swedish with English Summary) Technical Paper 25, The Swedish Power Board, Vallingby, Sweden, 36pp.

Barbarick, K.A., and Workman, S.M. (1987)Ammonium bicarbonate-DTPA and DTPA extractions of sludge-amended soils. *J of Envir Qual* **16**: 125–130.

Bardgett, R.D., and Saggar, S. (1994) Effects of heavy metal contamination on the short-term decomposition of labeled [^{14}C] glucose in a pasture soil. *Soil Biol and Biochem* **26**: 727–733.

Brazin, M.J., Markham, P., Scott, E.M., Lynch, J.M. (1990) Population dynamics and rhizosphere interactions. In *The rhizosphere*, Lynch, J.M. (Ed), John Wiley and Sons Ltd., Chichester, West Sussex, UK, pp 99–127.

Beckett, H.T. (1989) The use of extractants in studies on trace metals in soils, sewage sludges, and sludge-treated soils. *Adv in Soil Sci* **9**: 143–176.

Bidwell, A.M., and Dowdy, R.H. (1987). Cadmium and zinc availability to corn following termination of sewage sludge applications. *J of Envir Qual* **16**: 438–442.

Binder, S., Sokal, D., and Maughan, D. (1986) Estimating soil ingestion: the use of tracer elements in estimating the amount of soil ingested by young children. *Arch of Environ Hlth* **41**: 341–345.

Bingham, F.T., Strong, J.E., and Sposito, G. (1983) Influence of chloride salinity on cadmium uptake by swiss chard. *Soil Sci.*, **135**: 160–165.

Bingham, F.T., Page, A.L., Mahler, R.J., and Ganje, T.J. (1976) Cadmium availability to rice in sludge-amended soil under "flood" and "nonflood" culture. *Soil Sci. Soc. Amer. Proc.* **40**: 715–719.

Bramley, R.G.V. (1990) Cadmium in New Zealand agriculture. *N. Z. J. Agric. Res.* **33**: 505–519.

Brookes, P.C. (1993) The potential of microbiological properties as indicators in soil pollution monitoring. *Soil Monit*, 227–253.

Brookes, P.C., McGrath, S.P., and Heijne, C. (1986) Accelerated paper metal residues in soils previously treated with sewage-sludge and their effects on growth and nitrogen fixation by blue-green algae. *Soil Biol and Biochem* **18**(4) : 345–353.

Brookes, P.C. (1993) The potential of microbiological properties as indicators in soil pollution monitoring. In *Soil monitoring: early detection and surveying of soil contamination and degradation*. Schulin, R., Desaules, A., Webster, R., and Steiger, B., (Eds), Birkhauser Verlag AG, Basel, Switzerland, pp 229–254.

Brown, S.L., Chaney, R.L., Lloyd, C.A., Angle, J.S., and Ryan, J.A. (1996) Relative uptake of cadmium by garden vegetables and fruits grown on long-term biosolid-amended soils. *Environ Sci Technol* **30**: 3508–3511.

Brown, R.M., Pickford, C.J., and Davison, W.L. (1984) Speciation of metals in soils. *Inter J of Environ Analy Chem* **18**: 135–141.

Brown, AL., Quick, J., and Eddings, J.L. (1971) A comparison of analytical methods for soil zinc. *Soil Sci. Soc. Am. Proc.* **35**: 105–107.

Bruemmer, G.W., Gerth, J. and Tiller, K.G. (1988) Reaction kinetics of the adsorption and desorption of nickel, zinc and cadmium by goethite. I. Adsorption and diffusion of metals. *J. Soil Sci* **39**: 37–52.

Bruemmer, G.W., Gerth, J. and Herms, U. (1986) Heavy metal species, mobility and availability in soils. *Z. Pflanzenernaehr.Bodenk.* **149**: 382–398.

Calabrese, E.J., Barnes, E., Stanek, E.J., Pastides, H., Gilbert, C.E., Veneman, P., Wang, X., Lasztity, A., and Kostecki, P.T. (1989) How much soil do young children ingest – an epidemiologic study. *Regulat Toxico and Pharmacol* **10**: 123–137.

Calabrese, E.J., Stanek, E.J., Gilbert, C.E. (1991) Evidence of soil-pica behaviour and quantification of soil ingested. *Human Experiml Toxico* **10**: 245–249.

Calabrese, E.J., Stanek, E.J., Gilbert, C.E., and Barnes, R.M. (1990) Preliminary adult soil ingestion estimates – results of a pilot-study. *Regulat Toxicol and Pharmacol* **12**: 88–95.

Campbell, R., Greaves, M.P., and Lynch, J.M. (1990) Anatomy and community structure of rhizosphere. In *The Rhizosphere*, Lynch, J.M. (Ed), John Wiley and Sons Ltd., Chichester, West Sussex, UK, pp 11–34.

Cavallaro, N., and McBride, M.B. (1978) Copper and cadmium adsorption characteristics of selected acid and calcareous soils. *Soil Sci. Soc. Am. J.* **42**: 550–556.

Chandler, K., Brookes, P.C., and Harding, S.A. (1995) Microbial biomass dynamics following addition of metal-enriched sewage sludges to a sandy loam. *Soil Biol and Biochem* **27**: 1409–1421.

Chaney, R.L., and Hornick, S.B. (1978) Accumulation and effects of cadmium on crops. In *Proceedings of the First International Cadmium Conference*, San Francisco, Metals Bulletin Ltd., London, pp 124–140.

Chang, A.C., Warneke, J.E., Page, A.L., and Lund, L.J. (1984) Accumulation of heavy metals in sludge treated soils. *J Environ. Qual.* **13**: 87–91.

Clark, C.J., and McBride, M.B. (1987) Chemisorption of Cu(II) and Co(II) on allophane and imogolite. *Clays and Clay Miner* **32**: 300–310.

Curl, E.A., and Truelove, B. (1986) *The Rhizosphere.* Springer Verlag, Berlin.

Davies, D.J.A., Thornton, I., Watt, J.M. (1990) Lead intake and blood lead in two year old UK urban children. *Sci. Total Environ.* **90**: 13–29.

Dowdy, R.H., and Volk, V.V. (1984) Movement of heavy metals in soils. In *Chemical Mobility and Reactivity in Soil Systems*. SSSA, Madison, USA, pp. 229–240.

Dowdy, R.H., Larson, W.E., Titrud, J.M., and Latterell, J.J. (1978) Growth and metal uptake of snap beans grown on sewage sludge-amended soil: A four-year field study. *J. Environ. Qual.* **7**: 252–257.

EEC 1982. Directive 79/831, Annex V, Part C: Methods for the determination of ecotoxicity — level 1, Commission of European Communities, DG. X1/ 127–129/82.

EEC 1983. Directive 79/831, Annex V, Part C: Methods for the determination of ecotoxicity— level 1, Earthworms—artificial soil. Commission of European Communities, DG. X1/128/82. Rev. 5.

Ekelund, F., Ronn, R., and Christensen, S. (1994) The effect of three different pesticides on soil protozoan activity. *Pest Sci* **42**: 71–78.

Eriksson, J.E. (1989) The influence of pH, soil type and time on adsorption and uptake by plants of Cd added to the soil. *Water Air Soil Pollut* **48**: 317–335.

Fliessbach, A., Martens, R., and Reber, H.H. (1994) Soil microbial biomass and microbial activity in soils treated with heavy metal contaminated sewage sludge. *Soil Bio and Biochem* **26**: 1201–1205.

Forge, T.A., Berrow, M.L., Darbyshire, J.F., and Warren, A. (1993) Protozoan bioassays of soil amended with sewage sludge and heavy metals, using the common soil ciliate *Colpoda steinii*. *Biol and Fert of Soils* 16: 282–286.

Fotovat, A., and Naidu, R. (1997) Determination of Free hydrated Zn^{2+} and Cu^{2+} in soil-water system at low concentration by ion exchange resin and MINTEQA2. Presented at the fourth International Conference on the Biogeochemistry of Trace Elements. June 23–26, 1997. pp. 385–386.

Francis, A.J., Dodge, C.J., and Gillow, J.B. (1992) Biodegradation of metal citrate complexes and implications for toxic-metal mobility. *Nature* 356: 140–142.

Giller, K.E., Beare, M.H., Lavelle, P., Izac, A.M.N., and Swift, M.J. (1997) Agricultural intensification, soil biodiversity and ecosystem function. *App Soil Ecol* 6: 3–16.

Gough, LP., McNeal, J.M., and Severson, R.C. (1980) Predicting native plant copper, iron, manganese and zinc levels using DTPA and EDTA soil extractants, Northern Great Plains. *Soil Sci. Soc. Amer. J.* 44: 1030–1036.

Gupta, V.V.S.R., and Naidu, R. (1997) A protozoan bioassay to determine the bioavailability of heavy metals in soil. *Abstracts of the 10th International Congress of Protozoology*, Sydney, p. 94. Business Meetings and Incentives, Sydney, Australia.

Gupta, V.V.S.R., Dalby, P.R., Naidu, R., and Smith, L.H. (2000) Transformations of chromium by soil microorganisms and toxicity of chromium to earthworms. In *Towards better management of soils contaminated with tannery waste*. Naidu, R., Willett, I.R., Mahimairajah, S., Kookana, R. and Ramasamy, K. (Eds) Proceedings of a Workshop held in Coimbatore, India, 31 January to 4 February. 1998. ACIAR Proceedings No. 88, pp 112–124.

Hale, M.G., and Moore, L.D. (1979) Factors affecting root exudation II: 1970-1978. *Adv. Agron.* 31:93–124.

Haq, A.U., Bates, T.E., and Soon, Y.K. (1980) Comparison of extractants for plant-available zinc, cadmium, nickel, copper in contaminated soils. *Soil Sci. Soc. Am. J.* 44: 772–777.

Haq, A.U., and Miller, M.H. (1972) Predication of available Zn, Cu and Mn using chemical extractants. *Agron J.* 64: 779–782.

Harter, R.D., and Naidu, R. (1995) Role of metal-organic complexation on metal sorption by soils. *Adv in Agron* 55: 219–264.

Haynes, R.J., and Naidu, R. (1989) Soil fertility and management considerations for efficient crop production in the South Pacific. In Haynes, R.J. and Naidu, R. (Eds.). *Agricultural Development in the Pacific Island in the 90's*: Proceedings of an International Conference, and Workshop held in Suva, Fiji on 31 March to 5 April 1990. ISBN 0–477–03152–8. pp 21–47.

Heckman, J.R., Angle, J.S., and Chaney, R.L. (1987) Residual effects of sewage sludge on soybean: I. Accumulation of heavy metals. *J of Envir Qual* 16: 113–117.

Heitzer, A., Webb, O.F., Thonnard, J.E., and Sayler, G.S. (1992) Specific and quantitative assessment of naphthalene and salicylate bioavailability by using a bioluminescent catabolic reporter bacterium. *Appl and Environ Microb* 58(6): 1839–1846.

Helmke, P.A., and Naidu, R. (1996) Fate of contaminants in the soil environment: metal contaminants. In *Contaminants and the Soil Environment in the Australasia-Pacific Region*: Naidu, R., Kookuna, R.S., Oliver, D. P., Rogers, S., McLaughlin, M. J. (Eds) Kluwer Academic Publishers, Dordrecht, pp. 69–94.

Hodgson, J.F. (1963) Chemistry of micronutrient elements in soils. *Adv. Agron.* 15: 119–159.

Hortenstine, C.C., and Webber, L.R. (1981) Growth and cadmium uptake by spinach and corn in an acid typic Hapludalf as affected by cadmium, peat and lime additions. *Soil and Crop Sci Soc of Florida Proceed* 40: 40–44.

Hughes, M.N., and Poole, R. K. (1989) Metals and micro-organisms, London; New York: Chapman and Hall. p 400.

Hutton, M., and Symon, C.J. (1986) The quantities of cadmium, lead, mercury and arsenic entering the UK environment from human activities. *Sci. Tot. Environ.* **57**: 129–150.

Innis, M.A., Gelfand, D.H., and Sninsky, J.J. (1999) *PCR Applications, Protocols for Functional Genomics.* Academic Press, San Diego, pp 551.

Isenberg, D.L. (1993) The microtox toxicity test: a developers commentary. In *Ecotoxicol Monitor*, Richardson, M. (Ed), VCH, Weinham, pp 3–15.

Iyengar, S.S., Martens, D.C., Miller, W.P. (1981) Distribution and plant availability of soil zinc fractions. *Soil Sci. Soc. Am. J.* **45**: 735–739.

Jackson, A.P., and Alloway, B.J. (1991) The bioavailability of cadmium to lettuce and cabbage in soils previously treated with sewage sludges. *Pl and Soil* **132**: 179–186.

Jarvis, S.C., Jones, L.H.P., and Hopper, M.J. (1976) Cadmium uptake from solution by plants and its transport from roots to shoots. *Pl and Soil* **44**: 179–191.

John, M.K. (1972) Effect of lime on soil extraction and on the availability of soil applied cadmium to radish and leaf lettuce plants. *Sci. Total Environ.* **1**: 303–308.

Jolley, V.D., and Brown, J.C. (1989) Iron efficient and inefficient oats. I. Differences in phytosiderophore release. *J. Pl Nutr.* **12**: 423–435.

Kawai, S., Takagi, S., and Sato, Y. (1988) Mugineic acid family phytosiderophore in root secretions of barley, corn and sorghum varieties. *J. Pl Nutr.* **11**: 633–642.

Khasawneh, F.E. (1971) Solution ion activity and plant growth. *Soil Sci. Soc. Amer. Proc.* **35**: 426–436.

Kookana, R.S., and Naidu, R. (1997) Effect of soil solution composition on cadmium transport through variable charge soils. *Geod* **84**: 235–248.

Kookana, R.S., Naidu, R., and Tiller, K.G. (1997) Desorption of cadmium is determined by its adsorption affinity to soils. In *Proc. 4th International Conference on the Biogeochemistry of trace elements.* 23–26 June 1997, Berkeley, California.

Lagerwerff, J.V. (1971) Uptake of cadmium, lead and zinc by radish from soil and air. *Soil Sci.* **111**: 129–133.

Latterell, J.J., Dowdy, R.H., and Larson, W.E. (1978) Correlation of extractable metals and metal uptake of snap beans grown on soil amended with sewage sludge. *J of Envir Qual* **7**: 435–440.

Lee, K.E. (1985) Earthworms: Their ecology and relationships with soils and land use. Academic Press, Sydney, p. 411.

Lindsay, W.L. (1972) Zinc in soils and plant nutrition. *Adv in Agron* **24**:147–186.

Lindsay, W.L., and Norvell, W.A. (1978) Development of a DTPA soil test for zinc, iron, manganese and copper. *Soil Sci. Soc. Am. J.* **42**: 421–428.

Linehan, D.J., Sinclair, A.H., and Mitchell, M.C. (1989) Seasonal changes in Cu, Mn, Zn and Co concentration in soil in the root zone of barley (Hirdeum vulgare L). *J Soil Sci.* **40**: 103–115.

Marchner, H., Ramheld, V., Horst, W.J., and Martin, P. (1986) Root induced changes in the rhizosphere: Importance for the mineral nutrition of plants. *Z Pflazenernahr. Bodenk.* **149**: 441–456.

Martensson, A.M., and Witter, E. (1990) Influence of various soil amendments on nitrogen-fixing soil-microorganisms in a long-term field experiment, with special reference to sewage-sludge. *Soil Biol. and Biochem.* **22**: 977–982.

Martin, M.H., Duncan, E.M., and Coughtrey, P.J. (1982) The distribution of heavy metals in a contaminated woodland ecosystem. *Environ Pollu* B **3, (2)**: 147–157.

McBride, M.B. (1989) Reactions controlling heavy metal solubility in soils. *Adv. in Soil Sci.* **10**: 1–56.

McClean, A.J. (1976) Cadmium in different plant species and its availabilities in soils as influenced by organic matter and addition of lime, P, Cd and Zn. *Can J. Soil Sci.* **56**: 129–138.

McGrath, S.P. (1994) Effects of heavy metals from sewage sludge on soil microbes in agricultural ecosystems. In *Toxic Metals in Soil-Plant Systems*. Ross, S.M. (Ed.), John Wiley & Sons Ltd., Chichester pp 248–274.

McGrath, S.P., Chaudri, A.M., and Tiller, K.E. (1995) Long-term effects of metals in sewage sludge on soils, microorganisms and plants. *J. of Indusl. Microb.* **14**: 94–104.

McBride, M.B. (1995) Toxic metal accumulation form agricultural use of sludge: are U.S.E.P.A. regulations protective? *J of Envir Qual* **24**: 5–18.

McBride, M.B. (1989) Reactions controlling heavy metal solubility in soils. *Adv in Soil Sci* **10**: 1–56.

McLaren, R.G., Lawson, D.M., and Swift, R.S. (1986) The forms of cobalt in some Scottish soils as determined by extraction and isotopic exchange. *The J of Soil Sci* **37**: 223–234.

McLaren, R.G., and Smith, C. J. (1996) Issues in the disposal of industrial and urban wastes. In *Contaminants and the Soil Environment in the Australasia-Pacific Region:* Naidu, R., Kookuna, R.S., Oliver, D. P., Rogers, S., McLaughlin, M. J. (Eds), Kluwer Academic Publishers, Dordrecht. pp 183–212.

Mench, M., and Martin, E. (1991) Mobilization of cadmium and other metals from two soils by root exudates of Zea mays L., Nicotiana L., and Nicotiana rustica L. *Pl and Soil* **132**: 187–196.

Minnich, M. M., McBride, M.B., and Chaney, R.L. (1987) Copper activity in soil solution: II. Relation to copper accumulation in young snapbeans. *Soil Sci. Soc. Amer. J.* **51**: 573–578.

Murthy, A.S.P. (1982) Zinc fractions in wetland rice soils and their availability to rice. *Soil Sci* **133**(3): 150–154.

Naidu, R., and Harter, R.D. (1998) Effectiveness of different organic ligands on sorption and extractability of cadmium by soils. *Soil Sci Soc Amer J.* **62**: 644–650.

Naidu, R., Kookana, R.S., Sumner, M.E., Harter, R.D and the late Tiller, K.G. (1996) Cadmium sorption and transport in Variable Charge Soil: A review. *J of Envir Qual* **26**: 602–617

Naidu, R., Sumner, M.E. and Harter, R.D. (1998) Sorption of heavy metals in strongly weathered soils: An Overview. *J Environ Geochem and Hlth* **20**: 5–9.

Neilson, D., Hoyt, P.B., and MacKenzie, A.F. (1987) Measurement of plant-available zinc in British Columbia orchard soils. *Commun in Soil Sci. and Pl Anal.* **18** (2): 161–186.

Newman, E.I. (1985) The rhizosphere: carbon sources and microbial populations. In *Ecological interaction in soil: plants, microbes and animals*. Fitter, A.H., Atkinson, D., Read, D.J., and Usher, M.B. (Eds), Oxford, UK Blackwell Scientific Publications.

Novozamsky, I., Lexmond, T.M., Houba, V.J.G. (1993) A single extraction procedure of soil for evaluation of uptake of some heavy metals by plants. *Int. J Environ. Anal. Chem.* **55**: 47–58.

O'Connor, G.A. (1988) Use and misuse of the DTPA soil test. *J of Envir Qual* **17**: 715–718.

Oliver, D., Gartrel, J.W., Tiller, K.G., Correll, R., Cozens, G.D., and Youngberg, B.L. (1995) Differential responses of Australian wheat cultivars to cadmium concentration in wheat grain. *Aust. J. Agric. Res.* **46**: 873–886.

Otte, M.L., Rozema, J., Koster, L., Haarsma, M.S., and Broekman, R.A. (1989) Iron plaque on roots of Aster tripolium L.: interaction with zinc uptake. *New Phyto* **111**: 309–317.

Page, AL., Bingham, F.T., and Chang, A.C. (1981) In *Effect of Heavy Metal Pollution on Plants*, Vol. 1, Lepp, N.W. (Ed.). Applied Science Publishers, London and New Jersey, p. 77.

Paton, G.I., Palmer, G., Kindness, A., Campbell, C., Glover, L.A., and Killham, K. (1995a) Use of luminescence-marked bacteria to assess copper bioavailability in malt whisky distillery effluent. *Chemosphere* **31**: 3271–3224.

Paton, G.I., Campbell, C.D., Glover, L.A., and Killham, K. (1995b) Assessment of bioavailability of heavy metals using lux modified constructs of Pseudomonas fluorescens. *Lett in Appl Micro* **20**: 52–56.

Pankhurst, C.E., Rogers, S., and Gupta, V.V.S.R. (1998) Microbial parameters for monitoring soil pollution. In *The Biotechnology-Ecotoxicology Interface*. Lynch, J., and Wiseman, A. (Eds), Cambridge University Press, Cambridge, UK, pp 46–69.

Peijnenburg, W.J.G.M., Posthuma, L., Eijsackers, H.J.P., and Allen, H.E. (1997) A conceptual framework for implementation of bioavailability of metals for environmental management processes. *Ecotoxicol and Environ Safety* **37**: 163–172.

Pepper, I.L., Bezdicek, D.F., Baker, A.S., and Sims, J.M. (1983) Silage corn uptake of sludge-applied zinc and cadmium as affected by soil pH. *J Environ. Qual.* **12**: 270–275.

Puls, R.W., and Powell, R.M. (1992) Transport of inorganic colloids through natural aquifer material: Implications for contaminant transport. *Env. Sci. Tech.* **26**: 614–621.

Pratt, J.R., Mochan, D., and Xu, Z. (1997) Rapid toxicity estimation using soil ciliates: sensitivity and bioavailability. *Bull of Environ Contam and Toxicol* **58**: 387–393.

Pratt, J.R., McCormick, P.V., Pontasch, K.W., and Cairns, J. (1998) Evaluating soluble toxicants in contaminated soils. *Water, Air, and Soil Poll* **37**: 293–307.

Rapparport, B.D., Marten, D.C., Reneau, R.B. Jr., and Simpson, T.W. (1988) Metal availability in sludge-amended soils with elevated metal levels. *J. Environ. Qual.* **17**: 42–47.

Rogers, S.L., and Colloff, M. (1999) How functionally resilient are Australian production systems? Future concepts for the study of functional biology and functional resilience of soil systems. In *Fixing the Foundations: National Symposium on the Role of Soil Science in Sustainable Land and Water Management*. 11–12 November 1999, South Australian Research and Development Institute, Adelaide.

Ross, A.D., Lawrie, R.A., Whatmuff, M.S., Keneally, J.P., and Awad, A.S. (1991) Guidelines for the use of sewage sludge on Agricultural Land. NSW Agriculture, Sydney, Australia. p. 17.

Rovira, A.D., Bowen, G.D., and Forster, R.C. (1983) The significance of rhizosphere microflora and mycorrhizas in plant nutrition. In *Inorganic plant nutrition*. Lauchli, A., and Bieleski, R.L. (Eds) Encyclopedia of Plant Physiology, New Series, Vol. 15A, Springer Verlag, Berlin, pp 61–93.

Rundgren, S., and Nilsson, P. (1997) Sublethal effects of aluminum on earthworms in acid soil: the usefulness of *Dendrodrilus rubidus* (Sav.) in a laboratory test system. *Pedobiologia* **41**: 417–436.

Ryan, J.A. and Chaney, R.L. (1994) Development of limits for land application of municipal sewage sludge: risk assessment. In *15th World Congress of Soil Science*, Acapulco, Mexico, 10-16 July 1994. Transactions, Volume 3a: commission II symposia, pp 534–553.

Sandaa, R.A., Torsvik, V., Enger, O., Daae, F.L., Castberg, T., and Hahn, D. (1999). Analysis of bacterial communities in heavy metal-contaminated soils at different levels of resolution. *FEMS Microbio Ecol.* **30**: 237–251.

Sarkar, A.N., and Wyn Jones, R.G. (1982) Effect of rhizosphere pH on the availability and uptake of Fe, Mn and Zn. *Pl and Soil* **66**: 361–372.

Scott-Fordsmand, J.J., Weeks, J.M., and Hopkin, S.P. (1998) Toxicity of nickel to the earthworm *Eisenia veneta* (Oligochaeta: annelida) and the applicability of the neutral-red retention assay to indicate nickel toxicity. *Ecotoxicol* **7**: 291–295.

Scott-Fordsmand, J.J., Weeks, J.M., and Hopkin, S.P. (2000) Importance of contamination history for understanding toxicity of copper to earthworm *Eisenia fetica* (Oligochaeta: Annelida), using neutral-red retention assay. *Environ. Toxicol. Chem.* **19**: 1774–1780.

Sheppard, S.C. (1995) A model to predict concentration enrichment of contaminants on soil adhering to plants and skin. *Environ. Geochem. and Hlth.* **17**: 13–20.

Shuman, M.S. (1992) Dissociation pathways and species distribution of Al bound to an aquatic fulvic acid. *Envir. Sci. Tech.* **6**: 1033–1035.

Singh, B.R., and Narwal, R.P. (1984) Plant availability of heavy metals in a sludge-treated soil: II. Metal extractability compared with plant metal uptake. *J of Envir Qual* **13**: 344–349.

Siriratpuriya, O., Vigerust, E., and Selmer-Olsen, A.R. (1985) Effect of temperature and heavy metal application on metal content in lettuce. Meldinger fra Norges Land-brukshogskole 64, Agricultural University of Norway, p. 29.

Smith, S.R. (1994) Effect of soils-pH on availability to crops of metals in sewage sludge-treated soils II. Cadmium uptake by corps and implications for human dietary-intake. *Environ Poll* **86**: 5–13.

Symeonides, C., and McRae, S.G. (1977) The assessment of plant-available cadmium in soils. *J of Envir Qual* **6** (2): 120–123.

Tanji, K., Lauchli., A., and Meyer., J. (1986) Selenium in the San Joaquin Valley. *Environ* **286**: 6-11, 34–39.

Tiller, K.G. (1996) Soil contamination issues: past, present and future, a personal perspective. In *Contaminants and the Soil Environment in the Australasia-Pacific Region:* Naidu, R., Kookana, R. S., Oliver, D. P., Rogers, S., and McLaughlin, M. J. (Eds), Kluwer Academic Publishers, Dordrecht, pp 1–28.

Tiller, K.G., Gerth, J., and Bruemmer, G. (1984) The relative affinities of Cd, Ni and Zn for different soil clay fractions and goethite. *Geod* **34**: 17–35.

Tyler, L.D., and McBride, M.B. (1982) Influence of Ca, pH and humic acid on Cd uptake. *Pl and Soil* **64**: 259–262.

Uren, N.C., and Reisenauer, H.M. (1988) The role of root exudates in nutrient acquisition. *Adv. Plant Nutr.* **3**: 79–114.

van Straalen, N.M., and Krivolutsky, D.A. (1996) Bioindicator Systems for Soil Pollution. Kluwer Academic Publishers, London, UK.

van Straalen, N.M., and Lokke, H. (1997) Ecological Risk Assessment of Contaminants in Soil. Chapman and Hall, NY, USA.

van Wersem, J., Vegter, J.J., and Van Straalen, N. M. (1994) Soil quality criteria derived from critical body concentrations of metals in soil invertebrates. *App Soil Eco.* **1**: 185–191.

van Wijnen, J.H., Clausing, P., and Brunekreef, B. (1990) Estimated soil ingestion by children. *Enviorn. Res.* **51**: 147–162.

Viets, F.G.Jr. (1962) Chemistry and availability of micronutrients in soils. *J Agric. Food Chem.* **10**: 165–178.

Wear, J.I., and Evans, C.E. (1968) Relationship of zinc uptake by corn and sorghum with soil zinc measured by three extractants. *Proc. Soil Sci. Soc. Am.* **32**: 543–546.

Weeks, J.M., and Svendsen, C. (1996) Neutral red retention by lysosomes from earthworm (Lumbricus rubellus) coelomocytes: A simple biomarker of exposure to soil copper. *Envrion. Toxicol. Chem.* **15**: 1801–1805.

Wenzel, W.W., and Blum, W.E.H. (1995) Assessment of metal mobility in soil-methodological problems. In *Metal Speciation and Contamination of Soil.* Allen, H.E., Huang, C.P., Bailey, G.W., and Bowers, A.R. (Eds), Lewis, London, pp 227–236.

Whipps, J.M. (1990) Carbon economy. In *The rhizosphere*. Lynch, J.M. (Ed), John Wiley and Sons Ltd., Chichester, West Sussex, UK, pp 59–97.

Will, M.E. and Suter, G.W. (1995) Toxicological benchmarks for screening potential contaminants of concern for effects on soil and litter invertebrates and heterotrophic process. Report ES/ER/TM-126/R2, Oak Ridge National Laboratory: Oak Ridge, Tennessee.

Williams, C.H., and David, D.J. (1976) The accumulation in soil of cadmium residues from phosphate fertilizers and their effect on the cadmium content of plants. *Soil Science* **121**: 86–93.

Winistörfer, D. (1995) Speciation of heavy metals in extracted soil solutions by a cation exchange batch equilibrium method. *Commun. Soil Sci. Plant Anal*. **26**: 1073–1093.

Wong, M.S., Bundy, D.A.P., and Golden, M.H.N. (1991) The rate of ingestion of Ascaris lumbricoides and Trichuris trichiura eggs in soil and its relationship to infection in two children's homes in Jamaica. *Transactions of the Royal Society of Tropical Medicine and Hygiene* **85**: 89–91.

Wolt, J.D. (1994) *Soil Solution Chemistry: Application to Environmental Science and Agriculture*. John Wiley and Sons, New York, USA, p. 345.

3

The Role of Chemical Speciation in Bioavailability

Sébastien Sauvé

INTRODUCTION

The role of chemical speciation in metal bioavailability can be interpreted in many different ways. The term *speciation* has been used for many varied meanings. *Bioavailability* is even more contentious; different fields have precise but different definitions for it. The link between speciation and bioavailability has long been recognized and many earlier texts have summarized some of the important aspects (Allen et al 1980, Florence 1982, Langston and Bryan 1984, Bernhard et al. 1986, Bruemmer et al. 1986, Allen 1993, Luoma 1995, Tack and Verloo 1995, Allen and Hansen 1996). This chapter interprets bioavailability within the context of ecotoxicological risk assessment (as opposed to anthropocentric *health* assessment) as seen by the author. For human-based risk assessment, readers are referred to Chapter 1. Various approaches for chemical speciation will be explored in more detail, from simple separation in a contaminant pool to the free ion activity model.

Bioavailability

In the broadest sense, bioavailability has its etymological source as "biological availability". Strictly speaking, this simply means "available to living organisms". It may refer to actual ingestion or intake of a substance by certain biological organisms, or it might also refer to effects on the biological organism with or without concomitant uptake. For clarity, one should be certain to clarify as to which "biological" component the use of bioavailability is actually referring to. The vision of bioavailability from a pharmacologist's view of gastro-intestinal absorption (see Chapter 1) is drastically different from that of a microbiologist studying impacts

Department of Chemistry, Université de Montréal, 2900 Édouard-Montpetit, Montréal, QC, Canada, H3C 3J7.

on soil respiration. The environment, the biological components, the media, the contaminants, their chemical states, and the actual processes being studied are very different. To accommodate the various disciplines using "bioavailability", it is necessary to keep a rather loose or at least a flexible definition.

In reality, the term "bioavailability" is used by many people to contrast with that which is non-bioavailable. For example, bioavailability can differentiate the non plant-available Pb within Pb shoots in a calcareous clayey soil from the mostly reactive dissolved Pb pool in an acidic sandy soil exposed to Pb-based chemical sprays. It can also differentiate between a contaminant that will be taken up through gastro-intestinal transit from a non-bioavailable portion which is excreted without uptake.

Bioavailability, within a loose context, reflects the relative proportion of a contaminant that is sufficiently biologically reactive to effect a biological response (beneficial or detrimental) or to be taken up by an organism. Tissue accumulation and biological effects may or may not be related. Certain organisms may respond to changes in environmental availability without any changes in their tissues burden; alternatively, others organisms may accumulate contaminants without inducing any physiological changes. There is also a certain homeostatic range where both biological response and tissue accumulation can be independent of given variations of environmental bioavailability. Similarly, above a high environmental threshold, most organisms are overwhelmed and may take up drastically higher levels of contaminants or stop uptake because of disruption of normal physiological activities.

To further define bioavailability within the context of actual food ingestion, some more specific definitions are helpful (Penry 1998): "Assimilation efficiency is the fraction of the ingested elements or compounds that is incorporated into biological tissue, whereas absorption efficiency is the fraction of ingested material that is taken up across the gut lining. Assimilation thus equals absorption minus excretion."

Speciation

Etymologically, chemical speciation refers to the actual identification of chemical species. This has been interpreted in different ways and has been implemented and reported using various techniques. Speciation in this chapter is restricted to the solution phase. For example, for a divalent metal in a given medium, chemical speciation refers to identifying and quantifying the proportion of the metal (Me) present as divalent free Me^{2+} or as inorganic chemical ion-pairs (i.e. $Me(OH)_2^0$, $MeCl_3^-$, etc.), as well as the determination of metal bound to dissolved organic ligands (including humic and fulvic acids and other natural or synthetic ligands).

Speciation, which is based on the solution phase, contrasts with fractionation, which is related to the differentiation among various solid-phase fractions. Fractionation is used to differentiate between various minerals as well as to identify the association with various solid-phase components. The various sequential extraction schemes used to identify associations of trace elements with various soil or sediment solid phases is an obvious example. There has been much discussion and argument over these schemes that are *not* able to segregate the specific solid phases they were conceptually designed to identify and there has been no conclusive linking of soil fractionation results with bioavailability tests. A more elaborate discussion and a generous reference list can be found in Chapter 2 (Naidu et al 2002) and in (Sauvé 2001).

In a simpler approach, it is possible to use a partitioning or compartmentation approach based on the differentiation among different physicochemical states (i.e. partitioning of soil total metal vs. dissolved metal vs. free metal). Although partitioning has been mostly used to segregate total solid metals from total dissolved metal, the same concept can be applied to other compartments that are not related to specific chemical species but rather to distinct pools.

Linking Speciation to Bioavailability

From a purely chemical perspective, chemical reactions are controlled by speciation. For example, metal reactions in soils are mostly mediated through divalent free metal ions. Metal sorption is directly proportional to free-metal activities in solutions; similarly, ligand exchange reactions are also dependent on free-metal activity and not on total dissolved concentrations.

Metal complexation dominates the chemistry of metals in most natural systems; nevertheless, it remains critical to understand the species that is driving the study (whether the study is a chemical sorption, a bioassay, or some other test). Within a regulatory framework, it is still quite difficult to link speciation results quantitatively to bioavailability. Most risk assessors and environmental legislators are willing to recognize that bioavailability should *eventually* be considered and that certain physicochemical-dependent speciation characteristics actually control bioavailability determinations. For example, soil acidity is known to exacerbate metal toxicity and bioavailability; similarly, dissolved organic matter usually reduces the bioavailability of most contaminants in aquatic systems. Nevertheless, although the importance of pH in modulating bioavailability is well recognized, it is rarely quantitatively integrated within the regulatory framework for the determination of contamination thresholds.

CONTAMINANTS

Natural vs. Anthropogenic Inputs

Metals and metalloids are inorganic elements that occur naturally at various levels. Higher local background levels of certain trace elements are often observed, and concentrations are dependent on the geochemical origins of the parent materials studied. On the other hand, most organic contaminants are xenobiotics and even trace levels do not occur naturally. Also, the concentrations at which inorganic elements or organic compounds occur in the environment have no direct relationships with the availability from either a deficiency or a toxicity perspective. Similar naturally occurring levels may prove toxic to certain organisms under certain circumstances, and different organisms may experience deficiency under the same environmental conditions. Also, bioavailability relationships may be species-specific, element- or compound-specific, site-specific, etc. Although one could consider that above-normal concentrations of a substance represent a contamination, evaluating if that "increased" level actually represents a real toxicity risk is not straightforward.

Although some natural sources may occasionally generate relatively large metal concentrations or fluxes (e.g. methylmercury in aquatic sediments, volcanic dust deposition, mineral weathering, etc.), in many circumstances elevated metals concentrations have an anthropogenic origin.

Applying a chemical speciation approach to pollution inputs (fluxes) represents a special situation. Whether one considers anthropogenic inputs originating from atmospheric deposition or from the land application of sewage sludge, one must not confuse the chemical speciation of the material being added (the new input—the *contaminant*) from that which occurs in that particular environment. Biological organisms are not affected by the bioavailability of a contaminant in the physical form it is being supplied. What is critical is the actual chemical speciation form that the contaminant will take in the environment of the relevant biota. Also, biological organisms do not distinguish among the origins of a contaminant; they are sensitive to the relative distribution among inert and reactive fractions, independent of their origin.

In this sense, certain chemical species emitted to the environment (e.g. dust emission of metal oxides) are by themselves quite unreactive. The actual bioavailability of such compounds as they reach a particular environment largely depends on the relative rate at which they react to reach a chemical equilibrium which reflects the actual physicochemical properties of the environment they have settled in. Some materials and some compounds react in a matter of hours; in such situations, the chemical speciation is totally dependent on environmental properties. On the other hand, some other contaminant (e.g. solid metal pieces) may be very slow to react with their environment and may reach a pseudo-equilibrium only over decades or centuries.

In a risk assessment of the long-term quality and sustainability of an ecosystem, it is mostly irrelevant to consider the time needed to reach "speciation equilibrium". In that situation one is interested in an hypothetical pseudo-equilibrium into the future where only the actual fate and speciation of the contaminant matters. This "final status" is independent of the means through which that ecosystem evolved to get there. Neglecting the eventual transformation of slowly reactive compounds to potentially available forms will create a "time-bomb" scenario. From a sustainability perspective, critical load regulations should aim at predicting the long-term fate and speciation of contaminants and must consider the natural physicochemical variability of the environment, especially with regards to large-scale trends such as acidification and global warming.

Metals and Metalloids

Metals and metalloids are inorganic elements whose chemical availability is closely related to reactions with the mineral phase. In theory, at excessively high concentrations, mineral equilibrium will control the solubility and dissolved concentrations in solution. An element may precipitate as a discrete mineral phase if it exceeds its mineral solubility or may dissolve and be released if the solution concentration goes below the corresponding mineral solubility equilibrium. Some soil solubility data are illustrated in Fig. 3.1 for Cd, Cu, and Pb. The units on the y-axis are p(activity) of the appropriate chemical species (much like pH, a pCd^{2+} of 6 is equivalent to $10^{-6} M$ divalent free Cd^{2+}). The experimental speciation data in Fig. 3.1 suggests that for high soil total concentrations (above 5000 mg Pb kg^{-1} or 15 000 mg Cu kg^{-1}), mineral solubility may actually control solution metal concentrations. Nevertheless, under most conditions in oxic soils, solution free-metal activities in contaminated soil solutions will be below the level necessary for mineral solubility control. This means that mineral equilibrium rarely controls metal solubility. For the most part, adsorption on mineral, organic, and mixed surfaces is the determinant factor. Mineral solubility still needs to be considered in risk assessment since it represents a potential "ceiling" for metal activities in solution in intimate contact with a solid phase (e.g. soil solutions and sediment pore water). But it is even more critical to have an excellent understanding of the retention and sorption processes that truly control metal availability. This also shows the complexity of predictive modeling for metal availability in natural systems. Solid phases interact with the solution but in most situations it is the adsorption properties of the surficial environment that controls the solution and not mineral solubility equilibrium.

The same retention processes is applicable in the surficial environment of aquatic sediments. Allowing for variations in physicochemical parameters such as pH, organic matter, particle size, etc., it is expected that speciation in aerobic sediments should be qualitatively similar to that in oxic soils. But, aquatic sediments are often partly anaerobic. In such situations, the exceedingly low solubility of metal sulfides may effectively control pore water free-metal activities below the levels expected from simple retention mechanisms. Nevertheless, the mineral solubility approach presented in Fig. 3.1 is applicable; it only requires the addition of the metal-sulfide solubility lines, which would fall below

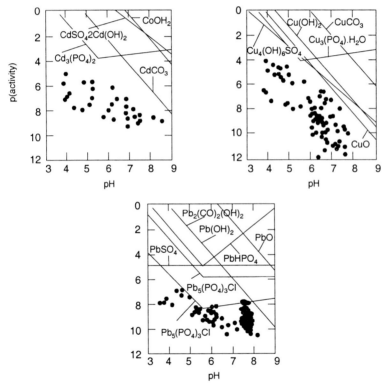

Fig. 3.1:   Soil solution mineral solubility equilibria of Cd, Cu, and Pb. The solid lines illustrate the theoretical free-metal activity maintained at equilibrium by various mineral phases. The computations assume 0.005 M Cl⁻, 0.001 M SO_4^{2-} and NO_3^-, atmospheric CO_2, 0.01 M Ca and PO_4^{3-} solubility controlled by $FePO_4$ below pH 5.6 and by $Ca(PO_4)_3OH$ at higher pH values. The dots represent actual electrochemical speciation measurements in field-collected contaminated soils (Sauvé et al., 1997a, 1997b, 2000b, Sauvé, 1999).

those already illustrated. The extent of solubility control by sulfides then depends on the stochiometric ratio of sulfides to metals. This approach will be discussed later in the section on aquatic sediments partitioning, but the principles remain valid and perfectly applicable for anaerobic, flooded soils.

Organic Contaminants

Organic contaminants are different from metals and metalloids (Lee and Allen, 1998), in that the latter are not volatile under standard environmental conditions (with the most notable exception of mercury). This simplifies some of the multimedia, multicomponent modeling often used to predict the environmental fate of contaminants. On the other hand, many quantitative structure-activity relationships developed to predict properties and affinities of organic compounds are totally inapplicable to metals and metalloids. In this case, alternative models are necessary, as will be discussed later.

Organic contaminants do not possess a mineral equilibrium constant that effectively generates a maximum solubility but, in a similar fashion, they have an intrinsic water solubility maximum that, although related to hydrophobicity, effectively works in the same way. Again, as was the case for metals, solubility processes such as these are not always effective within the contamination ranges usually encountered in contaminated sites. In many situations, partitioning of organic contaminants between solid and liquid phases is dependent on sorption processes.

Organic contaminants are usually hydrophobic and lipophilic to varying degrees. Such properties are often quantified using the octanol-water partition coefficient (K_{ow}), which is a measure of the partitioning of the chemical between water and octanol at equilibrium. Higher K_{ow} values represent a chemical with a higher affinity for octanol than for water. Failing the availability of experimental data for K_{ow}, some models are available to provide default K_{ow} values using comparison of molecular structure to that of chemicals and functional groups with known properties (see further details in Chapter 1).

Sorption of organic contaminants is presumed to be proportional to the organic carbon content of the solid phase. Some empirical estimations of organic-carbon partition coefficients (K_{oc}) have been proposed to obtain default K_{oc} from the measure of the hydrophobic character of the material (K_{ow}).

Risk assessment of soil contamination with petroleum hydrocarbons is a specific problem. The Total Petroleum Hydrocarbon Working Group (Total Petroleum Hydrocarbon Working Group, 1997b) proposed using the extractable petroleum hydrocarbon (EPH) fraction as opposed to the total petroleum hydrocarbon (TPH) content. Various analytical procedures are available to quantify the EPH fraction (Total Petroleum Hydrocarbon Working Group, 1998). Although this is strictly speaking not a *real* chemical speciation, the concept is similar. It is an attempt to differentiate the proportion of the organic contaminant that is potentially mobile or bioavailable pool from that which is inert and unreactive. Since in this case the attempt at quantifying EPH is not really directed towards identifying specific chemical species (although it usually segregates among equivalent carbon numbers), it truly is more analogous to partitioning. We envision that fraction-specific reference dose can be used to propose EPH-dependent soil quality criteria (Total Petroleum Hydrocarbon Working Group, 1997a; Loranger et al. 2002).

PARTITIONING OF METALS

Water Columns vs. Soil Profiles

The most obvious difference between soil and aquatic systems is the relative importance of the solid phase over the solution phase. Within the water column, the buffering processes occurring in free water are limited to reactions with dissolved and colloidal particulate substances. Aquatic sediments resemble soils more closely, the pore water is in intimate contact with the solid phase and its chemistry is controlled by sorption/desorption and precipitation/dissolution processes. Even then, aquatic sediments are almost never dry and often exposed to anaerobic conditions. Although surface soils may become anaerobic, in many situations they are in an oxidized state. In both soils and sediments, it is usual to consider that for most organisms, contaminant exposure occurs through pore water or soil solution. This assumption is more complex for soil-eating and bottom-feeding organisms (e.g. earthworms, mussels, etc.), but the dissolved phase remains a significant pathway to consider in most situations.

Besides differences between soil and sediments, the paradigm for assessing potential risk is similar, evaluating whether the dissolved reactive contaminant fraction (in the soil solution or sediment pore water) exceeds the recognized threshold for toxic effects. For metals, in most situations, the *reactive* dissolved fraction seems to be the divalent free metal species (Me^{2+}). In some situations, other chemical species are also bioavailable (e.g. phytoavailability of cadmium chloride, hydroxide species ($MeOH^+$) in aquatic systems, and other exceptions (Campbell, 1995)). The actual means used to evaluate soils and sediments differ, but both aims are the same, evaluating whether the chemical speciation of the bioavailable species exceeds the threshold for toxic effects.

Soil Metal Partitioning

An approximate approach to the determination of bioavailability is a simple separation among total, dissolved, and free/reactive pools, although according to the earlier definitions this is not strictly speaking speciation. When chemical speciation data is unavailable or difficult to implement, the partitioning approach is definitely more discriminate than simply soil total and allows an estimation of bioavailability and potential mobility. The huge differences among the relative sizes of these compartments are illustrated for soil Cu in Fig. 3.2. Five soils are illustrated with total soil metal ranging from 22 to 640 mg Cu kg^{-1} and with soil solution pH ranging from 3.9 to 7.4. The corresponding dissolved Cu concentrations in the extracted soil solutions vary from 7 to 645 µg Cu L^{-1} and free Cu^{2+} determined using an ion-selective electrode varies from $10^{-10.6}$ to $10^{-5.4}$ M. Other details on this dataset can be found in Sauvé et al. (1997a). The first three soils have near-neutral pH values and the first two show relatively similar dissolved Cu levels for a similar total Cu. The second one shows a lower free Cu^{2+} activity, despite a similar total and a lower pH. The third soil shows a lower dissolved Cu and a lower free Cu^{2+}, reflecting the lower level of total Cu contamination. On the other hand, the last two soils are acidic and despite total Cu and dissolved Cu concentrations within a similar range, their free solution Cu^{2+} activities are 2–5 orders of magnitude higher than that found in the soils at neutral pH. Soil solubility studies on Cu have shown that soil solution dissolved Cu concentrations are mostly dependent on the total soil concentrations and free Cu^{2+} speciation in soil solution is highly dependent on both total Cu and soil solution pH (Sauvé et al., 1997a; Yin et al., 2000). Also, a part of the variability in dissolved and free metal relationships is not explained by simple regression with pH or total soil metal. This is illustrated in the difference in the order of magnitude in free Cu^{2+} between the first two soils in Fig. 3.2 despite similar total Cu and solution pH. This discrepancy can be partly attributed to other physicochemical characteristics: soil organic matter, mineralogy, dissolved organic ligands, concentrations of competing cations and complexing anions, oxide content, etc.

It is critical to realize that those relationships could be different using another dataset consisting of soils from different origins. It is also impossible to extrapolate results of a particular element to another. Even the qualitative trends could be drastically different and experimental validation is required for element-specific and soil-specific relationships.

The partitioning between the solid and the solution phase is often represented using the partition coefficient (K_d). This coefficient is usually described as:

$$K_d = \left(\frac{\text{Total adsorbed metal}}{\text{Dissolved metal}} \right) \qquad (1)$$

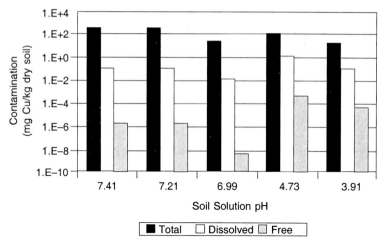

Fig. 3.2: Relative proportion of total, dissolved, and free copper in contaminated soils. Data adapted from Sauvé, 1999.

where total adsorbed metal is in mg kg^{-1} and dissolved metal in mg L^{-1}; hence, the units for K_d's are L kg^{-1}. Since K_d values are usually very large, it usually does not make a significant difference whether one uses "total" or adsorbed metal. A compilation of K_d values determined using varied experimental protocols for various trace elements reported in Chapter 2 with reference to sewage sludge loading thresholds are given in Table 3.1.

The background total values are an estimate of what is expected from normal, non-contaminated soils. The corresponding dissolved concentration results from applying Eq. (1) to the appropriate K_d. The threshold values correspond to the EEC limits for sewage sludge application and the corresponding dissolved concentration is calculated as done previously. The total levels in groundwater correspond to a back-calculation of the soil concentration level that would generate the criteria concentration in groundwater. It is interesting to note that for Cd and Hg, both independently determined soil and groundwater criteria concur well with the suggested K_d values.

Aquatic sediments

Many, but not all, aquatic sediments are partly anaerobic; hence, risk assessment of metals in sediments is partly based on acid volatile sulfide (AVS) determinations. This AVS approach is limited to sediments with significant AVS (>1 µmol/g), which excludes oxidized sediments (Ankley

Table 3.1: Soil partition coefficients and resulting soil and solution concentrations*.

Metal	K_d^a	Background[b] Total (mg kg^{-1})	Dissolved (μg L^{-1})	Threshold[c] Total (mg kg^{-1})	Dissolved (μg L^{-1})	Groundwater[d] Criteria (μg L^{-1})	Total (mg kg^{-1})
B	160		0.00	3	19		0.0
Cd	2869	0.5	0.17	3	1.0	1	2.9
Cr	14920	25	1.7	600	40	15	224
Cu	4799	30	6.3	130	27	25	120
Hg	8946	0.1	0.011	1	0.11	0.1	0.89
Ni	16761	25	1.5	70	4.2	10	168
Pb	171214	20	0.12	500	2.9	10	1712
Se	43937	1	0.023	3	0.068	1	44
Zn	11615	50	4.3	300	26	50	581

[a] K_d values were obtained from a compilation of various soil sorption studies (Sauvé et al., 2000a).

[b] The background values were obtained from a survey of 3045 Agricultural U.S. soils (Holmgren et al., 1993) and from values compiled by the Canadian Council of Ministers of the Environment (CCME, 1991).

[c] The threshold values correspond to the EEC limits for sewage sludge application, as discussed in detail in Table 1, Chapter 3.

[d] The groundwater criteria were obtained from values compiled by the Canadian Council of Ministers of the Environment (CCME, 1991).

et al., 1996). The pore water solubility of metals that have sparingly soluble sulfides (Cd, Cu, Hg, Ni, Pb and Zn) is controlled by the ratio of available sulfides to that of available metals. If enough sulfides are available to immobilize the metals, none will be left in a bioavailable form. If the capacity of the AVS is exceeded, that portion of the metals is possibly reactive and could potentially become bioavailable. In practice, the AVS is evaluated concurrently with the simultaneously extracted metals (SEM). The evaluation is then simple. If the SEM from the pore water is equal or lower than AVS, no toxic effects should occur.

Using this approach does not mean that a level of SEM exceeding the ASV represents a toxic sediment. In such situations, it is still possible for other processes to further reduce the availability of metals. For example, organic carbon and oxides (Fe, Mn, Al) may also adsorb metals and reduce their chemical and biological availability. Those approaches can be further refined (Ankley et al., 1996), but the concept remains the same: comparing a bioavailability measurement against some defined thresholds.

---

* The dissolved concentration for background and threshold levels are calculated using Eq. 1 and the appropriate K_d. The corresponding total element concentration for the groundwater is similarly calculated using dissolved concentrations to predict total ones, instead of the reverse.

FREE ION ACTIVITY MODEL (FIAM)

Aquatic Bioassays

The validity of the free ion activity model (FIAM) is well recognized in aquatic ecotoxicology. The FIAM is relatively tightly defined (Campbell, 1995). It assumes that biological effects are dependent on the free metal species in solution and that the mode of action is mediated through the amount of membrane-bound metals. Certain assumptions need to be made for metal transport and transfer: the relative size of the free metal activity should permit a pseudoequilibrium with respect to exposure. Also, biological organisms should only respond to free-metal activities, and the response itself should not be influenced by the metal (Campbell, 1995). The FIAM then becomes equivalent to considering that the metal-to-biota relationships are somewhat similar to a chemical equilibrium reaction between a metal and a given complexing agent.

This approach is mostly applicable to aquatic systems, where metal reactions with a fish gill can be modeled, accounting for pH, dissolved organic matter, contaminant levels, competing ions, etc. (Di Toro et al. 2001). In soils, the FIAM framework needs to be loosened. Many of the assumptions and conditions described by Campbell (1995) are difficult or impossible to confirm. The solid-phase components in soils further complicate the chemistry and increase the potential exposure pathways.

There are certainly exceptions even in aquatic toxicology, but the bulk of the data available demonstrate that the free metal activity and not the pool of dissolved metal controls toxicity. Figure 3.3 illustrates this using a survival assay for *Ceriodaphnia dubia* in solutions of humic acids and Cu (Ma et al., 1999). It is clear that little if any of the variability is related to the total dissolved Cu concentration and that, although not perfect, the fit is much improved by the use of measured free Cu^{2+} activities. Even though this is well recognized, implementing it in a regulatory framework

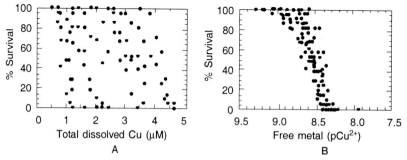

Fig. 3.3:   Survival of *Ceriodaphnia dubia* to a Cu stress, expressed as a function of dissolved Cu (A), or measured free Cu^{2+} (B). Data adapted from (Ma et al., 1999) with permission from SETAC.

is rendered difficult by the analytical challenges surrounding free-metal activity measurements. Often, the result is that criteria are categorized into classes based on pH or water hardness. In other instances, some effluent-specific bioassays may be used to assess the relative toxicity of organic-rich waters, which are difficult to categorize within some of the simpler schemes.

Even the acid volatile sulfide (AVS) approach for sediments is partly based on a loose FIAM. In this case, instead of directly measuring pore water free-metal activities, the AVS approach estimates the possible excess of the binding/precipitation capacity of sediment sulfides. When sulfides control metal solubility, it is not necessary to determine the exact level of the free-metal activity as it is excessively low. When sulfides no longer control free-metal activity (because of oxidation or saturation), it becomes necessary to quantify pore water free-metal activity. It is then also possible to use another model that attempts to quantify whether there is enough organic carbon in the sediments to actually bind the metals and reduce the free metal activity in the pore water below the critical threshold (Ankley et al., 1996). Even an approach based on AVS or organic carbon ratios is conceptually dependent on estimating whether free-metal activities are above or below a given toxicity threshold.

Soils

Free-metal activities in soil solutions drive chemical reactions. Independent of bioavailability, chemical equilibrium reactions of complexation, mineral solubility, and binding are based on free-metal activities. From an environmental fate modeling perspective, it is important to know the chemical speciation of the metals of interest, or at least recognize the potential errors that might be propagated because of the assumptions used to circumvent the absence of experimental data. Figure 3.2 already illustrated that for soils, orders of magnitude separate total content from solution dissolved concentrations and free-metal activities. Figure 3.4 illustrates the extent of the complexation of three metals by dissolved organic matter in extracted soil solutions. Wide variations in the level of complexation are clearly evident. For Cu, most of the samples show more than 98% complexation; on the other hand, Cd and Pb show a more homogenous distribution. For Cu, there seems to be a clear effect of pH, whereas for Cd and Pb there is no obvious pH trend. A constant ratio is sometimes used to predict metal complexation, i.e. presuming that 98% of dissolved Cu, 50% of dissolved Pb, or negligible amounts of Cd are complexed. Although Fig. 3.4 illustrates data only for Cd, Cu, and Pb, similarly variable trends are expected for other metals. A simple dissolved/complexed ratio approach is therefore not appropriate to describe metal complexation in soil solutions.

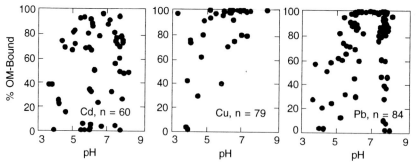

Fig. 3.4: Complexation of Cd, Cu, and Pb by dissolved organic matter in soil solution extracts. The data is calculated from actual electrochemical speciation measurements of free-metal activities in field-collected contaminated soils (Sauvé et al., 1997a, 1997b, 2000a, Sauvé, 1999).

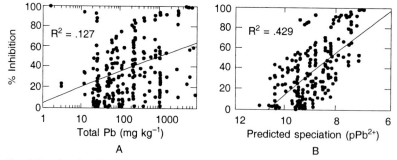

Fig. 3.5: Contrast between inhibition effects predicted from total Pb with an R^2 of 0.127 (A) or from predicted free Pb^{2+} activity with an R^2 of 0.409 (B). Data adapted from (Sauvé et al., 1998) with permission from SETAC.

Despite the high variability illustrated in Fig. 3.4, there is some evidence that soil solution free-metal activities are better predictors of toxic effects than soil total metal measurements are. To evaluate this, various bioassays were compiled from literature (using plant, soil microbial processes, and soil invertebrates) and related to predicted free-metal activities (Sauvé et al., 1998). The results show that even using free-metal activity predictions (as opposed to actual measurements), greatly improve the predictive power over the "standard" soil total metal (Fig. 3.5). It is also critical to note that Fig. 3.5(A) is the approach currently in use by most environmental protection agencies, which usually presume that a level of soil total Pb between 100 and 1000 mg kg^{-1} is protective of the environment. It must be recognized that the variability of inhibition effects against free Pb^{2+} predictions is still excessively high and needs to be improved. Nevertheless, in this case, predicted free-metal activities are better predictors of toxic effects than is total soil metal.

Speciation Techniques

The relative lack of soil data on free-metal speciation or extractable hydrocarbons is due partly to experimental difficulties. There are many problems inherent to soil extractions, detection limits, analytical interferences and a variety of potential artifacts. Reviews of electrochemical and other exchange resin- or membrane-based metal speciation techniques can be found in (Florence 1982; Florence 1986; Gulens 1987; Apte and Natley 1995; Mota and Correia dos Santos 1995; Tack and Verloo 1995; Sauvé 2001). Also, some innovative experimental techniques for extractable petroleum hydrocarbons can be found in Total Petroleum Hydrocarbon Working Group (1998).

It is worth noting that few speciation techniques have been compared and validated concurrently (Pickering 1986; Waller and Pickering 1993; Holm et al. 1995; Sauvé 2001) and some efforts are definitely necessary to standardize or at least ease comparability of some of the speciation methodologies used as well as the methods used to obtain soil solutions.

Chemical Equilibrium Models

Chemical equilibrium models can be used to predict chemical conditions at equilibrium. They are most appropriate for the inorganic components in solution. For example, Fig. 3.6 illustrates the distribution of various

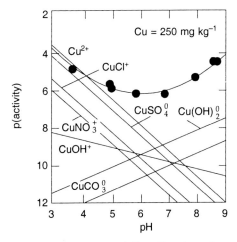

Fig. 3.6: Soil solution chemical speciation. The free Cu activities are predicted using the competitive adsorption model (Equation 25 Sauvé, 2001) and the corresponding inorganic ion-pairs are computed using standard equilibrium constants (such as in (Smith and Martell (1974–1989) and Sauvé (2001)). The total dissolved metal was obtained using experimental measurements of a spiked-soil containing the reported amount of total soil metal (and total soil organic matter of 5.46% C—(Sauvé, 1999)).

inorganic ion-pairs for Cu in contaminated soil solutions. The divalent free Cu activity in Fig. 3.6 is calculated using the semimechanistic predictive model presented later. The relative importance of complexation by dissolved organic matter is further contrasted using experimental data from a dataset of pH-adjusted, metal salt-spiked soil extractions (Sauvé, 1999). Similar to what was shown by Fig. 4, much of the total dissolved Cu is bound to organic matter, the next most predominant species being free metal (Cu^{2+}), sulfate ($CuSO_4^0$) or hydroxide ($Cu(OH)_2^0$) species. Again, the relative importance of various inorganic species is dependent on the affinity constants for each metal-inorganic ligand. Other metals in different systems will emphasize different chemical species.

Figure 3.6 clearly illustrates the importance of chemical speciation. It shows that the presumably bioavailable free Cu^{2+} can be predicted as a function of pH, but the chemistry of the soil solution system is dominated by Cu bound to dissolved organic matter. Also, some inorganic chemical species may become significant and possibly exceed the relative abundance of free Cu^{2+} (e.g. $Cu(OH)_2^0$ at alkaline pH). The bioavailability contribution of those easily dissociated (relative to OM complexes) ion pairs is not clear and needs further study. The necessary computations to describe the inorganic chemical reactions depicted by Fig. 3.6 are straightforward for synthetic systems of controlled composition. The exact chemistry of the solution is known and the corresponding equilibrium constants are clear and relatively undisputed. On the other hand, the description of speciation in natural systems is much more complex, as formulated by Turner (1995):

The overall recipe sounds like a chemist's nightmare: all the stable elements of the periodic table at a wide range of concentrations, plus several radiogenic elements; dissolved organic matter of largely unknown structure; suspended particles comprising many different solid phases and often organic coated; and inorganic and organic colloidal material. In addition, a healthy variety of microorganisms are busy consuming and repackaging particulate matter.

Therefore, although the chemical speciation of inorganic solutions is simple, natural solutions are much more difficult to model. Various computation models are available, but there is some controversy over which model is more appropriate to describe metal binding by dissolved organic ligands and how one should also consider the pool of metal bound to colloidal materials.

Using free metal and OM complexation data, it is possible to use various geochemical models for a complete speciation of the potential inorganic species. Those models also compute the solution mineral equilibrium potentially generated by different solid phases (similarly to the solubility lines shown in Fig. 3.1). The main difficulties are explaining the variability observed in mixed and heterogeneous natural systems. In those cases, some of the model results will differ and it is still often necessary to experimentally measure the free-metal speciation to validate or adjust the speciation model to the particular system's conditions. The most sophisticated fully mechanistic models used to describe solution metal speciation and solid-phase sorption are often difficult to parameterize. They often require complex mathematical treatment and data processing and, surprisingly, they have not been shown to be significantly better tools to model metal speciation and partitioning as compared with some simple multisite binding approaches (Buffle 1980; Cabaniss et al. 1984; Dzombak et al. 1986; Fish et al. 1986; Martell et al. 1988).

Semi-mechanistic Predictive Approach

It may not be necessary, or even desirable, to use the most sophisticated and the most thorough chemical modeling approaches to link chemical speciation and bioavailability within a risk assessment framework. More qualitative models that reflect the most important actual trends may prove more useful. These would also be more intuitive and transparent and permit nonspecialists to grasp the most fundamental principles involved. There are many applications of such models (Benedetti et al. 1995; McBride et al. 1997; Rieuwerts et al. 1998; Egli et al. 1999; Sauvé 2001). From a risk assessment perspective, we should prone the simplest approach that will explain the observed variability. Some specific sub-models may prove quite accurate but their overall importance does not justify their inclusion in the risk assessment process. On the other hand, some semiempirical regression models may improve normalization of chemical speciation across a wide range of physicochemical conditions. Although such an approach is not perfectly representative of the real systems, it is nevertheless useful for risk quantifications.

One such semimechanistic and partly empirical approach relates soil solution free-metal to soil solution pH and the total soil metal (McBride, et al. 1997; Sauvé, 1999; Sauvé, 2001).

$$pMe^{2+} = a + b \cdot pH + c \cdot \log_{10} \text{(Total Metal)} \qquad (2)$$

This regression was used for Pb^{2+} speciation data in Fig. 5(B), the predictions of free Cu^{2+} activities in Fig. 3.6, and the relationships among its three components (free Cu^{2+}, pH and total metal) is illustrated for Cu in Fig. 3.7. Although such mechanistic simplifications forfeit some of the details, they allow to better grasp some of the underlying environmental processes.

LINKING BIOASSAYS TO SPECIATION

Speciation Versus FIAM

There are some situations where FIAM fails, but chemical speciation can

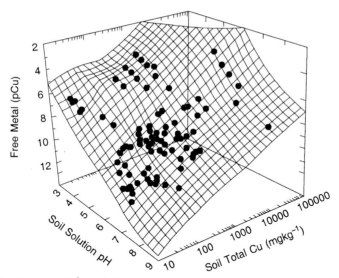

Fig. 3.7: The free Cu^{2+} activity measured by ion selective electrode is illustrated as a function of solution pH and total metal loading (log_{10}). Data taken from Sauvé et al. (1997a) and Sauvé (1999). The graph surface is obtained using a distance-weighted least square smoothing (Wilkinson, 1992).

still control and explain bioavailability. It is important to differentiate FIAM from chemical speciation. For example, in some agricultural systems, chloride is the main physicochemical parameter controlling plant uptake of cadmium. This reflects the soil solution predominance of cadmium chloride complexes, although it is a situation where free Cd^{2+} does not control or predict phytoavailability, chemical speciation encompasses $CdCl^+$ and $CdCl_2^0$ and demonstrates its critical importance for understanding metal bioavailability. It is emphasised again that if free-metal activity is known along with pH and inorganic ligand concentrations, it is relatively straightforward to estimate the distribution of various inorganic species in solution, and this estimate of Cd complexation by chloride allows an evaluation of Cd phytoavailability (McLaughlin et al. 1999).

There are also other situations where metal bioavailability is mediated through free-metal activities, but where actual bioavailability is dependent on another factor that modulates free-metal activities. It is important to recognize that a lack of correlation between biological effects and free-metal activities or chemical speciation results does not necessarily mean that it does not still mediate bioavailability. Similarly, the reverse is possible: an apparent concordance of bioavailability to free-metal activities does not preclude a biological response actually dependent on another unknown or unobserved process.

Some of the exceptions to FIAM can be explained using chemistry and the relative bioavailability of other chemical species. In some situations, bioavailability depends on FIAM or chemical speciation through specific chemical species; in other cases it may depend mostly on an ingestion pathway that is much more independent of environmental speciation (e.g. trophic bioaccumulation of methylmercury in predator species).

Kinetics and Intensity/Capacity Relationships

As Figs. 3.2 and 3.6 illustrate, only a very small proportion of dissolved metal is actually present in a free divalent form. This means that in most situations, the pool of accessible free metal is quickly depleted. The bulk of the metal in the environment is present as slowly accessible or unreactive forms. Bioavailability then depends not only on the actual intensity of the free-metal activity but as much, or even more so, on the system's capacity to supply free metal and maintain a constant activity. Similarly, there are also some experimental results for plant metal uptake. For an identical free-metal activity (*intensity*), the presence of organic ligands increases metal bioavailability and uptake (through an increase in supply *capacity*) (McLaughlin et al., 2000).

The preceding examples referred to the dissociation kinetics of metal desorption rates; interestingly, Fig. 3.3 showed a reverse example. Figure 3(A) shows the wide spread of toxicity generated by Cu/humic solutions. The actual toxic free Cu^{2+} activity depicted in Fig. 3.3(B) was obtained by varying the reaction time of the humic solution with the Cu salt solution. For a given dissolved Cu concentration, increasing reaction time allowed a higher proportion of Cu to bind with the relatively bulky humic acids. This decreases the free Cu^{2+} activity and results in a lower toxicity (Kim et al., 1999; Ma et al., 1999). Therefore, in this case, bioavailability depends on the slow kinetics of forward binding of Cu to humic acid. In a similar fashion, the kinetics of the backward reaction, i.e., releasing metal to the solution, may also control bioavailability in other circumstances. This shows that a simple FIAM approach may not be sufficient to explain metal bioavailability in complex systems. The dynamic components of most ecosystems need to be recognized and should help identify the limiting or critical steps to understand metal bioavailability.

Spiked vs. Field-collected Samples

Much of the soil testing data has been derived from an approach based on the potential hazard of specific substances. Therefore, much of the quality criteria derivation data was obtained using synthetic salt-spiked solutions or freshly contaminated soils. Laboratory water or synthetic soils are very different from natural media. Furthermore, the chemical state and bioavailability of freshly added compounds is drastically different from that of the same material after it has had time to react in the environmental medium. Bioavailability determinations are relevant when the contaminant has reached pseudoequilibrium with respect to chemical speciation and environmental conditions. In theory, spiked samples can be representative of environmental samples if, and only if, the chemical speciation is similar to field conditions. Nevertheless, it is critical to use real environmental matrices, even if one is using spikes. For example, artificial soil media or synthetic waters will never be equivalent to a real environment, regardless of the reaction and equilibration time they are given. If one needs to revert to using spiked samples, it is important to begin with field-collected samples, which contain a plethora of heterogeneous components, all of which are important to be representative of real systems. Even then, it will be important to understand the dynamics of chemical speciation in that system and insure that an appropriate and sufficient equilibration period was included. Unfortunately, chemical speciation and especially its kinetics are not well understood and too often, toxicological assays do not give it enough consideration, if any. Even with a proper equilibration period, the sample might not be representative of environmental media due to accumulation of counter ions or other unforeseen physicochemical modifications.

Real field-collected samples are ideal but unfortunately difficult to obtain. Often, the samples or sites available have multiple contaminants, complicating or preventing a clear interpretation of the toxicological significance of the bioassay results. Spiked samples are interesting for establishing a toxicological dose-response to a single contaminant or to evaluate the influence of a specific environmental variable. Unamended field-collected sets of data are restricted to the variability encountered in the available samples. It may sometimes be extremely difficult to dissociate and partition the influence of various parameters that have a high correlation (Basta et al., 1993).

Bioavailability is directly dependent on chemical speciation, which is controlled by environmental physicochemical conditions. The first step in evaluating the validity and usefulness of a particular is to assess to what extent the environmental conditions (pH, organic matter, colloids, competing ions, complexing agents, contaminant aging, environmentally relevant concentrations, etc.) are actually representative of the relevant environmental problem. It is therefore critical to carefully evaluate and report all the necessary physicochemical characteristics. Obtaining experimental chemical speciation data would be ideal, but at least if the relevant physicochemical data is available, some attempt can be made to estimate chemical speciation.

CONCLUSIONS

The role of chemical speciation is critical in mediating the bioavailability of most contaminants in most environments. Toxicological studies should measure the chemical speciation and all the relevant physicochemical parameters. If chemical speciation is not an alternative (because of financial, technical, or other difficulties), at the very least, the relevant physicochemical parameters should be reported to allow some approximations of the chemical speciation.

Chemical speciation is often explored using a chemical equilibrium approach, but the dynamic aspects of many kinetically limited reactions need further study. The research aimed at improving our understanding of chemical speciation also needs to be coupled with toxicological assays so that we may better understand the links and feedback mechanisms between the chemical speciation dynamics of contaminants and their toxicological effects.

REFERENCES

Allen, H.E. (1993) The significance of metal speciation for water, sediment and soil quality standards. Sci Tot Environ 134: 23–45.

Allen, H.E., Hall, R.H., Brisbin, T.D. (1980) Metal speciation effects on aquatic toxicity. Environ Sci Technol 14: 441–443.

Allen, H.E., Hansen, D.J. (1996) The importance of trace metal speciation to water quality criteria. Water Environ Res 68: 42–53.

Ankley, G.T., Di Toro, D.M., Hansen, D.J., Berry, W.J. (1996) Technical basis and proposal for deriving sediment quality criteria for metals. Environ Toxicol Chem 15: 2056–2066.

Apte, S.C., Natley, G.E. (1995) Trace metal speciation of labile chemical species in natural waters and sediments: Non-electrochemical approaches. In: Tessier, A., Turner, D.R. (eds) Metal Speciation and Bioavailability in Aquatic Systems. John Wiley & Sons, Ltd, Chichester/New York, pp 259–306.

Basta, N.T., Pantone, D.J., Tabatabai, M.A. (1993) Path analysis of heavy metal adsorption by soil. Agron J 85: 1054–1057.

Benedetti, M.F., Milne, C.J., Kinniburgh, D.G., van Riemdijk, W.H., Koopal, L.K. (1995) Metal ion binding substances: application of the non-ideal competitive adsorption model. Environ Sci Technol 29: 446–457.

Bernhard, M., Brinckman, F.E., Irgolic, K.J. (1986) Why "Speciation". In: Bernhard, M,, Brinckman, F.E., Sadler, P.J. (eds) The importance of chemical "speciation" in environmental processes. Springer-Verlag, Berlin, pp 7–14.

Bruemmer, G., W., Gerth, J., Herms, U. (1986) Heavy metals species, mobility and availability in soils. Z., Pflanzenernaehr Bodenkd 149: 382-398.

Buffle, J. (1980) A critical comparison of studies of complex formation between copper(II) and fulvic substances of natural waters. Anal Chim Acta 118: 29–44.

Cabaniss, S.E., Shuman, M.S., Collins, B.J. (1984) Metal-organic binding: A comparison of models. In: Kremer, C.J.M., Duinker, J.C. (eds) Complexation of trace metals in natural waters. Martinus Nijhoff/W. Junk Publishers, The Hague, pp 165–179.

Campbell, P.G.C. (1995) Interactions between trace metals and aquatic organisms: A critique of the free-ion activity model. In: Tessier, A., Turner, D.R. (eds) Metal speciation and bioavailability in aquatic systems. John Wiley & Sons, Ltd, New York, pp 45–102.

CCME, Canadian Council of Ministers of the Environment (1991) Review and Recommendation for Canadian interim Canadian environmental quality criteria for contaminated sites, Environment Canada—Conservation and Protection: Publication No: CCME EPC-CS34, Winnipeg, Manitoba, Canada.

Di Toro, D.M., Allen, H.E., Bergman, H., Meyer, J., Paquin, P.R., Santore, R.C. (2001) A biotic ligand model of the acute toxicity of metals. I. Technical basis. Environ Toxicol Chem 20: 2383–2396.

Dzombak, D.A., Fish, W., Morel, F.M.M. (1986) Metal-humate interactions. 1. Discrete ligand and continuous distribution models. Environ Sci Technol 20: 669–675.

Egli, M., Fitze, P., Oswald, M. (1999) Changes in heavy metal contents in an acidic forest soil affected by depletion of soil organic matter within the time span 1969–1993. Environ Pollut 105: 367–379.

Fish, W., Dzombak, D.A., Morel, F.M.M. (1986) Metal-humate interactions. 2. Application and comparison of models. Environ Sci Technol 20: 676–683.

Florence, T.M. (1982) The speciation of trace elements in waters. Talanta 29: 345–364.

Florence, T.M. (1986) Electrochemical approaches to trace element speciation in waters. A review. Analyst 111: 489–505.

Gulens, J. (1987) Assessment of research on the preparation, response and application of solid-state copper ion-selective electrodes. Ion-Sel Electrode Rev 9: 127–171.

Holm, P.E., Andersen, S., Christensen, T.H. (1995) Speciation of dissolved cadmium: Interpretation of dialysis, ion exchange and computer (GEOCHEM) methods. Wat Res 29: 803–809.

Holmgren, G.G.S., Meyer, M.W., Chaney, R.L., Daniels, R.B. (1993) Cadmium, lead, zinc, and nickel in agricultural soils of the United States of America. J Environ Qual 22: 335–348.

Kim, S., Ma, H., Allen, H., Cha, D. (1999) influence of dissolved organic matter on the toxicity of copper to Ceriodaphnia dubia: Effect of complexation kinetics. Environ Toxicol Chem 18: 2433–2437.

Langston, W.J., Bryan, G.W. (1984) The relationships between metal speciation in the environment and bioaccumulation in aquatic organisms. In: Kramer, C.J.M., Duinker, J.C. (eds) Complexation of trace metals in natural waters. Martinus Nijhoff/W. Junk Publishers, The Hague/Boston, pp 375–392.

Lee, C.M., Allen, H.E. (1998) The ecological risk assessment of copper differs from that of hydrophobic organic chemicals. Hum Ecol Risk Assess 4: 605–617.

Loranger, S., S. Sauvé, Y. Pouliot, L. Dussault, and Y. Courchesne. 2002. Environmental fate and human exposure modeling of the residual TPH contamination in a bioremediated petroleum storage site. In: Sunahara, G.I., Renoux, A.Y., Thellen, C., Gaudet, C.L., Pilon, A. Bioremediation and land reclamation: Tools to measure success or failure. John Wiley & Sons, New York, pp. 411–431.

Luoma, S.N. (1995) Prediction of metal toxicity in nature from bioassays: Limitations and research needs. In: Tessier, A., Turner, D.R. (eds) Metal speciation and bioavailability in aquatic systems. John Wiley & Sons, Ltd, New York, pp 609–659.

Ma, H., Kim, S., Cha, D., Allen, H. (1999) Effect of kinetics of complexation by humic acid on toxicity of copper to Ceriodaphnia dubia. Environ Toxicol Chem 18: 828–837.

Martell, A.E. and Smith, R.M. (1974–1989) Critical stability constants, Plenum Press: New York, NY. c1974–c1989

Martell, A.E., Motekaitis, J., Smith, R.M. (1988) Structure-stability relationships of metal complexes and metal speciation in environmental aqueous solutions. Environ Toxicol Chem 7: 417–434.

McBride, M.B., Sauvé, S., Hendershot, W. (1997) Solubility control of Cu, Zn, Cd and Pb in contaminated soils. Europ J Soil Sci 48: 337–346.

McLaughlin, M.J., Maier, N.A., Correl, R.L., Smart, M.K., Sparrow, L.A., McKay, A. (1999) Prediction of cadmium concentrations in potato tubers (Solanum tuberosum L.) by pre-plant soil and irrigation water analyses. Aust J Soil Res 37: 191–207.

McLaughlin, M., Zarcinas, B., Stevens, D., Cook, N. (2000) Soil testing for heavy metals. Commun Soil Sci Plant Anal 31: 1661–1700.

Mota, A.M., Correia dos Santos, M.M. (1995) Trace metal speciation of labile chemical species in natural waters: Electrochemical methods. In: Tessier, A., Turner, D.R. (eds) Metal speciation and bioavailability in aquatic systems. John Wiley & Sons, Ltd, New York, pp 205–257.

Naidu, R., Rogers, S., Gupta, V.V.S.R., Kookana, R.S., Bolan, N.S., Adriano, D. (2002) Bioavailability of metals in the soil-plant environment and its potential role in risk assessment: An overview. In: Naidu, R., Gupta, V.V.S.R., Kookana, R.S., Rogers, S., Adriano, D. (eds) Bioavailability, toxicity and risk relationships in ecosystems. Oxford & IBH Publishing Co. Pvt. Ltd. Chapter 2 this book.

Penry, D.L. (1998) Applications of efficiency measurements in bioaccumulation studies: Definitions, clarifications, and a critique of methods. Environ Toxicol Chem 17: 1633–1639.

Pickering, W., F. (1986) Metal ion speciation-soils and sediments. Ore Geol Rev 1: 83–146.

Rieuwerts, J., Thornton, I., Farago, M., Ashmore, M. (1998) Quantifying the influence of soil properties on the solubility of metals by predictive modelling of secondary data. Chem Speciation Bioavailability 10: 83–94.

Sauvé, S. (1999) Chemical speciation, solubility and bioavailability of lead, copper and cadmium in contaminated soils. Ph. D. Dissertation, Cornell University, Ithaca, N.Y., 174 p.

Sauvé, S. (2001) Speciation of metals in soils. In: Allen, H.E. (ed) Bioavailability of Metals in Terrestrial Ecosystems: Importance of Partitioning for Bioavailability to Invertebrates, Microbes and Plants. Society for Environmental Toxicology and Chemistry, Pensacola, F.L., USA, pp 7–58.

Sauvé, S., Dumestre, A., McBride, M., Hendershot, W. (1998) Derivation of soil quality criteria using predicted chemical speciation of Pb^{2+} and Cu^{2+}. Environ Toxicol Chem 17: 1481–1489.

Sauvé, S., McBride, M., Norvell, W.A., Hendershot, W. (1997a) Copper solubility and speciation of in situ contaminated soils: Effects of copper level, pH and organic matter. Water Air Soil Pollut 100: 133–149.

Sauvé, S., McBride, M.B., Hendershot, W.H. (1997b) Speciation of lead in contaminated soils. Environ Pollut 98: 149–155.

Sauvé, S., Hendershot, W., Allen, H.E. (2000a) Solid-Solution Partitioning of Metals in Contaminated Soils: Dependence on pH and Total Metal Burden. 34:1125–1131.

Sauvé, S., Norvell, W.A., McBride, M., Hendershot, W. (2000b) Speciation and complexation of cadmium in extracted soil solutions. Environ Sci Technol 34:291–296.

Tack, F.M.G., Verloo, M.G. (1995) Chemical speciation and fractionation in soil and sediment heavy metal analysis: A review. Intern J Environ Anal Chem 59: 225–238.

Total Petroleum Hydrocarbon Working Group (1997a) Development of Fraction-Specific Reference Dose (RfDs) and Reference Concentration (RfCs) for Total Petroleum—Volume 4, Amherst Scientific Publ.: Amherst , Mass, 137 p.

Total Petroleum Hydrocarbon Working Group (1997b) Selection of Representative TPH Fractions Based on Fate and Transport Consideration—Volume 3, Amherst Scientific Publ.: Amherst, MA, 102 p.

Total Petroleum Hydrocarbon Working Group (1998) Analysis of petroleum hydrocarbons in environmental media - Volume 1, Amherst Scientific Publishers: Amherst, MA, 98 p.

Turner, D.R. (1995) Problems in trace metal speciation modeling. In: Tessier, A., Turner, D.R. (eds) Metal speciation and bioavailability in aquatic systems. John Wiley & Sons, Ltd, Chichester/New York, pp 149–203.

Waller, P.A., Pickering, W.F. (1993) The effect of pH on the lability of lead and cadmium sorbed on humic acid particles. Chem Speciation Bioavailability 5: 11–22.

Wang, W-X., Fisher, N.S. (1999) Assimilation efficiencies of chemical contaminants in aquatic invertebrates: A synthesis. Environ Toxicol Chem 18: 2034–2045.

Wilkinson, L. (1992) SYSTAT for Windows: Statistics Version 5, SYSTAT Inc.: Evanston, IL, 636 p.

Yin, Y., Lee, S-Z., You, S-J., Allen, H. 2000. Determinants of metal retention to and release from soils, In: Iskandar I (ed) Environmental restoration of metals-contaminated soils. CRC Press LLC, Boca Raton, FL, pp. 77–91.

SECTION B

INDICATORS OF BIOAVAILABILITY

4

Microbial Parameters as Indicators of Toxic Effects of Heavy Metals on the Soil Ecosystem

P.C. Brookes

INTRODUCTION

In high-productivity agricultural ecosystems, natural soil fertility is commonly supplemented by applications of nutrients, either as inorganic fertilizer or organic manures, and occasionally both. However, the activities of the soil micro-organisms (collectively the soil microbial biomass) in decomposing plant and animal residues and in the formation and mineralization of soil organic matter still underpins the fertility of these managed systems. In natural ecosystems these natural processes determine, almost entirely, the fertility of their soils. Any decline in natural soil fertility will therefore have disproportionately large effects in natural systems but still cannot be ignored in managed ones. The soil-plant ecosystem may be damaged, either in the long- or short-term, by agents that inhibit or stop the natural functioning of the soil micro-organisms.

The heavy metals, e.g. Cu, Ni, Cd, Cr, Zn, Pb, are by far the most important inorganic pollutants of managed soil ecosystem. They differ from organic pollutants in that, once they have entered soil they persist, for all practical purposes, indefinitely. Currently, the only practical means for their removal is to remove the soil itself, hardly a practical proposition in most cases.

Mandatory government limits e.g. European Union (EU) and United States Government Protection Agency (USEPA) are designed to stop the accumulation of heavy metals above 'safe' soil metal concentrations. The limits are based upon known effects of heavy metals on plant and animal health. Until recently, they took no account of possible effects on soil micro-organisms or microbial activities despite their essential role in maintaining soil fertility.

Agriculture and Environment Division, IACR-Rothamsted, Harpenden, Herts., AL5 2JQ, UK.

Brookes and McGrath (1984) reported decreased total amounts of soil microbial biomasss in soils contaminated with heavy metals more than 20 years previously from past sewage-sludge applications (reviewed elsewhere in this paper). Since then a great deal of further research in this area has been carried out and new UK legislation has been recently introduced decreasing the current limit for Zn from 300 to 200 µg Zn g^{-1} soil, based on observed adverse effects of Zn on *Rhizobium* in agricultural soils (MAFF/DOE, 1993a and b). Even so, amounts of metals permitted in agricultural soils in Europe are still set at concentrations at, or only just below, those at which effects on the soil microbial ecosystem can be detected. This is in marked contrast to limits for some other pollutants (pesticides in drinking water, for example) which are many times lower than concentrations at which adverse biological effects have been demonstrated.

In the USA, maximum permitted metal concentrations in agricultural soils are, depending upon the metal, 3 to 10 times larger than in the EU, which involves markedly different philosophies (Table 4.1). Irreversible effects upon the soil microbial ecosystem have been consistently demonstrated at soil metal concentrations well below the USA limits (Brookes 1994). In Table 4.2 are given heavy metal concentrations at which significant (commonly 50% decreases) changes in soil ecosystem functioning can be demonstrated.

Table 4.1: Maximum concentrations of metals allowed in agricultural soils treated with sewage sludge.

Country	Year	Total soil metal concentration, mg kg^{-1} soil						
		Cd	Cu	Cr	Ni	Pb	Zn	Hg
European Community	1986	1-3	50-140	100-150[a]	30-75	50-300	150-300	1-1.5
US[b] 1993		20	750	1500	210	150	1400	8

[a] Now withdrawn
[b] Calculated from maximum cumulative pollutant loading limits, assuming incorporation to 15 cm depth and average soil bulk density of 1.33 g cm^{-3}, but not including the background concentration of these elements in soils.
(From McGrath *et al.*, 1995).

The European approach is based upon the view that, while there is inevitably some escape of metals into the environment in industrial societies, it is best to operate in ways which cause the minimal contamination that is compatible with modern life.

Here I review recent findings of the effects of heavy metals on the functioning of the soil microbial ecosystem. Several 'levels' of microbial ecosystem system structure are considered. These are (1) Total microbial communities and activities, (2) Specific micro-organisms or functional

Table 4.2: Soil heavy metal concentrations at which significant effects (1-5) on the microbial ecosystem were detected.

Microbial indicator	Soil total metal concentration, mg kg^{-1} soil			
	Cu	Ni	Cd	Zn
Total microbial biomass[1]	45	nd	3–5	160
Heterotrophic N$_2$-fixation[2]	37	21	3.4	127
Autotrophic N$_2$-fixation[3]	20	2.5	3	50
Symbiotic *Rhizobium*-legume N$_2$-fixation[4]	99	27	10	334
Rhizobium leguminosarum bv. *trifolium*	27–48	11–15	0.8–1	130–200

nd—not determined
Effect:
[1] 50% decrease (Brookes *et al.*, 1997)
[2] Significant decrease (McGrath *et al.*, 1995)
[3] 50% decrease (Brookes *et al.*, 1986)
[4] 50% decrease (McGrath *et al.*, 1988)
[5] Several orders of magnitude (McGrath *et al.*, 1995)

groups, (3) Specific biochemical markers, and (4) Novel approaches. These include DNA technology and the use of *lux* genes. Where appropriate, the different methodologies are evaluated for their suitability as diagnostic tests for evaluating heavy metal effects on the soil microbial ecosystem.

EFFECTS OF HEAVY METALS ON MICROBIAL COMMUNITIES AND THEIR ACTIVITIES

Soil Microbial Biomass

A full description of the soil microbial biomass concept, the method of measurement and its limitations, were first presented by Jenkinson and Powlson (1976). Instead of considering soil micro-organisms as separate species or even classes (e.g. fungi and bacteria) the biomass was measured as a single unit, or pool, of the total mass of micro-organisms or the nutrients, initially C or N, immobilized within the microbial cells. This permitted, for the first time, the measurement of a single discreet pool of soil organic matter, the micro-organisms themselves. It is this pool that is responsible for the decomposition of plant and animal residues, the immobilization and mineralization of the major plant nutrients (C, N, P, and S) and for the formation and degradation of soil organic matter. It is thus ultimately responsible for the maintenance of soil fertility and is indeed, as eloquently described by Jenkinson (1977), "the eye of the needle through which all organic matter must pass" as it is broken down into simple inorganic components, including water, carbon dioxide, nitrate, phosphate and sulphate, that plants can use again. By treating the micro-organisms as a defined and measureable pool, and by appropriate use of isotopically labelled substrates, fluxes of C and N (and later P and S) through the microbial population could be measured, leading to a new understanding of the importance and role of the biomass in the maintenance and regulation of soil fertility.

From this pioneering work was also developed the concept of the biomass as an 'early warning' of changing soil conditions and as an indicator of the direction of change. For example, on changing from forest or grassland to arable, microbial biomass decreased much more rapidly than total soil organic matter (Ayanaba et al., 1976). Similarly, Powlson *et al.* (1987) found no significant increase in total soil organic matter following 18 years of straw incorporation in two Danish soils. In contrast, the total amount of microbial biomass had increased by nearly 50% over the same period, compared to soils where the straw had been burnt.

Brookes and McGrath (1984) used the biomass concept to investigate the residual effects of heavy metals from past applications of sewage-sludge on microbial and soil organic matter dynamics. The experiment they studied was the Market Garden Experiment at Woburn, UK, a sandy loam of about 10% clay and a pH of 6.5 which had received annual applications of sewage sludge or sludge compost (high-metal soils) from 1942 to 1961 or farmyard manure or inorganic fertilizer (low-metal soils) from 1942 to 1967. All plots received inorganic fertilizer annually since the applications of organic manures ceased. In 1984, the high-metal soils contained Cu, Ni and Zn at up to about current European Union (EU) permitted limits and Cd at up to three times the current permitted limit.

In the low-metal soils there was a reasonably close linear relationship between soil biomass content and total soil organic matter, with the biomass-C comprising about 1 to 2% of total soil organic C (Brookes and McGrath, 1984). This was within the range typically reported for sandy soils in temperate regions (Jenkinson and Ladd, 1981). In contrast, in the sludge-treated high-metal soils the amounts of biomass were only about half those in the low-metal manured soils and some were lower than in the soils given inorganic fertilizer. Equally surprisingly, there was no relationship between amounts of biomass and amounts of organic matter in the high-metal soils, unlike the low-metal soils. This was despite (1) the comparatively small amounts of heavy metals in the sludge-treated soils and (2) at that time, the last sludge applications were more than 20 years ago. Because total concentration of Cu, Ni, Cd and Zn were very closely correlated it was not possible to determine which metals or combinations of metals were producing these effects.

Brookes et al. (1997) measured biomass and heavy metals along a gradient obtained by sampling along the middle of adjacent plots of the Market Garden Experiment which had previously received continuous inorganic fertilizer (NPK), farmyard manure (FYM) or sewage-sludge, as described above. A very smooth gradient of heavy metals was measured and biomass increased linearly between the NPK and FYM plots, in line with increasing soil organic matter concentrations. However, after 120 μg

Zn g^{-1} soil, obtained from the second sludge soil sample, there was a smooth decline in biomass nearly to the level in the soil given inorganic fertilizer. I attribute this to heavy metals. It is exceedingly unlikely that organic pollutants, if present in the sludges initially, would have persisted for so long.

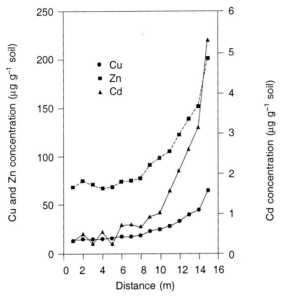

Fig. 4.1: Heavy metals along a soil transect of Wobum Market Garden Experiment.

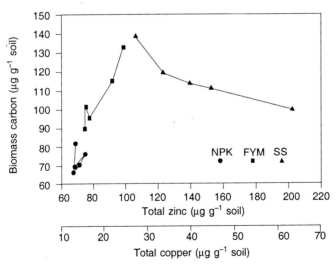

Fig. 4.2: Microbial biomass along a soil transect of Wobum Market Garden Experiment.

Chander and Brookes (1991a) reported biomass and organic C measurements in field experiments at Luddington (sandy loam, 15% clay) and Lee Valley (silt loam, 21% clay) experimental farms. Both soils received half- and full-rate dressings of metal-contaminated sludge and full rates of uncontaminated sludges. Both soils had similar pH's (5.6 to 5.9) and were under grass. The major findings, summarized, were:

1. Both Cu and Zn at about 2–3 times current EU permitted limits decreased soil microbial biomass in both the sandy loam (15 %) clay and silt loam (21% clay) soil.
2. At these soil concentrations, Cu at both sites decreased the biomass by about 40% compared to soils given uncontaminated sludge. Zinc decreased the biomass by about 40% in the sandy loam and 30% in the silty loam soils.
3. Nickel at about 2 to 3 times current EU limits did not affect microbial biomass in either soil. Similarly, Cd at twice current limits had no effect in the silt loam soil.

In a further field experiment, sewage-sludges, enriched with different rates of single metals (Zn, Cu and Ni) and metal combinations, were applied to a sandy loam soil (9% clay, pH around 6.5) in 1982 at Gleadthorpe Experimental Husbandry Farm. This experiment was used to investigate effects of single metals and combinations of metals on the biomass and on relationships between biomass and soil organic matter (Chander and Brookes, 1993). Main findings included:

1. Zinc at about 2.3 and Cu at about 4.9 times current permitted EU limits decreased soil biomass by about 40 and 50% respectively compared to soils receiving uncontaminated sludges.
2. A combination of Cu and Zn, each at about 1.5 times current EU limits decreased biomass by about 60% compared to soils given sludge. This suggests synergistic interactions between metals and biomass.

Other work has shown decreased microbial biomass formed during sludge decomposition after incorporation to soil. Two field experiments were established at Braunschweig, Germany with (in brief) the following treatments: inorganic fertilizer; uncontaminated sludge; metal-contaminated sludge - both at two rates (5 or 16 t ha^{-1} dry weight). Two sites were used, an old arable (pH 6.1 to 6.8) or ex-Woodland (pH 5.3 to 5.7) of the same soil type. After 9 years, total metal concentrations in some of the sludge-amended soils exceeded the German limits (mg kg^{-1}) of 300 Zn, 100 Cu, and 3 Cd, but not for Ni. On both soils, the biomass was increased each year with either rate of uncontaminated sludge. However there was less, or even no, increase in biomass following addition of metal-contaminated sludges, even at the higher rate (Fliessbach et al., 1994).

In a laboratory experiment, Chander *et al.* (1995) added sewage sludges enriched, or not, with the single metals Cu, Ni, Cd or Zn, to a sandy loam soil of the Woburn Market Garden Experiment. The unenriched (low metal) and metal-enriched (high metal) soils were added separately at four rates so that total soil metal concentrations were between 1 and 4 times the European Union (EU) current permitted limits. The main aim was to determine the maximum individual soil metal concentrations which decreased either the amount or activity of the biomass.

Sludge addition increased biomass C by about 30% at the lowest rate of application (40 t ha^{-1} dry weight) and by about 4.5 fold at the highest (160 t ha^{-1} dry weight) rate after four weeks, with no effects attributable to the metals. (It should be noted that these are much higher sludge rates than in the previous experiment.) However, during the longer 64 week incubation, the biomass declined exponentially in all treatments. Larger applications of high metal sludges caused final biomasses which were smaller than those given low metal sludge and no sludge. None of the single metals at the current EU permitted limits showed any adverse effects on biomass. However, Zn, Cu, or Cd, individually at about twice the EU limit, decreased biomass C by about 20% whereas Ni at four times the limit decreased the biomass by about 15%. Thus this suggests that although toxic effects of heavy metals may be delayed by the antagonistic effect of enhanced substrate availability via the sludge in the short-term, once the sludge has been decomposed the metals exert their negative effects upon the soil microbial biomass. In a review of the long-term implications of sludge metal bioavailability over time, McBride, (1995), postulated the 'time bomb theory', suggesting that as the organic component of sludge degrades over time, organically complexed metals are released into the soil matrix and thus become available to soil biota.

Microbial Activity Measurements

The fertility of all natural ecosystems depends upon the mineralization and immobilization of soil organic C, N, P and S and on the decomposition of plant and animal residues that enter soil. These processes are all mediated by a suite of complex metabolic process provided by the soil microbial biomass and higher soil organisms (micro-, meso-, and macrofauna). Since heavy metals are proven inhibitors of most enzymic reaction in soil (e.g. Tyler, 1981) it is essential that heavy metals are not permitted to accumulate in soils to concentrations at which these processes are inhibited or suppressed.

Microbial activities (e.g. respiration and N mineralization) can fluctuate enormously, even over a few days (temporal fluctuations), under field conditions, even in pristine soils—see Brookes (1994) for a discussion of this. However, under controlled laboratory conditions of suitable moisture (usually between 40 and 50% water holding capacity) and temperature (usually between 15 and 25 °C) the microbial mineralization of both C and N in sieved (2 to 6.25 mm) soil proceed practically linearly for long periods and can be determined accurately and precisely (Jenkinson and Powlson, 1976).

Most work indicates that mineralisation of both C and N are little, if at all, affected at soil metal concentrations at around maximum current EU limits. Thus, Tyler (1981) reported that microbial respiration was not depressed below about 1000 µg Cu or Zn g^{-1} soil in forest soils. Similarly, there was no apparent change in soil C or N mineralisation in arable soils containing heavy metals from past sewage-sludge applications compared to similar 'low-metal' control soils (Brookes and McGrath, 1984; Brookes et al., 1984). This was despite the fact that the biomasses in the high-metal soils were often only half the size of the biomasses in the low-metal soils. Thus these much smaller microbial populations in the high-metal soils were able to mineralize soil organic matter and decompose crop and animal residues to the same extent and at the same rate as the much larger populations in the low-metal soils.

Microbial Specific Activity Measurements

From the above, measurements of microbial biomass appear useful indicators of environmental stress due to heavy metals, while microbial activity measurements at soil metal concentrations commonly occurring in agricultural soils may not. ED: change is ok. However, combining the two measurements, to give rates of activities per unit of biomass (biomass specific activities), has been shown to be a much more subtle indicator of environmental stress. In summary, evidence is accumulating that environmental stress such as that produced by heavy metals causes a diversion of energy from biosynthesis to microbial activity in the soil microbial biomass.

Thus, rates of CO_2 evolution (µg CO_2-C g^{-1} soil d^{-1}) OK now Edfrom both low- and high-metal soils from the Woburn Market Garden Experiment were not significantly different during laboratory incubations. However, rates of biomass specific respiration (measured as µg C respired g^{-1} biomass C d^{-1}) was twice the rate in the high-metal soils than in the low-metal ones (Brookes and McGrath, 1984).

From such observations, Killham (1985) developed a simple bioassay procedure based on proportionating ^{14}C-labelled glucose between biomass-^{14}C and ^{14}C evolved. He showed, for a given increase in stress, that the ratio: [(respired ^{14}C):(biomass ^{14}C)] was, on average, twice as great as the magnitude of the decrease in either respiration or dehydrogenase activity.

Similarly, about 10% more total and 20% more ^{14}C-labelled CO_2 were evolved from a high-metal than a low-metal Woburn soil during the first five days following addition of ^{14}C-labelled glucose and maize (Chander and Brookes, 1991b). In contrast, about 30% less ^{14}C-labelled biomass was synthesised per unit of added substrate, which is in line with the findings of Killham (1985). Similarly, Chander and Brookes (1991c) showed that plant-derived inputs of organic ^{14}C were about 20% less in the high-metal than low-metal soil. Also, the biomasses in the high-metal soil contained about 30% less of this ^{14}C-labelled organic C than in the low-metal soil. These results suggest that two mechanisms operate in causing smaller biomasses in metal-contaminated soils. These are (1) decreased C inputs from growing plants and (2) decreased efficiency of conversion of this C into new biomass. The latter mechanism appears to be the more important.

Results from our laboratory and elsewhere indicate that measurements of linked parameters such as microbial biomass and soil respiration, giving microbial specific respiration, are much more useful than either measurement standing alone. Indeed, it could be argued that 'stand alone' measurements can only really be interpreted when dealing with well-designed field experiments with proper 'control' plots. Non-experimental field data, be it from agricultural or unmanaged ecosystems is usually difficult to interpret because of lack of suitable 'controls' with which to compare it. To overcome this problem, Brookes (1994) suggested that linked parameters such as microbial specific respiration (or the link between biomass C and total soil organic C, as discussed above) may itself constitute an internal control. Thus, when soils deviate much from biomass specific activity or biomass specific C ratios perceived as normal for the particular management, soil type or climate, it may provide an 'early warning' that the soil ecosystem is under stress and that more research is needed.

EFFECTS OF HEAVY METALS ON SPECIFIC MICRO-ORGANISMS OR FUNCTIONAL GROUPS

Soil is a complex material with, for example, cation exchange, buffering properties and chelation reactions which may be quite different to those in simple media. These soil properties will have very different effects on metal bioavailability than would occur in less complex systems. Thus, while there has been a great deal of research into the effects of heavy metals on soil micro-organisms grown in aqueous or solid culture media, it is impossible to extrapolate the results to the soil environment with any certainty. In most cases, it is equally difficult to study single species in vivo and their use in this field is therefore limited. In addition most important soil processes, for example the mineralization or formation of soil organic matter, depend upon the functioning of the microbial community as a whole rather than the activity of individual species. At present, there seem few possibilities to use single microbial genus or species as biological indicators, as discussed below.

N_2-Fixation (Legume-Rhizobium Symbiosis)

Biological N_2-fixation requires the presence of the enzyme nitrogenase, which occurs only in micro-organisms. The three main types of N_2-fixing micro-organisms, autotrophs, heterotrophs and symbionts, vary in the amounts of N_2 they can fix in temperate regions. In order to directly convert the measurements on nitrogenase activity to N_2-fixed requires ^{15}N analytical techniques which may not always be feasible. However, the conversion of acetylene to ethylene and subsequent chromatographic analysis provides a simple and sensitive test for nitrogenase activity provided the results are not extrapolated too far (Giller and Day, 1985). There is certainly some data to suggest that both heterotropic and non-symbiotic N_2-fixation could be suitable tests for soil pollution by heavy metals. This is discussed below.

Heterotrophic N_2-fixation

Free-living heterotrophs have very slow rates of N_2-fixation in most soils which are difficult to measure. Fixation can also be variable, developing measurably in some but not in other apparently similar soils. This lack of reproducibility convinced Lorenz et al. (1992) that difficulties in optimizing and standardizing incubation conditions presently effectively prevents the use of heterotrophic N_2-fixation measurements as indicators of effects of heavy metals on soil ecosystems.

However, this conclusion was based on unamended soils. Following amendment with up to 5000 µg glucose g^{-1} soil, both Brookes et al. (1984) and Lorenz et al. (1992) found that heavy metals close to, or less than, current EU permitted limits decreased heterotrophic N_2-fixation by up to 90%. More work is certainly required to standardize operating conditions if heterotrophic N_2-fixation is to have any value as a biological indicator in metal-contaminated soils.

Autotrophic N_2-fixation

Field measurements of autotrophic N_2-fixation by Cyanobacteria are extremely variable. For example, Witty et al. (1979) estimated that up to 28 kg N ha^{-1} y^{-1} could be fixed on soils of the Broadbalk Continuous Wheat Experiment. However, amounts varied five-fold in their work. In contrast, when I tried to measure nitrogenase activity in situ over a year, I did not find measurable amounts of nitrogenase activity at any time.

However, measurement of nitrogenase activity as an indicator of N_2-fixation by Cyanobacteria under standard laboratory conditions could be a possible indicator of heavy metal effects. However, the measurements would reflect the potential of contaminated and uncontaminated soils for N_2-fixation rather than fixation in the field.

Brookes et al. (1986a incubated moist, fresh low- and high-metal soil from the Woburn Market Garden Experiment under laboratory conditions of 20 °C day, 16 °C night, 16 h day and 50% water holding capacity. In the low-metal soil there was an initial lag-period of about 14 d, then the rate of acetylene reduction increased rapidly, peaking on day 28 then declining slowly until day 118. In contrast, acetylene reduction had barely commenced by day 50 in the high-metal soil. It then increased regularly but much more slowly than in the low-metal soil, and until the experiment ended. There was about three times more acetylene reduction in the low- than high-metal soil by day 118. Similarly, the low-metal soil fixed about ten times more $^{15}N_2$ in 24 h than did the high-metal soil. A thick crust of Cyanobacteria formed on the surface of the low-metal soil by day 25 but hardly any was obvious on the high-metal soil by day 118.

In a further experiment (Brookes *et al.*, 1986a), soil was sampled at 40 cm intervals along the middle of a low- and high-metal plot. Concentrations of EDTA-extractable Zn, Cu, Ni and Cd increased in a curvilinear manner between the low- and high-metal plots. In contrast, total acetylene-reduction decreased linearly with increasing soil metal concentration during the 60 d experimental period. It was halved at about 50 µg total Zn, 20 µg Cu, 2.5 µg Ni and 3 µg Cd. Because the soils contained all these metals, in closely correlated concentrations, it is not known which metal or combination of metals induced these effects. However, apart from Cd, the maximum concentrations of individual metals were well within individual EU maximum permitted limits (about 30% of the soils contained above 3 µg Cd g^{-1} soil).

Lorenz *et al.* (1992) obtained a metal gradient by mixing different proportions of a low- and a high-metal Woburn soil. They also reported inhibition of the growth of Cyanobacteria and decreased acetylene reduction in the high-metal soils but only at the maximum concentrations (a mixture of 83% sludge and 17% FYM soil or 100 sludge soil.

The lag phase of 14 d that Lorenz *et al.* (1992) determined for the high-metal soil was much shorter than the 50 d reported by Brookes et al. (1986a). It seems likely that this was because, however carefully the mixing was done, particles—or islands—of uncontaminated soils would exist side by side with contaminated soils. In contrast the metals would be very much more homogenously distributed when sampled along a natural gradient in the field.

Cyanobacteria also appear very sensitive to incubation conditions in other ways. It is interesting that Lorenz *et al.* (1992) failed to get them to grow even on uncontaminated Luddington (UK) soil (similar to Woburn soil in most respects). They also reported that in other Swedish experiments Cyanobacteria failed to grow even when uncontaminated soils were incubated under apparently ideal conditions. If Cyanobacteria and autotrophic N_2-fixation are to be used as indicators of metal-contamination in soil we need to know more about how to culture them on soils under laboratory conditions.

Symbiotic N_2-fixation

The ability of various *Rhizobium* species to infect legumes and to fix atmospheric nitrogen is well known. Symbiotic N_2 fixation by *Rhizobium leguminosarum* biovar *trifoli* in symbiotic association with *Trifolium repens* (white clover), along a transect of soils of the Woburn Market Garden Experiment, was decreased by 50% or more in pots of soil containing above 334 μg Zn, 99 μg Cu, 27 μg Ni and 10 μg Cd g^{-1} soil (McGrath *et al.*, 1988). The nodules were small and white and easily distinguishable from the much larger, pink nodules found on clover grown in low-metal soils, which were activity-fixing N_2. Yields of clover of the high-metal soils were restored to those of the low-metal soils by applying inorganic N, i.e. the effects were not caused by phytotoxicity. Rather, McGrath et al. (1998) proved that the decreased clover yields and N_2-fixation were because the clover root nodules were ineffective in fixing N_2, although nodulation did occur in the high-metal soil. Free-living *Rhizobium* sp. added to soil were also less able to survive in high-metal than low-metal soils from the same experiment (Giller et al., 1993). Giller et al. (1989) showed that the ineffectiveness of the *Rhizobium* sp. in fixing N_2 was not directly due to metal toxicity. Instead, the metals had selected for the survival of a single *Rhizobium* sp. genotype, which was ineffective in N_2-fixation.

An understanding of the interactions between heavy metals and the legume-Rhizobium symbiosis is clearly important. However, due to the long bioassay periods (more than 2 weeks) required, it seems unlikely that the legume-*Rhizobium* symbiosis will be developed as a routine indicator of soil pollution by heavy metals 'for non-specified use'. While the sampling, preparation and measurement of clover dry matter yields, total %N and total plant ^{15}N are feasible, the work involved is considerable.

Mycorrhizae

There is evidence for depressed rates of mycorrhizal infection in metal-contaminated soils at 'agricultural' concentrations (e.g. Koomen et al., 1990). However they also suggested that the metals may have encouraged the proliferation of metal-resistant mychorrhizae already present. As McGrath et al. (1995) pointed out, sludge applications invariably increase soluble P concentrations in soil which may confound the results by also suppressing mycorrhizal infection. The analytical difficulties and skill required in measuring mycorrhizal infection in roots, both in pots and field soils would probably preclude mycorrhizae as a bioindicator. It can also take a considerable time for measurable mycorrhizal infection to occur in pot experiments, which may also cause analytical problems.

In summary, there are problems in monitoring and interpreting effects of heavy metals on single microbial species or microbial groups and their use as bioindicators of heavy metal pollution appears negligible at this stage.

BIOCHEMICAL MARKERS AS BIOLOGICAL INDICATORS OF EFFECTS OF HEAVY METALS

Markers for Whole Biomass

Specific biochemicals extracted from the cells of the soil micro-organisms can be used theoretically to investigate the effects of heavy metals on soil microbial ecosystem functioning. Some constituents can be used to gauge heavy metal effects upon the whole biomass. Jenkinson and Ladd (1981) laid down stringent criteria to be met:

1. The constituent must be present in the same concentration in the biomass in different soils.
2. It must be present only in living organisms, i.e. it must not occur exocellularly.
3. It must be capable of being extracted quantitatively from soil.
4. There must be an accurate and precise method(s) to estimate it.

While no cell constituent fully meets these conditions, several have been tried. These include adenosine 5′ mono-, di- and triphosphate, muramic acid, N-acetylyglucosamine and the nucleic acids. The above, with the exception of adenosine 5′ triphosphate (ATP), all occur exocellularly in soil in sufficient quantities to violate the above criteria (Jenkinson and Ladd, 1981). Soil ATP analyses however have proved very useful in assessing heavy metal effects on the soil ecosystem.

Adenosine 5′ Triphosphate

Adenosine 5′ triphosphate occurs in all living cells but has only a transitory existence in dead cells or exocellularly. Once extracted from soil it may be analysed with high accuracy and sensitivity by the fire-fly luciferin-luciferase enzyme system. A successful method of extracting ATP from soil must (1) release all ATP from the microbial cells, and (2) inactivate ATPases and phosphatases, so preventing ATP hydrolyis. We use a reagent based on a mixture of trichloroacetic acid (TCA), paraquat and phosphate (P), developed and described by Jenkinson and Oades (1979). The phosphate and paraquat prevent sorption of the ATP by positive and negative sites respectively on soil surfaces and the TCA provides maximum inhibition of the enzymic hydrolysis of ATP. Other extraction reagents have been proposed, for example the sulphuric acid-phosphate based reagent of Eiland (1983) which gives similar results. However, it is essential that neutral or alkaline reagents are not used under any circumstances, (for example the sodium bicarbonate-chloroform based reagent of Paul and Johnston, 1977) as they do not inhibit enzymic dephosphorylation of ATP and give incorrect, low values (Brookes et al. 1987a).

Due to its unique role in cellular energetics, it would seem likely that ATP would provide a better indication of microbial activity than total biomass. However, for reasons that we do not yet understand, soil ATP content is very closely correlated with total soil biomass content in soils, not with activity (Jenkinson, 1988), in soils incubated with and without substrates such as straw (Ocio and Brookes, 1990) or glucose and ryegrass (Chander and Brookes 1991b). In the latter two cases the addition of the substrates caused several-fold increases in rates of CO_2 evolution, a reliable indicator of metabolic activity. Addition of the substrates also caused both biomass and ATP to increase by up to two-fold but the concentrations of ATP in the biomass remained constant and at the same concentration as in unamended soils (about 11 μmol ATP g^{-1} soil).

Brookes and McGrath (1984) reported that heavy metals from past sludge additions in the Woburn Market Garden Experiment caused biomass decreases of up to 50% compared to similar soils given FYM (see above). Similar decreases in ATP in the same experiment were also measured so that the biomasses in the low- and high-metal soils had the same ATP concentrations (again around 10 to 12 μmol ATP g^{-1} soil) despite decreases in biomass and increases in the specific respiration of this biomass in the high-metal soils.

Chander and Brookes (1991b) incubated both low- and high-metal soils from the same experiment at 25 °C and 40% WHC with and without separate additions of 5000 μg C g^{-1} soil as ryegrass or glucose for up to 50 days. Again, remarkably close linear corrrelations between biomass and ATP were found, irrespective of metal concentrations or amendment with the two very different substrates, ryegrass and glucose.

The value of ATP analyses in this work is that it is a completely independent estimate of biomass yet correlates remarkably closely with biomass C and other biomass estimates by fumigation-extraction (see above). This gives considerable confidence in both types of analyses in research into effects of heavy metals on soil microbial ecosystem functioning.

The finding of identical biomass ATP concentrations in unamended and substrate-amended soils, despite huge differences in activities, 'biomass standing crops' and soil metal concentrations ranging from background to above current EU permitted limits was unexpected. The true biological significance is currently unknown. What these methods cannot do is to shed light upon possible differences in the community structure of the biomasses in high- and low-metal soils.

Adenine Nucleotide and Adenylate Energy Charge in Metal-contaminated Soils

The adenylate energy charge {AEC= [(ATP) + (0.5ADP)] /[(ATP) + (0.5ADP) + (AMP)]} is defined as a linear measure of the metabolic energy stored in the adenine nucleotide pool of ATP, adenosine 5′ diphosphate (ADP) and adenosine 5′ monophosphate (AMP) (Atkinson 1977). Most data has come from estimates obtained in vitro and AEC's between about 0.95 and 0.80 indicate a highly metabolically-active population undergoing rapid cellular division and biosynthesis. Values of AEC between about 0.8 to 0.4 indicate a stressed population with a low metabolic rate and incapable of much cellular biosynthesis. Adenylate energy charges lower than about 0.4 indicate a moribund or dying population, although microbial spores may have an AEC lower than 0.1. The potential value of AEC measurements in the work described here is that it is a ratio and thus detailed knowledge of past site history may not be critical, which is a limitation of many other approaches.

So far, most AEC measurements have been made on organisms grown in vitro. Brookes et al. 1987b) measured the AEC in a low- and high-metal soil of the Woburn Market Garden Experiment. The soil ATP concentration, and total adenine nucleotide pool were significantly lower in the high-metal soil, indicating, as found previously, a smaller total soil microbial biomass. However, the AEC's of the low- and high-metal soil were both high (0.85 and 0.89 respectively) and comparable to others reported previously for moist soils extracted with acidic reagents (e.g. Tateno 1985). Therefore, although, as discussed previously, several indices of microbial activity are considerably decreased in high-metal soils, this is not reflected in a lower AEC. This suggests that the magnitude of soil AEC's may not be a valid indicator of environmental stress.

Markers for Bacteria and Fungal Biomass

The above methods refer to biological markers for the entire microbial biomass. Some progress has also been made in splitting the biomass at least into its fungal and bacterial components. Some approaches relevant to determining the effects of heavy metals on the soil fungal and bacterial communities are discussed below.

Direct Microscopy

Total microbial biomass may be differentiated into spherical and cylindrical forms (fungi and actinomycetes) and the spherical further subdivided into a bacterial size class and an above-bacterial size class by visual or automated counting of suitably stained organisms in appropriately prepared soil suspensions in agar (e.g. Jenkinson *et al.*, 1976). The technique is tedious, requires considerable skill and (often subjective) judgement and is not generally very popular. Nevertheless it does have the huge advantage that it is a direct measurement of the entire biomass and also reveals something of its complexity, unlike most indirect methods. Brookes et al. (1986a) found, as previously, about twice as much total biomass in a low- than high-metal Woburn soil measured by microscopy. However the ratio of fungal to bacterial biomass were very similar in both cases (6.4 and 5.4 respectively). Therefore, although the heavy metal decreased the biomass 'standing crop' they did not alter bacterial/fungal ratios. Of course it is quite conceivable that the metals caused other changes to the microbial community structure which were not detected.

Phospholipid Fatty Acids

The ester-linked fatty acids in the phospholipids (PLFAs) are considered the most sensitive and useful chemical measures of microbial community structure. The fungal and bacterial components of the microbial biomass can be determined by specific 'signature' PFLAs. For example, bacteria characteristically contain odd-chain, methyl-branched and cyclopropane fatty acids. The PLFAs in fungi are typically saturated, even-chained, polyenoic fatty acids. Many actinomycetes contain methyl-branched tuberculostearic acid (Tunlid and White, 1992).

In both a forest and arable soil, the double-unsaturated 18.2ω6 PFLA increased proportionately, indicating a shift to fungi two weeks following Zn addition. There were also indications of changes in the proportions of several individual bacterial PFLAs, indicating shifts within the bacterial communities following Zn amendment of the soils (Frostegård et al., 1996).

Much more complex changes in PLFAs were found by Frostegård et al. (1993) at six months after separate additions of Cd, Cu, Ni or Zn at different concentrations to a forest humus soil and an arable soil. In summary, PFLAs indicative of actinomycetes increased in the forest soil but tended to decrease in the arable soil. Various types of bacterial PLFAs increased in all metal-contaminated arable soils but were unaffected by metals in the forest soil. The fungal fatty acid, 18:2ω6, generally increased in response to increasing metal concentrations in the arable soils except following Cu amendment, where it decreased. Effects on PFLA patterns occurred at metal concentrations similar to, or lower than, those at which effects on ATP, respiration or total PFLAs occurred.

Changes in PLFA profiles have now been observed in soils containing heavy metals at below current EU permitted limits (Abaye et al., 2000). For example, such soils contained both a larger number and greater concentrations of hydroxy- and cyclo-propyl fatty acids than similar butuncontaminated soils. The contaminated soils also contained larger concentrations of lipopolysaccharide hydroxy-fatty acids. These findings indicate a larger proportion of gram-negative bacteria in the metal-contaminated soils.

Soil Ergosterol Content

Ergosterol (ergosta-5,7,22-trien-3B-ol) is the predominant sterol in most fungi (Tundlin and White, 1992). Methods to measure soil ergosterol have been developed (e.g. Grant and West, 1986) and proposed as a way to estimate the soil fungal biomass content. The basic procedure involves extraction of the ergosterol from soil with methanol, followed by saponification and then re-extraction with hexane. The ergosterol is then determined by HPLC using a UV detector.

It is not currently known if the fungal biomass in different soils has a very constant ergosterol content, as with ATP. Certainly, in vitro, ergosterol contents can vary at least three-fold, depending upon species and growth conditions. West et al. (1987) considered that ergosterol analyses were most useful "to quantify *changes* in the fungal populations of soils". Frostegård and Båth (1996), however, showed that the PLFA 18:2ω6 (see above) was closely correlated (r=0.92) with soil ergosterol content, which indicates that both components are measuring the soil fungal biomass. Three independent biomass measurements (biomass C by fumigation-extraction, substrate-induced respiration and ATP) closely followed decreases in soil ergosterol content along a heavy metal gradient from a Finnish Cu-Ni smelter (Fritze *et al.*, 1989). Soil ergosterol may therefore have potential as an indicator of fungal biomass in metal-contaminated soils but this requires further evaluation.

Soil Enzymes

If enzymes are to be used as bioindicators it may be important to differentiate between exocellular and endocellular enzymes. For example, Brookes (1994) reported less dehydrogenase activity in metal-contaminated soils of the Woburn experiment than in similar uncontaminated soils. In contrast, soil phosphatase activity was unaffected by the metals. Soil phosphatase activity can occur exo- and endocellularly while dehydrogenase only functions within the living cell in soil. These results, at face value, suggest therefore that dehydrogenase is a more reliable indicator than phosphatase of effects of metals on soil microbial activity. However, Chander and Brookes (1991d) showed that the dehydrogenase assay is sensitive to interference from Cu in soil because Cu stops the red colour developing of the artificial end-product (triphenyl formazan). Thus, when Cu is added to the soil in sludges, or in ionic form in solution, this abiological reaction can be incorrectly interpreted as decreased dehydrogenase activity caused by Cu. Other common heavy metals e.g. Ni, Cd or Zn do not cause this effect. As the interference is specific to Cu among the metals tested, dehydrogenase activity may be useful as an indicator of other heavy metals.

NOVEL APPROACHES

DNA Technology

DNA, or gene, technology is proving to be an awesomely powerful new science with vast and unknown potential. The ability to transfer genetic material between species, even between the plant and animal kingdoms, gives rise to enormous potential benefits in many areas such as agriculture, drug manufacture and medicine. Many would also argue that the risks that accompany such procedures are as massive as the potential benefits.

Recently, Dolly the sheep, the world's first mammal to be cloned from a somatic cell, was announced by the Roslin Institute, UK, and many countries are currently drafting legislation to prevent similar technology being used on humans. However, the progress with these methodologies in soil science has been desperately slow by comparison. This is not merely a reflection of too little research money being available. In fact, research into soil applications of DNA technology, for example to detect different genetic diversities in soil, has attracted a large amount of funding, sometimes by diverting money from other, apparently less exciting but potentially more productive, areas of soil science.

The major problems which are currently limiting progress are the nature both of soil and of the soil microbial biomass itself. Soil is an extremely heterogenous material, containing both negatively and positively charged surfaces capable of absorbing DNA, enzymes and co-factors. Soil also contains large amounts of humic substances which are powerful inhibitors of many enzymes such as restriction enymes and polymerases and of nucleic acid hybridisation reactions, all of which are required to function at high efficiency.

Moreover, soil microbial populations are also relatively large and heterogenous. Most of the new methods only work well with cultures of single species, or, at best, simple mixtures of species. The base sequence ratio of guanosine to cytosine in different soil organisms also varies widely, from about 40 to 80% which causes problems in optimizing analytical procedures (P. R. Hirsch, pers. comm.). However, given some potential limitations of molecular based approaches, these techniques are now being successfully applied to the study of soil metal impacts on microbial populations. Techniques such as denaturing gradient gel electrophoresis of 16S, PCR amplified rDNA (Muyzer et al. 1996), and 16S rDNA sequencing (Stackerbrand et al., 1996) are now being successfully applied to the study of heavy metal impacts on soil microbial communities. Sandaa et al. (1999), reported the application of 16S rDNA molecular techniques to study microbial population changes following sewage-sludge application to soils, and demonstrated significant changes in microbial communities based on the diversity of the 16S and 23S gene sequences.

Other molecular approaches to the study of metal impacts on microbial communities are based on the study of important 'functional' genes. Stephen et al. (1999) studied the impact of heavy metals on the diversity of the *amo*A gene, that codes for the production of the ammonium monoxgenase enzyme, in chemoautotrophic ammonia oxidiser bacteria such as *Nitrosomonas* sp. This is an important rate limiting step in the mineralization of soil N. Techniques based on the extraction of community DNA from soil, and PCR of a 490bp *amo*A fragment with a set of degenerate primers Rotthauwe et al. (1997), demonstrated significant changes in the community structure of ammonia oxidizing populations, whilst traditional techniques used to study the total number of these organisms showed no changes.

Whilst evidence for the successful application of molecular approaches to the study of metal impacts on soil micro-organisms have only started appearing in the scientific literature in the last couple of years, it is clear that the advent and continued rapid development of such technologies will provide additional tools for the study of metal impacts on population diversity and functionally important gene sequences.

Lux Genes

Bioluminescence-based biosensors are being developed as indicators of soil pollution. Paton et al. (1997) described an experiment where a soil isolate of *Rhizobium leguminosarum* bv. *trifoili* was marked with a *lux* gene cassette to enable the expression of bioluminescence. They found that bacterial bioluminescence responded sensitively and negatively to increasing heavy metal concentration in solution, in the order Cd > Ni > = Zn > Cu in an acute test and in the order Cd > Ni = Zn = Cu in a chronic test. On the basis of this work they considered that it may be possible to develop a 'microbial battery test system' to assess both chronic and acute responses of bacteria from different ecological niches to pollution in the soil system. So far, however, the difficulty of detecting bacterial bioluminescence in intact soil has not been overcome.

Biolog Assessments of Substrate Utilization

Microbial communities which differ phenotypically may also differ in the range of carbon substrates which they are capable of utilizing (Garland and Mills, 1991). The Biolog microtitre plate system for identifying micro-organisms offers a simple and fast potential method to face the soil heterotrophic microbial community with up to 95 separate substrates. Knight et al. (1997) tested if the Biolog approach could show differences in the ranges of substrates which could be metabolized by the micro-organisms from uncontaminated and metal-contaminated soils. The soils were obtained by adding Cu, Cd or Zn, in simple salt forms, at around current maximum permitted EU concentrations for agricultural soils. After a 3-year equilibration, microbial biomass was measured and compared with substrate utilization patterns of micro-organisms using the Biolog approach. The metabolic potential of the extracted microbial populations were decreased by both Cu and Zn, and also generally in lower pH soils. In contrast, total microbial biomass was unaffected except for a significant decrease at the lowest pH (4.1) with Cu.

This work needs to be repeated with freshly sampled field soils given long-term sludge applications rather than single doses of simple metallic salts. In principle, the use of Biolog microtitre plates is rapid and could offer a fast screening technique to detect stressed populations in situations where an appropriate control soil is available. However the relations between Biolog plate results and microbial ecosystem functioning in metal-contaminated and uncontaminated soils awaits evaluation. In addition the interference of metal such as Cu on the colour development reaction in the Biolog plates should not be confused to its impact on the microbial community itself.

CONCLUSIONS

1. There is no single microbiological property that is ideal for monitoring soil pollution caused by different types of heavy metals.
2. Problems in interpretation of environmental measurements are common because of lack of suitable control, or baseline, measurements.
3. There are advantages in using measurements that have some form of internal control e.g. microbial biomass as a percentage of total soil organic matter, as it helps side-step the lack of environmental control data.
4. The rapid development of molecular techniques and their continued successful application to the study of soil microbial ecology and function, provides significant future potential for the monitor-

ing of soil pollution impacts. However, molecular tools are comparatively expensive, the level of information these techniques provide, over and above more traditional microbial indicator techniques, needs to be clearly demonstrated.

ACKNOWLEDGEMENTS

I thank A. Chaudri, P. Hirsch and B. Knight for helpful discussions and C. Grace and H. Richardson for help in preparing the manuscript. I also thank C. Grace and B. Tiwari (North Eastern Hill University, Shillong, India) for allowing me to publish data in Figs. 4.1 and 4.2.

REFERENCES

Abaye D. (2000) Lipid derivatives and the relationships between microbial biomass, community structure and activity in soils. PhD thesis. University of Nottingham.

Atkinson, D.E. (1977) *Cellular energy metabolism and its regulation.* Academic Press, New York.

Ayanaba, A., Tuckwell, S.B. and Jenkinson, D.S. (1976) The effects of clearing and cropping on the organic reserves and biomass of tropical forest soils. *Soil Biology and Biochemistry* 8, 519–525.

Brookes, P.C. (1994) The use of microbial parameters in monitoring soil pollution by heavy metals. *Biology and Fertility of Soils* **19**, 269–279.

Brookes, P. and McGrath, S.P. (1984) Effects of metal toxicity on the size of the soil microbial biomass. *Journal of Soil Science* **35**, 341–346.

Brookes, P.C., McGrath, S.P., Klein, D.A. and Elliott, E.T. (1984) Effects of heavy metals on microbial activity and biomass in field soils treated with sewage sludge. In: *Environmental contamination.* CEP Consultants Ltd, Edinburgh, pp. 574–583.

Brookes, P.C., McGrath, S.P. and Heijnen, C. (1986a) Metal residues in soils and their effects on growth and nitrogen fixation by blue-green algae. *Soil Biology and Biochemistry* **18**, 345–353.

Brookes, P.C., Heijnen, C.E., McGrath, S.P. and Vance, E.D. (1986) Soil microbial biomass estimates in soils contaminated with metals. *Soil Biology and Biochemistry* **18**, 383–388.

Brookes, P.C., Newcombe, A.D. and McGrath, S.P. (1987) Adenylate energy charge in metal-contaminated soil. *Soil Biology and Biochemistry* **19**, 219–220.

Brookes, P.C., Newcombe, A.D. and Jenkinson, D.S. (1987) Adenylate energy charge measurements in soil. *Soil Biology and Biochemistry* **19**, 211–217.

Brookes, P.C., Tiwari, B.K. and Grace, C.A. (1997) The role of microbial parameters in monitoring soil pollution by chromium. In: *Chromium Environmental Issues.* (Eds. S. Canali, F. Tittarelli, P. Sequ). Franco Angeli. pp. 40–59.Chander, K. and Brookes, P.C. (1991a) Effects of heavy metals from past applications of sewage sludge on microbial biomass and organic matter accumulation in a sandy loam and a silty loam UK soil. *Soil Biology and Biochemistry* **23**, 927–932.

Chander, K. and Brookes, P.C. (1991b) Microbial biomass dynamics during the decomposition of glucose and maize in metal-contaminated and non-contaminated soils. *Soil Biology and Biochemistry* **23**, 917–925.

Chander, K. and Brookes, P.C. (1991c) Plant inputs of carbon to metal-contaminated soil and effects on the soil microbial biomass. *Soil Biology and Biochemistry* **23**, 1169–1177.

Chander, K. and Brookes, P.C. (1991d) Is the dehydrogenase assay invalid as a method to estimate microbial activity in Cu contaminated soils? *Soil Biology and Biochemistry* **23**, 901–915.

Chander, K. and Brookes, P.C. (1993) Effects of Zn, Cu and Ni in sewage sludge on microbial biomass in a sandy loam soil. *Soil Biology and Biochemistry* **25**, 1231–1239.

Chander, K., Brookes, P.C. and Harding, S.A. (1995) Microbial biomass dynamics following addition of metal-enriched sewage sludges to a sandy loam. *Soil Biology and Biochemistry* **27**, 1409–1421.

Eiland, F. (1983) A simple method for quantitative determination of ATP in soil. *Soil Biology and Biochemistry* **15**, 665–670.

Fliessbach, A., Martens, R. and Neber, H.H. (1994) Soil microbial biomass and activity in soils treated with heavy metal contaminated sewage sludge. *Soil Biology and Biochemistry* **26**, 1201–1205.

Fritze, H., Niini, S., Mikkola, K. and Mäkinen, A. (1989) Soil microbial effects of a Cu-Ni smelter in south western Finland. *Biology and Fertility of Soils* **8**, 87–94.

Frostegård, Å., Tunlid, A. amd Bååth, E. (1993) Phospholipid fatty acid composition ,biomass and activity of microbial communities from two soil types experimentally exposed to different heavy metals. *Applied and Environmental Microbiology* **59**, 3605–3617.

Frostegård, A. and Bååth, E. (1996) The use of phospholipid fatty acid analysis to estimate bacterial and fungal biomass in soil. *Biology and Fertility of Soils* **22**, 59–65.

Frostegård, Å, Tunlid, A. and Bååth, E. (1996) Changes in microbial community structures during long-term incubation in two soils experimentally contaminated with metals. *Soil Biology and Biochemistry* **28**, 55–63.

Garland J. L. and Mills A. L. (1991) Classification and characterisation of heterotrophic microbial communities on the basis of patterns of communty-level sole-carbon-source utilization. *Applied and Environmental Microbology* **57**, 2351–2359.

Giller, K.E. and Day, J.M. (1985) Nitrogen fixation in the rhizosphere. Significance in natural and agricultural systems. In: Fitter, A.H., Atkinson, D., Read, D.J. and Busher, M.B. (eds) *Ecological interactions in soil*. Spec. Publ. no 4, British Ecological Society. Blackwell Scientific Publications, Oxford, pp. 127–147.

Giller, K.E., McGrath, S.P. and Hirsch, P.R. (1989) Absence of nitrogen fixation in clover grown in soil subject to long-term contamination with heavy metals is due to the survival of only ineffective *Rhizobium*. *Soil Biology and Biochemistry* **21**, 841–848.

Giller, K.E., Nussbaum, A.M., Chaudri, A.M. and McGrath, S.P. (1993) *Rhizobium meliloti* is less sensitive to heavy metal contamination in soil than *R. leguminosarum* bv. *trifoli* or *R. loti*. *Soil Biology and Biochemistry* **25**, 273–278.

Grant, W.D. and West, A.W. (1986) Measurement of ergosterol, diaminopimelic acid and glucosamine in soil: evaluation as indicators of microbial biomass. *Journal of Microbial Methods* **6**, 47–53.

Hirsch, P.R. and Skinner, F.A. (1992) The identification and classification of *Rhizobium* and *Bradyrhizobium*. In: *Identification methods in applied and environmental microbiology*. Eds. R.G. Board, D. Jones and F.A. Skinner. Blackwell Scientific Publications, Oxford, pp. 45–65.

Jenkinson, D.S. and Powlson, D.S. (1976) The effects of biocidal treatments on metabolism in soil. V. A method for measuring soil biomass. *Soil Biology and Biochemistry* **8**, 209–213.

Jenkinson, D.S., Powlson, D.S. and Wedderburn, R.W.M. (1976) The effects of biocidal treatments on metabolism in soil. III. The relationship between soil biovolume, measured by optical microscopy, and the flush of decomposition caused by fumigation. *Soil Biology and Biochemistry* **8**, 189–202.

Jenkinson, D.S. and Oades, J.N. (1979) A method for measuring adenosine triphosphate in soil. *Soil Biology and Biochemistry* **11**, 193–199.

Jenkinson, D.S. and Ladd, J.N. (1981) Microbial biomass in soil: measurement and turnover. In: Paul, E.A. and Ladd, J.N. (eds) *Soil Biochemistry*, Vol 5, Marcel Dekker, New York, pp. 415–471.

Jenkinson, D.S. (1988) Determination of microbial biomass carbon and nitrogen in soil. In: Wilson, J.T. (ed) *Advances in nitrogen cycling in agricultural ecosystems*. CAB International, Wallingford, pp. 368–386.

Killham, K. (1985) A physiological determination of the impact of environmental stress on the activity of microbial biomass. *Environmental Pollution (Series A)* **38**, 204–283.

Knight, B.P., McGrath, S.P. and Chaudri, A.M. (1997) Biomass carbon measurements and substrate utilization patterns of microbial populations from soils amended with Cd, Cu or Zn. *Applied and Environmental Microbiology* **63**, 39–43.

McBride, M. B. (1999). Toxic metal accumulation from agricultural use of sludge: are the USEPA regulations protective? *Journal of Environmental Quality*. 24, 5–18.

Koomen, L., McGrath, S.P. and Giller, K.E. (1990) Mycorrhizal infection of clover is delayed in soils contaminated with heavy metals from past sewage sludge applications. *Soil Biology and Biochemistry* **22**, 871–873.

Lorenz, S.E., McGrath, S.P. and Giller, K.E. (1992) Assessment of free-living nitrogen fixation activity as a biological indicator or heavy metal toxicity in soil. *Soil Biology and Biochemistry* **24**, 601–606.

McGrath, S.P., Brookes, P.C. and Giller, K.E. (1988) Effects of potentially toxic metals in soil derived from past applications of sewage sludge on nitrogen fixation by *Trifolium repens* L. *Soil Biology and Biochemistry* **20**, 415–424.

McGrath, S.P., Chaudri, A.M. and Giller, K.E. (1995) Long-term effects of metals in sewage sludge on soils, micro-organisms and plants. *Journal of Environmental Microbiology* **14**, 94–104.

Ministry of Agriculture, Fisheries and Food and Department of Environment (1993a) Review of the rules for sewage-sludge application to agricultural land. Soil fertility aspects of potentially toxic elements. *Report of the Independent Scientific Committee*, MAFF Publications, London.

Ministry of Agriculture, Fisheries and Food and Department of Environment (1993b) Review of the rules for sewage-sludge application to agricultural land. Food safety and relevant animal health aspects of potentially toxic elements. *Report of the Independent Scientific Committee*, MAFF Publications, London.

Muyzer G. and Smalla K (1998) Application of denaturing gradient gel electrophoresis (DGGE) and temp[erature gradient gel electrophoresis (TGGE) in microbial ecology. *Antoine Van Leeuwenhoek International Journal of General and Molecular Microbiology*. **73**, 127–141.

Ocio, J.A. and Brookes, P.C. (1990) An evaluation of methods for measuring the microbial biomass in soils following recent additions of wheat straw and the characterization of the biomass that develops. *Soil Biology and Biochemistry* **22**, 685–694.

Paton, G.I., Palmer, G., Burton, M., Rattray, E.A.S., McGrath, S.P., Glover, L.A. and Killham, K. (1997) Development of an acute and chronic exotoxicity assay using *lux*-marked *Rhizobium leguminosarum* biovar *trifoli*. *Letters in Applied Microbiology* **24**, 296–300.

Paul, E.A. and Johnston, R.L. (1977) Microscopic counting and adenosine 5′ triphosphate measurements in determining microbial growth in soil. *Soil Biology and Biochemistry* **34**, 163–269.

Powlson, D.S., Brookes, P.C. and Christensen, B.T. (1987) Measurement of soil microbial biomass provides an early indication of changes in total soil organic matter due to straw incorporation. *Soil Biology and Biochemistry* **19**, 159–164.

Rotthauwe J. H., Witzel K. P. and Liesack W. (!997). The ammonium monooxygenase structural gene *amo*A as a functional marker: Molecular fine-scale analysis of natural ammonia oxidizing populations. *Applied and Environmental Microbiology.* **63**, 4704–4712.

Sandaa R. A., Enger O and Torsvik V. (1999) Abundance and diversity of Archaea in heavy-meal contaminated soil. *Applied and Environmental Microbiology.* **65**, 3293–3297.

Stackebrant E., Rainey F. A. and Ward Rainey N. (1996) Anoxygenic phototrophy across the phylogenic spectrum: current understanding and future perspectives. *Archives of Microbiology* **166**, 211–223.

Stephen J. R., Chang Y. J., MacNaughton S. J. Kowalchuk G. A., Leung K. T., Flemming C.A. and White D. C. (1999). Effect of toxic metals on indigenous soil beta-Subgroup proteobacterium oxidizer community structure and protection against toxicity by innoculated metal-resistant bacteria. *Applied and Environmental Microbiology* **65**, 95–101

Tateno, M. (1985) Adenylate energy charge in glucose amended soil. *Soil Biology and Biochemistry* **14**, 331–336.

Tunlid, A. and White, D.C. (1992) Biochemical analysis of biomass, community structure, nutritional status and metabolic activity of microbial communities in soil. *Soil Biochemistry,* Vol 7, pp. 229–262.

Tyler, G. (1981) Heavy metals in soil biology and biochemistry. In: Paul, E.A. and Ladd, J.N. (eds) *Soil Biochemistry,* Vol 5. Marcel Dekker, New York, pp. 371–414.

West, A.W., Grant, W.D. and Sparling, P. (1987) Use of ergosterol, diaminopimelic acid and glucosamine contents of soils to monitor changes in microbiological populations. *Soil Biology and Biochemistry* **19**, 607–612.

Whitty J. F., Keay P. J., Frogatt P. J. and Dart P.J. (1979) Algal nitrogen fixation on temperate arable fields-the Broadbalk experiment. *Plant and Soil.* **52**, 151–164.

5

Metal-Algae Interactions: Implications of Bioavailability

M. Megharaj, S.R. Ragusa and R. Naidu

INTRODUCTION

Algae are a large diverse assemblage of ubiquitous oxygen-evolving phototrophic organisms. They are represented by cyanobacteria (blue-green algae) and eukaryotic organisms belonging to Chlorophyta, Rhodophyta, Phaeophyta, Chrysophyta, Cryptophyta, Dinophyta, Euglenophyta, and Haptophyta. Cyanobacteria are the oldest group of oxygen-evolving photosynthetic prokaryotic organisms and are very important in both aquatic and terrestrial ecosystems because of their ability to fix carbon and dinitrogen (Dohler, 1986). Microalgae form an important component of soil microflora and are ubiquitous, probably accounting for up to 27% of the total microbial biomass in the soil (McCann and Cullimore, 1979). These organisms are involved in maintaining soil fertility besides oxygen production (Bold and Wynne, 1978). The important functional attributes of soil algae are addition of organic matter via carbon fixation, N_2- fixation, and surface consolidation of soil besides being primary colonizers of barren areas (Metting, 1981). Also, cyanobacteria are more widespread than the other free-living micro-organisms capable of dinitrogen fixation (Burns and Hardy, 1975) and are thus very important for the nitrogen economy of soils. In addition, free-living and symbiotic cyanobacteria are commonly the principal agents for input of N as components of biological crusts in semi-arid lands and deserts (Metting, 1981).

The surface of algae is a mosaic of metal-ion-binding sites. Algae have high surface-to-volume ratio, which is ideal for interaction with metals. The binding sites are made up of lipid, carbohydrate, and protein components, which create a spectrum of distinct binding sites that differ in affinity and specificity. Both anions and cations may be bound. Evaluation of the effect that metals have on algae in aquatic and terrestrial

CSIRO Land and Water, Private Mail Bag No. 2, Glen Osmond, Adelaide, SA 5064, Australia habitats is based on chemical analyses of the sample, toxicity tests, and bioaccumulation of metals by algae. Chemical analyses are the first step in establishing a relationship between the contaminant and any negative effect on an organism. Data on the total and bioavailable fraction of a contaminant and its speciation are critical factors that need to be determined. Advances in analytical capabilities have enabled scientists to rapidly estimate low levels of metal contaminants in the environmental samples on a routine basis. This information is invaluable in ascertaining what portion of the metal is bioavailable.

In the initial stages, interaction of metals with biota occurs at the cell wall or with membranes. In lakes and oceans, biological surfaces (e.g. algae, biological debris, and biologically coated surfaces) play a dominating role in binding heavy metals and regulating the residual concentration (and so the bioavailability) of heavy metals (Sigg, 1985, 1987). Interestingly, biological surfaces can exhibit more affinity than inorganic surfaces in scavenging metals in oceans, lakes, and rivers. Particulate materials (organic and inorganic) are important in heavy-metal interactions in soil and water systems. Algae in such systems exhibit large surface areas (see Table 5.1) containing various functional groups that can act coordinately with heavy metals (e.g. carboxylic, amino, thio, hydroxo and hydroxy-carboxylic). The typical surface area and total surface area of biological material in settling particles of lakes is of the same order as iron oxide material (Table 5.1). This is especially true, in oceans, where most of the particulate material is of biological origin. With this in mind, it is easy to envisage that algae play an important role in regulating the bioavailability of metals in soils and water. The ability of algae to adsorb heavy metals from contaminated water has been exploited in the clean-up of heavy-metal contaminated water.

Table 5.1: Properties of organic and inorganic surfaces in relation to their importance in metal binding

Material	Typical surface area (m^2 g^{-1})	Content of material in settling particle (mg g^{-1})	Total surface area of settling particle (m^2 g^{-1})
Iron oxide	200-500	5-10	1-10
Organic material	107	200-400	2.3-4.5
Manganese oxides	NA	NA	200

NA: not available

In addition to the major elements building up biological compounds (C, N, O, H, S, and P), a number of other elements are required in trace amounts. Trace elements are vital constituents of enzymes and perform many vital functions. When a certain requirement for a trace element is exceeded, the element may become toxic, whereas inadequate concentration may be growth limiting (see Fig. 5.1). For some non-essential elements, no beneficial effects on the organisms are observed and adverse effects increase with increasing metal concentration (see Fig. 5.1). Toxic effects are dependent on the species of metallic ion, the pH, and complexing agents found in natural and impacted water and soil. The toxicity effects of different metal ions are closely related to their coordination chemical properties. Mechanisms for toxicity often include binding of metals to sensitive cellular compounds, often enzymes. An example is Hg that has a very high affinity for sulfur ligands, which means that it interacts with -SH and -S- S- groups of enzymes and other proteins, rendering them inoperative.

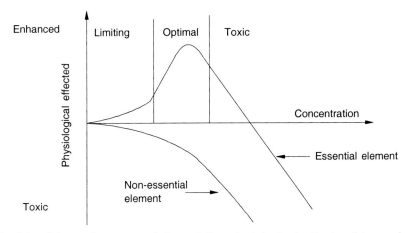

Fig. 5.1: Schematic representation of the physiological effects of increasing concentrations of essential and non-essential elements on organisms.

Environmental pollution by heavy metals is a serious global problem. Contrary to organic pollutants, heavy metals are not biodegradable and are more persistent in the environment. Algae are an ecologically beneficial group in both soil and water in addition to their role as primary producers representing the base of the food chain. Any interference of metals with normal functioning of these organisms could potentially lead to serious consequences to the ecosystem. Bioavailability is one of the key factors influencing the metal toxicity to algae. This chapter extends microbe-metal interactions reported in Chapters 4 and 5 by focusing on heavy metal-algae interactions in soil and water, especially the implications of such interactions to metal bioavailability.

TOXICITY TESTS

Toxicity tests provide a direct measure of bioavailability of contaminants, and these in combination with chemical analyses help to establish the link between contamination and negative ecological effects. Algal-based toxicity tests are used to evaluate the effect of acute and chronic exposure to the heavy metal. These tests measure the growth response, behavioural changes, or the effects on specific physiological processes such as photosynthesis, nitrogen-fixation, etc. The advantage of using a toxicity test is that it measures the organisms' response to the relative toxicity of mixtures of chemicals in terms of additive or synergistic interactions. Although algal toxicity tests may be used in both aquatic and terrestrial habitats, aquatic toxicity tests are better developed compared to the terrestrial toxicity tests. These tests usually involve either single species or occasionally multiple species. Testing can be done in <94 h for acute toxicity and 7–30 days for chronic toxicity. Compared to batch static tests, continuous culture tests using a chemostat are few (Hall et al., 1989; Chen and Lin, 1997). Although some researchers reported that continuous culture tests are more sensitive than batch tests, Kayser (1976) found that the difference between these tests was not highly significant. The differential results obtained by various authors may have been due to different experimental conditions and growth media employed by them.

Aquatic algal-based toxicity tests have been extensively used for ecotoxicity assessment for the past 25 years. There are many standard protocols published by US Environmental Protection Agency (USEPA), Organization for Economic Cooperation and Development (OECD) and others (OECD, 1984; ISO, 1987; USEPA, 1989; ASTM, 1994). The frequently used toxicity tests employ the concentration of the contaminant to relate to the observed effect on the growth of an organism. One such example is the chronic toxicity test employing the growth of an alga, *Selenastrum capricornutum* in which case the test is conducted over 94 h at a temperature of 25 °C under static conditions (USEPA, 1989).

The toxicity of heavy metals toward freshwater algae has been the subject of a few reviews (Whitton, 1970; Rai et al., 1981a; Stokes, 1983). The use of algae as test organisms for bioassays for evaluation of toxicity in metal wastes and effluent has long been recognized. In general, all heavy metals are inhibitory to growth, pigments, macromolecules, ATP generation, photosynthesis, nitrogen fixation, nitrate reductase, glutamine synthetase, and nutrient uptake. Many metals interfere with cell membrane permeability and induce loss of potassium ions. A brief account of literature on metal toxicity to algae is presented in Table 5.2. Rana and Kumar (1974) have employed algae to evaluate toxicity in a zinc smelter and mine effluents. No algae were found in these effluents. By conducting laboratory bioassays on these effluents, using various green and blue-green algae, these authors have suggested that (a) green algae were more tolerant to copper and zinc than blue greens, (b) differential response by the algae in the same environment was due to the variation in their physiological and genetic attributes, (c) the observed tolerance in some algae isolated from unpolluted habitats to high concentrations of metals might be genetically controlled since there was little scope for natural selection to occur in these isolates because of their isolation from unpolluted habitats, and (d) the presence of these tolerant and sensitive organisms can be used as indicators of heavy-metal pollution.

Table 5.2: Toxicity of heavy metals to algae

Alga	Concentration	Toxicity	Reference
Mercury			
A. inaequalis	$8 \ \mu g \ l^{-1} >$	Complete inhibition of growth	Stratton et al., 1979
	$100 \ \mu g \ l^{-1}$	Photosynthesis and nitrogenase activity Growth,	
N. linckia	$0.2 \ \mu g \ ml^{-1}$	nitrogenase and glutamine synthetase activities	Kumar et al., 1985
C. vulgaris	$0.5 \ \mu g \ ml^{-1}$ (LC_{50} value)	Growth, CO_2- fixation, O_2- evolution, ATP content, nitrate reductase activity, nutrient uptake (NO_3^-, NH_4^+)	Rai et al., 1991
A. nidulans	$0.5 \ \mu g \ ml^{-1}$ (5.0 μg ml^{-1} is lethal)	Growth	Lee et al., 1992
N. calcicola	$0.02 \ \mu g \ ml^{-1}$ (LC_{50} value for Hg_2^+) $0.008 \ \mu g$ ml^{-1} (LC_{50} value for CH_3Hg^+)	Growth, NH_4^+ uptake, glutamine synthetase activity	Singh and Singh, 1992
C. reinhardtii	$1–10 \ \mu g \ ml^{-1}$	Growth	Weiss-Magasic et al., 1997
Phormidium fragile	$0.01–1.5 \ \mu g \ ml^{-1}$	Growth, protein synthesis	Khalil, 1997
Copper			
Chlorella sp.	$0.0063 \ \mu g \ ml^{-1}$ $6.3 \ \mu g \ ml^{-1}$	Growth Photosynthesis and respiration	Cedeno-Maldonado and Swader, 1974
Coastal phytoplankton	$1–2.5 \ \mu g \ ml^{-1}$	Carbon -fixation	Davies and Sleep, 1980
A. nidulans	$0.25 \ \mu g \ ml^{-1}$ (LC_{50})	Growth	Singh, 1985
N. linckia	$0.2–5 \ \mu g \ ml^{-1}$	Growth, photosynthesis and nitrogenase activity	Kumar et al., 1985
Anabaena sp.	$0.1–0.4 \ \mu g \ ml^{-1}$	Growth	Clark et al., 1987
Nile river algae	$>0.05 \ \mu g \ ml^{-1}$	Growth	Lasheen et al., 1990
Isochrysis galbana	$1–25 \ \mu g \ ml^{-1}$ 1.58 $\mu g \ ml^{-1}$ (72 h EC_{50})	Growth	Edding and Tala, 1996
Dunaliella tertiolecta	$>12.5 \ \mu g \ ml^{-1}$ 38.80 $\mu g \ ml^{-1}$ (EC_{50})	Growth	Edding and Tala, 1996
Cylindrotheca fuciformis	$0.2–1.0 \ \mu g \ ml^{-1}$	Growth	Pistocchi et al., 1997
Gymnodinium sp.	$0.03–0.05 \ \mu g \ ml^{-1}$	Growth	Pistocchi et al., 1997

Contd.

Table 5.2 Contd.

Alga	Concentration	Toxicity	Reference
Cadmium			
Chlorella sp.	0.001 µg ml^{-1}	Growth	Hart and Scaife, 1977
Cylindrotheca closterium	0.001 mg ml^{-1}	Growth, photosynthesis and respiration	Lehman and Vas Cancelos, 1979
A. inaequalis	1.0 µg ml^{-1} 0.05 µg ml^{-1}	Photosynthesis, N$_2$-fixation Induced morphological changes	Stratton and Corke, 1979
A. falcatus, S. obliques and Chlorococcum spp.	2 µg ml^{-1} (5 µg ml^{-1}is lethal)	Growth	Devi Prasad and Devi Prasad, 1982
A. nidulans	4 µg ml^{-1} (EC$_{50}$)	Inhibition of nitrate uptake	Singh and Yadav, 1983
A. cylindrica	1–2 µg ml^{-1}	Nitrogenase activity	Becker, 1983
Nile river algae	> 0.05 µg ml^{-1}	Growth	Lasheen et al., 1990
C. vulgaris	15.72 (48 h LC$_{50}$)	Growth	Rai et al., 1991
Cobalt			
N. linckia	> 0.2 µg ml^{-1}	Growth, N$_2$-fixation, glutamine synthetase	Kumar et al., 1985
Aluminium (AlCl$_3$)			
N. linckia	LC$_{50}$ values pH 7.5, 16.2 µg ml^{-1} pH 6.0, 1.48 µg ml^{-1} pH 4.5, 0.49 µg ml^{-1}	Growth, photosynthesis, electron transport, Nitrogen and phosphorus metabolism, ATPase activity	Rai et al., 1996
C. vulgaris	LC$_{50}$ values pH 6.8, 108 µg ml^{-1} pH 6.0, 5.9 µg ml^{-1} pH 4.5, 4.0 µg ml^{-1}	Growth	Rai et al., 1998
Aluminium (AlF$_3$)			
N. linckia	LC$_{50}$ values pH 7.5, 13.5 µg ml^{-1} pH 6.0, 1.08 µg ml^{-1} pH 4.5, 0.32 µg ml^{-1}	Growth, photosynthesis, electron transport, nitrogen and phosphorus metabolism, ATPase activity	Rai et al., 1996
C. vulgaris	LC$_{50}$ values pH 6.8, 40.5 µg ml^{-1} pH 6.0, 4.8 µg ml^{-1} pH 4.5, 2.7 µg ml^{-1}	Growth	Rai et al., 1998
Aluminium			
S. obliquus	0.7 µg ml^{-1} (EC$_{50}$)	Growth, Acid phosphatase, nitrate reductase	Kong et al., 1999
Nickel			
Scenedesmus sp.	>0.05 mg ml^{-1}	Growth	Hutchinson, 1973

Contd.

Contd.

Alga	Concentration	Toxicity	Reference
Chlorella sp.	>0.3 mg ml^{-1}	Growth	Hutchinson, 1973
Chloroglea fritschii	0.0125 µg ml^{-1}	Nitrogenase activity	Henriksson and DaSilva, 1978
Nostoc sp.	0.025 µg ml^{-1}	Nitrogenase activity	Henriksson and DaSilva, 1978
Westiellopsis sp.	0.005 µg ml^{-1}	Nitrogenase activity	Henriksson and DaSilva, 1978
A. inaequalis	>0.05 µg ml^{-1}	Growth, photosynthesis and N$_2$-fixation	Stratton and Corke, 1979 b
A. flosaquae and A. cylindrica	0.6 µg ml^{-1}	Growth	Spencer and Greene, 1981
A. falcatus, S. obliques and Chlorococcum spp.	>2 µg ml^{-1}	Growth	Devi Prasad and Devi Prasad, 1982
N. linckia	>0.2 µg ml^{-1}	Growth, N$_2$-fixation and glutamine synthetase	Kumar et al., 1985
A. variabilis	>2 µg ml^{-1} (20 µg ml^{-1}is lethal)	Growth (pigments)	Ahluwalia and Kaur, 1989
N. muscorum	0.29–0.88 µg ml^{-1}	Growth, N$_2$-fixation and photosynthesis	Asthana et al., 1990
Chromium (VI)			
2 strains of Chlorella	>0.5 µg ml^{-1}	Growth	Meisch and Beck-mann, 1979
Nile river algae	0.25 µg ml^{-1}	Growth	Lasheen et al., 1990
S. acutus	1-µg ml^{-1}	Growth and cell cycle	Corradi and Gorbi, 1993
N. muscorum	78 µg ml^{-1}	Growth	Singh et al., 1997
Chlorococcum sp.	1 µg ml^{-1} 2 µg ml^{-1} is lethal in the absence of EDTA)	Growth, protein, carbohy-drate content, potassium leakage	Megharaj et al., unpubl. work
Chromium (III)			
A. doliolum	40 µg ml^{-1}	Growth	Rai and Dubey, 1989
Lead			
S. capricornutum, C. pyrenoidosa C. ellipsoidea C. vulgaris	0.5 µg ml^{-1}	50 % growth inhibition	Monohan, 1976
Scenedesmus sp. S. obtusiusculus	0.5 µg ml^{-1}	Growth	Monohan, 1976
Ankistrodesmus sp. Scnedesmus sp. D Scenedesmus sp. C	1.0 µg ml^{-1}	Growth	Monohan, 1976
S. costatum	0.05–10 ng ml^{-1} 100 ng ml^{-1}	Growth and respiration Lethal to growth	Rivkin, 1979
A. falcatus, S. obliques and Chlorococcum spp.	>1 µg ml^{-1}	Growth	Devi Prasad and Devi prasad, 1982

Contd.

Alga	Concentration	Toxicity	Reference
Zinc			
C. vulgaris	2 µg ml^{-1}(LC$_{50}$)	Growth, CO$_2$-fixation, O$_2$-evolution, ATP content, nitrate reductase activity and nutrient uptake	Rai et al., 1991
A. nidulans	25 µg ml^{-1}(+ EDTA) 50 µg ml^{-1}is lethal (+EDTA) 10 µg ml^{-1} is inhibitory (– EDTA) 25 µg ml^{-1} is lethal (– EDTA)	Growth	Lee et al., 1994
Arsenate			
Melosira granulata Ochromonas vallesiaca	0.08 µg ml^{-1}	Growth and phosphate uptake	Planas and Healey, 1978
C. reinhardtii	0.75 µg ml^{-1}	Growth and phosphate uptake	Planas and Healey, 1978
Cryptomonas erosa A. variabilis	7.5 µg ml^{-1}	Growth and phosphate uptake	Planas and Healey, 1978
Fluoride			
N. linckia	LC$_{50}$ values pH 7.5, 0.19 µg ml^{-1} pH 6.0, 0.095 >µg ml^{-1} pH 4.5, 0.057 µg ml^{-1}	Growth, photosynthesis, electron transport, Nitrogen and Phosphorus metabolism, ATPase activity	Rai et al., 1996
Tin			
A. dolialum	50 µg ml^{-1}(LC$_{50}$)	Growth	Rai and Dubey, 1989
Synechocystis aquatilis	1–10 µg ml^{-1} (as Sn II) 5–10 µg ml^{-1}(Sn IV)	Growth, photosynthesis	Pawlik-Skowlonska et al., 1997
Zirconium			
S. capricornutum	2–6 µg ml^{-1} (96 h EC$_{50}$)	Growth	Couture et al., 1989

Shehata et al. (1999) have reported the effect of metal mixtures (0.05 mg L^{-1} of Cd, 0.1 mg L^{-1} of Ni, Zn, Cu, and Cr) on physiological and morphological characteristics of natural phytoplankton populations from water of the Nile river, consisting of three main groups of algae, cyanobacteria, green algae, and diatoms. These investigators noticed a substantial change in algal community structure with the disappearance of sensitive species, resulting in the selection of tolerant species. The most tolerant group was found to be cyanobacteria (with *Oscillatoria mougeotti* as the dominant species), followed by green algae (*Scenedesmus quadricauda*) with diatoms being the most sensitive group. In addition, a change in morphology of the tolerant algae was observed. However, these phytoplankton populations recovered from the metal stress once the metals were eliminated from the media and resulted in the production of biomass equivalent to the control without metals. Accumulation of metals by these algae from the Nile was shown to be influenced by the type of metals, algal community structure, and the ratio between different morphological forms of algae with high uptake of Zn observed in the algal biomass compared to other metals in the mixture (Shehata et al., 1999). Shift in the composition of natural assemblages of algae in response to toxic metals seems to be a common mechanism of adaptation to the toxicant. For example, phytoplankton from Sado river in many metal-polluted sites responded by shifts in their composition towards resistant species (Monterio et al., 1995). Also, *Skeletonema costatum* was found to be the dominant diatom when a natural assemblage of algae were exposed to the toxic levels of metals (Hollibaugh et al., 1980). A study conducted by Kumari et al. (1991) on the impact of heavy-metal pollution (Fe, Zn, Pb, Mn, Co, Cu, Ni) on phytoplankton composition of the river Moosi (India) revealed that Chlorophycea were more tolerant to metals followed by Bacillariophyceae, Cyanophyceae, and Euglenophyceae. Further, they observed *Chlorella vulgaris* to be the alga most tolerant to all metals and suggested the usefulness of certain algae in that river as metal-pollution indicators.

BIOAVAILABILITY vs. TOXICITY

Assessment of heavy-metal pollution necessitates the understanding of various processes that influence the bioavailability and toxicity of the pollutant to the organisms.

Bioavailability is related to the availability of the compound to the specific organism and thus depends on the nature of the organism. Bioavailability can be measured in terms of effects such as growth rate, photosynthesis, uptake, accumulation, etc. caused by a specific concentration of the compound to an organism. Bioavailability of the metal to an organism is also dependent on its speciation rather than total metal content itself. Thus, much of the available evidence suggests that total concentration of the metal is not a good predictor of its bioavailability, but metal speciation greatly influences the bioavailability.

Traditionally, bioavailability has been explained in terms of physicochemical properties of the metal. Metals exist in solution as chemical species, the proportion of which changes depending on their complexation with ligands. For example, the free Cd ion is believed to be the most bioavailable form and hence toxic to organisms. Many aquatic toxicological studies have shown that free-metal ion is the main factor influencing toxicity. Also, the metal dissolved in water but bound to dissolved organic matter is not readily bioavailable. According to the free-ion model theory of trace metal-aquatic organisms interactions, the biological response of an organism is proportional to the free ion in the solution phase. However, this is not true in all the cases as has been recently demonstrated by Parent et al. (1996) in case of aluminium toxicity to *Chlorella* sp. where toxicity was not proportional to the activity of Al^{3+} ion in the presence of soil fulvic acid. These authors have shown that fulvic acid adsorbed to the cell surface increased the membrane permeability of alga whereas Al decreased the membrane permeability. Furthermore, the total Al^{3+} associated with the alga was more in the presence of fulvic acid compared to the medium without fulvic acid, assumed to be due to the adsorption to algal cell surface of the Al-fulvic acid complex. Also, soil fulvic acid has been suggested as a possible source of phosphorus to P-deficient algae. In view of the common occurrence of fulvic acids in the natural dissolved organic material, these findings are of practical significance. For this reason, it is important to consider the physiological influence of natural organic matter on the organism while assessing metal toxicity.

TOXICITY TO NATIVE SOIL ALGAE

Most of the work published on metal toxicity to algae is as a result of studies conducted with aquatic algae, and information on terrestrial algae is limited. Cyanobacterial colonization and autotrophic nitrogen fixation have been shown to decrease in a metal-contaminated soil (EDTA-extractable soil metal concentrations of Zn, 44; Cu, 57; Ni, 7; Cd, 4.7 mg kg^{-1}) analysed even 20 years after the last metal inputs from sewage-sludge were made (Brooks et al., 1986). We observed a drastic reduction in the density and species composition of algal populations in a long-term tannery-contaminated soil where chromium was the major contaminant (Megharaj et al., unpublished results). In highly contaminated soil, only one type of alga (*Chlorococcum* type unicellular alga) occurred in contrast to the control soil where a total of eight algal species, including three cyanobacteria, were present.

MECHANISMS OF HEAVY METAL UPTAKE AND EXCLUSION BY ALGAE

Algal Resistance to Metal Ions

This requires the input of cellular energy and thus represents a non-equilibrium reaction. Biomethylation is a mechanism that is used to detoxify metals and it is a plasmid-borne trait. Mechanisms for vitamin B_{12}-dependent synthesis of metal alkyls have been discovered for the metals Hg, Pb, Tl, Pd, Pt, Au, Sn, and Cr and the metalloids As and Se. Pathways for the synthesis of organo-arsenic compounds have been shown to occur via a mechanism where S-adenosyl-methionine is the methylating coenzyme.

Intracellular traps are also used by cells to prevent metals from reaching toxic levels. This precedes mechanisms for expulsion by vacuoles. An example is the biosynthesis of metallothionen and removal of Cd and Cu by a sulfhydryl-containing protein. Cellular exclusion mechanisms include the ability of an organism to selectively remove toxic metals by energy-driven mechanisms. Another protection mechanism is the ability of an organism to synthesize extracellular ligands that complex metals to the cell surface and prevent cellular uptake. Metals may also be precipitated on the surface of cells indirectly via the activity of membrane bound sulfate reductases or through biosynthesis of oxidizing agents such as O_2 or H_2O_2. The reduction of sulfate to sulfide and the diffusion of O_2 or H_2O_2 through the cell membrane provide highly reactive means by which metals can be complexed or precipitated before entering the cell.

Surface Binding and Precipitation

Many workers have studied the mechanisms of metal binding to algal surfaces. This includes ion exchange, electrostatic attraction, coordination bonding, H_2S, O_2, and H_2O_2 production. Metal binding has been found to be dependent on pH, temperature, and the presence of competing ions.

These include the amino group, peptide bonds, carboxyl group, the imidizole group of histidine, and the nitrogen and oxygen of the peptide bond, which can be available for coordination bonding with metallic ions (Fig. 5.2). Bond formation could be accompanied by displacement of protons, which is dependent on protonation and pH. Metallic ions could also be electrostatically bonded to unprotonated carbonyl oxygen and sulfate and might be evident by the appearance of two slopes in Langmuir isotherms.

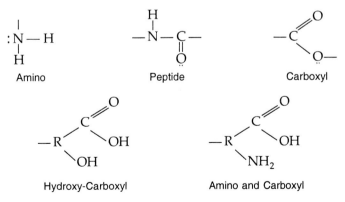

Fig 5.2: Examples of some functional groups.

Adsorption of metal ions was found to increase in the order Cu>Sr>Zn>Mg>Na (Xue et al., 1988). Sodium was found to decrease adsorption of positive metallic ions and enhance adsorption of negative ion complexes. Their results could be explained in light of protein, and polysaccharide composition of the cell wall and covalent bonding to amino and carbonyl groups with ionic charge bonding to carbonyls and sulfates. The extent of surface binding may be generalized by mass law equations:

$$\equiv SH + Cu^{2+} \leftrightarrow \equiv SCu^+ + H^+$$

Models describing metal binding to algal surfaces have been based on those developed for metal binding to hydrous-oxide surfaces.

$$\equiv SOH + Me^{2+} \leftrightarrow \equiv SOMe^+ + H^+ \quad and$$
$$2\equiv SOH + Me^{2+} \leftrightarrow (\equiv SO)_2 Me + 2H^+$$

The degree of surface protonation depends on the acid-base equilibria:

$$\equiv SOH_2^+ \leftrightarrow SOH + H^+ \quad and \equiv SOH \leftrightarrow SO^- + H^+$$

Similarly, algal cell surfaces contain various functional groups with particular acid base properties, which may be characterised by;

$$RH_2^+ \leftrightarrow RH + H^+ \text{ and } RH \leftrightarrow R^- + H^+$$

Complex formation of metals with the algal surface is similar to that with polyfunctional macromolecules and may be generalised as;

$$RiHx + Me^{2+} \leftrightarrow RiMe^{(2-x)} + xH^+$$

where Ri designates the deprotonated surface site. All surface groups are not identical, which means that it is very difficult, if not impossible, to determine all complexation constants; usually, an average constant is derived. Figure 5.3 shows some examples of metal reactions with functional groups.

Metal ion interactions with algae can be divided into three major categories depending on how binding is affected by pH. One group, which includes Hg(II), Au(III) as $AuCl_4^-$, Ag(I), Pd(II) and Au(I) thiomalate, bind to *Chlorella vulgaris* cells rather independently of pH between 2 and 7. According to Pearson (1973) these metals may be classified as "soft" and this behavior is consistent with the general coordination chemistry of metal ions. "Soft," metal ions undergo covalent binding to softer ligands such as sulfhydryl and amine groups, and these binding interactions are minimally affected by ionic interaction and pH.

A second group of metals termed the "hard" and "borderline soft" are bound more strongly as the pH is increased from 2 to 5. This group of metals included Cu(II), Ni(II), Zn(II), Co(II), Pb(II), Cr(III), Cd(II), U(VI), Be(II), and Al(III). When active metal ion binding sites such as carboxylate or amine groups that usually bind protons are present, then metal binding is pH dependent. In this instance, metal ions and protons compete for the same binding sites.

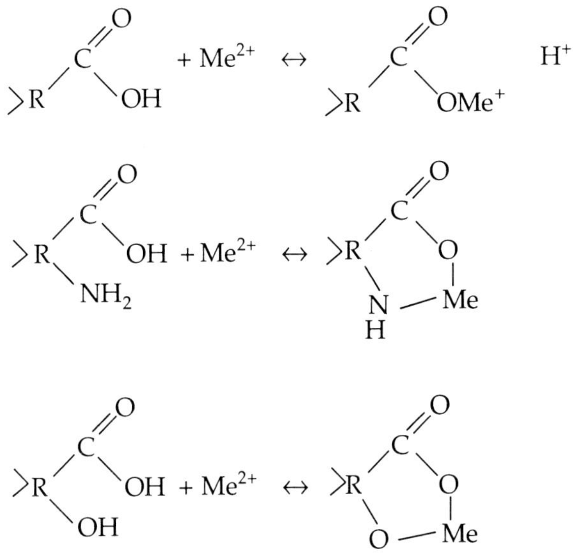

Fig. 5.3: Generalized examples of metal reactions with functional groups.

The third group of metal ions bind more strongly at pH 2 than at pH 5. This group is made up of oxoanions such as MoO_4^{2-}, SeO_4^{2-} and CrO_4^{2-} and other anionic metal complexes which include $PtCl_4^{2-}$ and $Au(CN)_2^-$. The increased binding at low pH is consistent with electrostatic binding to ligands such as amine and imidazoles that would be protonated (i.e. positively charged) at low pH. Although the above characteristics hold for *C. vulgaris*, differences in the binding capacity for certain metals by different algal species have been noted (Green and Darnall, 1990). These differences have led to the formulation of methods to selectively adsorb and desorb metals from algal biomass.

Uptake of Metals by Algae

Heavy-metal uptake by algal cells occurs in two steps. An initial rapid adsorption step followed by a slower step that is diffusion-controlled uptake into the cell. Binding of Cu(II) and Cd(II) is reduced in the presence of Ca^{2+} and Mg^{2+}. Adsorption is also affected by pH. The tendency to form surface complexes decreases with increasing metal loading. This is because metal ions bind first to the surface groups for which they have highest affinity and then to those with lower affinity. A two-site Langmuir isotherm can reproduce binding data very well. In living cells, transition metals bind better with sulphur-rich bases than they do with nitrogen and oxygen bases and finally by coordination to water molecules alone (Greene and Darnall, 1990). Such bases are readily available to metal ions because of the fast exchange of protons that occurs within cells in the pH range 7-7.2.

Algae possess the ability to take up toxic heavy metals from the environment, resulting in higher concentrations than that in the surrounding water. Bioaccumulation studies reveal the accumulation of contaminant in the organism via uptake of food or water containing the contaminant. Algae generally take up metals via membrane-transport proteins involved in regulation and intracellular transport of trace metals. These transport systems are not specific to any single metal, and hence these systems bind to essential and non-essential metals resulting in inhibition of essential metal uptake and intracellular transport of competing toxic metals. Usually, bioaccumulation occurs when the uptake rate of the pollutant exceeds the excretion rate. Several algae have been shown to be good accumulators of metals from water. Generally, bioaccumulation under laboratory conditions is measured by determining the residual contaminant in the organism after exposure to a contaminant-containing medium for a defined period of time. These assays provide a direct measure of the contaminant bioavailability whereas chemical assays do not.

The metal content of the algae depends on the metal concentrations in the ambient water as well as metal bioavailability (Davies, 1978; Canterford and Canterford, 1980). Uptake of copper in many algae was shown to be directly proportional to the metal concentration in the ambient medium (Fisher et al., 1981; Gowrinathan and Rao, 1991). The intracellular metal (Cu, Cd, Zn, Mn) content of phytoplankton originating from four different lakes in Switzerland was compared to the free metal ion concentration in the ambient waters (Knauer et al., 1998). In this study, cellular concentrations of Cu, Cd and Mn were related to the free metal ion concentration, and dissolved concentration under natural conditions despite high variability among the phytoplankton species composition. However, intracellular Zn content of the phytoplankton seems to be regulated differently, which could not be related to the free ion concentration in the lake water, a deviation from the free ion model theory (Knauer et al., 1998). Also, these authors have reported the production of phytochelatins by phytoplankton under natural conditions even at low levels of free Cu, Cd, and Zn.

In a carefully planned and controlled experiment conducted by Errecalde et al. (1998), it has been demonstrated that toxicity and uptake of Cd and Zn by the green alga *S. capricornutum* did not follow the free ion model in the presence of citrate. These investigators have demonstrated the enhanced uptake of Cd and toxicity of Cd and Zn in the presence of citrate over that with EDTA to the alga even by maintaining the free ion concentrations of Cd and Zn as well as the constituents of incubation media (H^+, Ca^+, $Mg_{,}^+$ and the cation trace nutrients). The uptake of the metal-ligand complex (Cd-H citrate, Cd-citrate$^-$, Cd-citrate$_2^+$) across the biological membrane has been suggested as the possible mechanism for the observed enhanced uptake of Cd by the alga. Thus, these results indicate that low-molecular-weight metabolites can affect the bioavailability of metals. Mercury and monomethyl mercury uptake by a marine diatom *Thalassiosira weissflogii* was shown to be by passive diffusion through the membrane rather than by facilitated transport (Mason et al., 1995).

METAL TOLERANCE

Tolerance to several toxic metals has been demonstrated in many algae (Table 5.3) and different mechanisms have been suggested for metal tolerance in algae (Stokes, 1983). These mechanisms include (1) physical exclusion of metal ions because of reduced membrane permeability. Examples include Cu tolerance in *C. vulgaris* (Foster, 1977), Hg tolerance in *D. tertiolecta* (Davies, 1976), Zn tolerance in *Chlorella* sp. (de Fillippis and Pallaghy, 1976b); (2) internal immobilization of the metal as observed by Daniel and Chamberlain (1981) in *Navicula* spp. and *Amphora* spp. where binding of Cu to polyphosphate bodies occurred, and in *Chlamydomonas variabilis* where Zn was sequestered by cellular polyphosphates (Bates et al., 1985); (3) external binding of metal ions by cell exudates as observed in many algal species (McKnight and Morel, 1979; Brown et al., 1988), protection of *Anabaena* sp. from Cu toxicity under siderophore production conditions (Clarke et al., 1987); (d) enzymatic reduction as in *Chlorella*, which produces glucose dehydrogenase, which in turn reduces Hg(II) ions to volatile metallic Hg at outer cell surfaces (de Filippis and Pallaghy, 1976a; Bartless et al., 1977).

Many algal species have been shown to develop tolerance to toxic metals by synthesising metal-binding proteins (ligands or siderophores) on their exposure to polluted environments. Siderophores are excreted into the external environment to bind essential metal ions and have an affinity for specific membrane-bound receptors. Specific metals may be transported into the cell by combining with a receptor or moving through special channels found in the membrane. Clarke et al. (1987) have demonstrated that siderophores produced in response to iron starvation were able to protect *Anabaena* spp. from Cu toxicity, presumably due to chelation of the Cu by the siderophore.

Table 5.3: Examples of metal tolerance in algae

Alga	Metal	Reference
Euglena gracilis	Cd	Bariaud and Mestre, 1984
Scenedesmus sp. B[4]	Cu	Silverberg et al., 1976
Selenastrum capricornutum	Cu	Kuwabara and Leland, 1986
Hormidium fluitans	Cu	Sorentino, 1985
Amphora coffeaeformis	Cu	Brown et al., 1988
Chlorella vulgaris	Cu	Butler et al., 1980; Foster 1977
Scenedesmus acutus	Cu	Twiss et al., 1993
Scenedesmus acutus	Cu and Ni	Stokes et al., 1973
Anacystis nidulans/Mn[9],Mn[10],Mn[14]	Cu	Singh,1985
Cylindrotheca fuciformis	Cu	Pistocchi et al., 1997
Chlorella sp.	Zn	De Filippis and Pallaghy, 1976
Anacystis nidulans	Zn	Shehata and Whitton, 1982
Anacystis nidulans	Co, Ni, Cu and Cd	Whitton and Shehata, 1982
Stigeoclonium tenue	Zn	Harding and Whitton, 1976
Phaeodactylium tricornutum,	Cd and Pb	Jennings 1979
Dunaliella tertiolecta	Cd	Nagel et al., 1996
Chlamydomonas		
Scenedesmus sp.	Pb	Silverberg et al., 1977

Chlorella ellipsoidea has been shown to develop tolerance to cadmium by synthesising Cd- binding proteins (Nagano et al., 1984). Phytochelatins have been found in the cytosol and chloroplast of Cd-tolerant cells of *Chlamydomonas reinhardtii*. However, these Cd-tolerant cells had impaired photosystem II activity, suggesting a mutation of a chloroplast or a nuclear gene (Nagel et al., 1996; Voigt et al., 1998). Cells of *C. vulgaris* previouly exposed to sublethal concentrations of Cd (9.5 μg L^{-1}) subsequently showed tolerance to high levels of Cd (14 μg L^{-1}) with less growth inhibition (Carr et al., 1998). This was attributed to the induction of intracellular metal-binding proteins that chelate the toxic metal thereby preventing it from damaging effects on cytoplasmic organelles. Kaplan et al. (1995) isolated a Cd-tolerant *Chlorella* strain by subjecting the cells to progressively higher concentrations of Cd. The LD$_{50}$ of the isolated strain increased tenfold over that of the original culture from which it was derived.

Twiss et al. (1993) gave experimental evidence for metal exclusion from cell interior due to high Cu-adsorptive capacity of cell surface in case of Cu-tolerant *S. acutus*. Usually, metal exclusion in algae may result from binding of the metal by cell wall or organic ligands so that metal is unavailable for interference with normal functioning of the organism. The importance of zeta potential in heavy-metal resistance of the two species of *Dunaliella*, acid-resistant *D. acidophila* and salt-resistant *D. parva* has been documented (Gimmler et al., 1991). These authors have studied the effects of Al (III), La (III), Cu (II), Cd (II), Hg (II), and W (IV) on the zeta potential of the cells and found that cells with a positive zeta potential (*D. acidophila*, pH 1-2) were extremely resistant to toxic di- and trivalent cations, but sensitive to toxic anions, whereas cells with negative zeta potential (*D. acidophila* and *D. parva*, pH 7.0) were resistant to toxic anions but sensitive to toxic cations. A study conducted on the response of 7 strains of eukaryotic algae and 11 strains of cyanobacteria towards Cu tolerance revealed that cyanobacteria lost their tolerance whereas eukaryotes did not change significantly, after these organisms had been subcultured in a medium without Cu for 2 years (Takamura et al., 1990). However, the tolerance of these cyanobacteria recovered on subculturing once in Cu-containing medium. This result shows that Cu tolerance in eukaryotic algae is genetically stable at least for the test period (2 years). The loss of tolerance in cyanobacteria can be attributed to loss of plasmids, given the fact that Cd tolerance in *Synechococcus* is believed to be controlled by plasmids (Olafson, 1986).

Extracellular carbohydrate is known to reduce Cu toxicity by possibly binding the Cu, as has been demonstrated in a Cu-tolerant diatom *Cylindrotheca fusiformis* which produced more carbohydrate in response to the presence of copper compared to the Cu-sensitive alga *Gymnodinium* sp. which only produced little carbohydrate (Pistocchi et al., 1997). The ability of polyphosphates in algal cells to mobilize/accumulate metals is considered to be a protective mechanism from metal toxicity. The accumulation of Pb and Zn in polyphosphate bodies of algae have been demonstrated using energy-dispersive X-rays (Jensen et al., 1982). The accumulation of metals in the nuclear region, thereby decreasing the metal content in the cytoplasm, is also considered to reduce metal toxicity. Accumulation of copper and lead as the intranuclear complexes was noticed in the copper-tolerant *Scenedesmus* species B_4 and in a lead supplemented *Scenedesmus* culture respectively (Silverberg et al., 1976, 1977). A gradual development of Zn tolerance upon 40 series of cell divisions has occurred in *Chlorella* spp. (de Filippis and Pallaghy, 1976b). Consequently, this Zn-tolerant strain had reduced number of Zn-binding sites accompanied by inhibition of the temperature-sensitive component of the Zn uptake system, which was suggested as the development of physiological exclusion mechanisms (de Filippis and Pallaghy, 1976b).

ALGAE AS INDICATORS OF METAL POLLUTION

Several workers have demonstrated the sensitivity or tolerance of algal species to metals, and this has led to the use of these species as indicator organisms for metal pollution. *Cladophora glomerata* has been found to be most sensitive to heavy-metal pollution. This alga is used as a monitor of metal pollution based on its presence or absence in water bodies (Whitton, 1970). Although algae have since long been used as monitors of heavy-metal pollution in fresh waters, unfortunately there is no universally applicable method developed so far. Whitton (1984) has reviewed the use of algae as monitors of metal pollution in fresh waters. Morphological and cytological changes in algae, such as the presence of metals in poly Pi granules, can also serve as an indication of metal pollution.

FACTORS AFFECTING THE BIOAVAILABILITY AND TOXICITY OF METALS TO ALGAE

The availability of metal ions depends not only on environmental factors (e.g. pH) but also on interaction among metal, ligands, and living cells. Metal ions compete for binding sites; they displace each other on particle and membrane surfaces. Although the presence of free metal ion is a good indicator of probable metal toxicity, it is difficult to determine which metal will enter the algal cells especially when a mixture of metals is present.

Heavy-metal toxicity and their bioavailability to algae is influenced by various environmental factors such as (1) composition of growth medium, (2) limiting nutrients, (3) population density (4) pH, (5) temperature, (6) salinity, (7) Hardness (8) presence of chelators, (9) redox transformations and (10) metal combinations, each of which are briefly discussed here.

Composition of Growth Medium

The composition of the growth medium was found to significantly influence the sensitivity of algal species (Millington et al., 1988; Stauber and Florence, 1989). Protective effects of metals such as Ca^{2+}, Mg^{2+}, Cu^{2+}, and Ni^{2+} on toxicity of Hg^{2+} and CH_3Hg^+ towards growth and nutrient uptake of the cyanobacterium N. calcicola has been reported (Singh and Singh, 1992). These metals effectively antagonized the Hg^{2+} and CH_3Hg^+ toxicities to the cyanobacterium with Ca^{2+} and Mg^{2+} being more effective than Cu^{2+} and Ni^{2+} ions. As these metals are essential components of algal growth media, their interaction with the other toxic metals in modifying the toxicity assumes importance. The toxicity of selenium to A. nidulans and A. variabilis has been shown to be reduced in the presence of sulfate in the growth medium (Kumar and Prakash, 1971). Presence of phosphate and carbonate reduced the uptake of uranyl by Chlorella regularis (Nakajima et al., 1979).

Presence of sulphur-containing compounds, amino acids, and other reducing agents in the water/medium can protect the organisms from heavy-metal toxicity. This assumes special importance as these compounds are likely to occur in natural habitats. Sulphur-containing compounds and amino acids have been shown to reduce toxicity against Cd, Co, Hg, Ni, Ag, Cr, and Pb in Chlorella, Anacystis nidulans, and Nostoc muscorum

(Rai et al., 1981b; Stokes, 1983; Whitton, 1984; Rai and Raizada, 1988. Rai and Raizada (1988) have demonstrated the protective effect of reduced glutathione and ascorbic acid against Cr and Pb toxicity towards *N. muscorum*. The suggested protective mechanisms include (1) ascorbic acid acting as a reducing agent thereby protecting the oxidation of –SH group by donating electrons or reducing power; (2) ascorbic acid and its products binding to metals and blocking their movement across biological membranes; (3) glutathione-induced reduction of Cr(VI) to Cr(III); and (4) sulphur-containing amino acids (L-methionine and L-cysteine) acting as metal-binding agents (e.g. metals bind to –SH groups), thereby affecting cell membrane permeability.

Limiting Nutrients

The nature of limiting nutrient is also believed to be a possible factor influencing metal toxicity. For example, Hall et al. (1989) reported that algae were more sensitive under P limitation than under NO_3^- limitation. However, Li (1978) reported a reduction in Cd toxicity to a diatom *Thalassiosira fluviatilis* in the presence of NO_3^-. A decrease in P concentration increased the Cd-induced inhibition of alga *Selenastrum capricornutum* in a chemostat culture indicating the importance of limiting nutrient in tolerance of algal cells to the metal (Chen and Lin, 1997). Aluminium uptake by *C. vulgaris* decreased in the presence of PO_4^{3-} due to interaction of PO_4^{3-} with Al (Rai et al., 1998). Several unicellular green algae (*Selenastrum capricornutum*, three species of *Chlorella* and two species of *Scenedesmus*) have been shown to be sensitive to 0.5 µg ml^{-1} of Pb at pH 6.2 under PO_4^{3-}-limiting growth conditions (Monohan, 1976). Reduction of heavy metal-toxicity by PO_4^{3-} has been demonstrated in many algae (Harding and Whitton, 1977; Say and Whitton, 1977; Say et al., 1977; Rai and Kumar, 1980; Rai et al., 1981b). The reason for this reduced toxicity by PO_4^{3-} is due to the formation of insoluble precipitates with heavy metals, resulting in decreased metal bioavailability and toxicity to the organisms. For example, the reduction in Fe toxicity in *C. pyrenoidosa* by P has been shown to be due to oxidation and precipitation of Fe (Zarnowski, 1972). However, Skaar et al. (1974) reported that the addition of P increased the Ni-binding capacity of the diatom *P. tricornutum* resulting in an increased bioavailability and toxicity of metal to the alga.

Population Density

An increase in the amount of algal inoculum has been shown to reduce metal toxicity (Vasseur and Pandard, (1988). Also, Nyhlom and Kallqvist (1989) reported that high biomass density could lead to elevated pH levels due to poor gas-exchange, which in turn can influence the toxicity.

The rate of inhibition in the growth of the alga, *Selenastrum capricornutum* increased with decrease in cell density in the presence of 0.075 µg Cd ml^{-1} in a chemostat culture, which means with high cell density the toxicant concentration per cell is reduced resulting in lower toxicity (Chen and Lin, 1997). Copper toxicity to *Chlorella* sp. has been shown to be dependent on the ratio of cells to cupric ions present in the growth medium (Cedeno-Maldonado and Swader, 1974). An increase in the population of *Chlamydomonas variabilis* caused a decrease in the toxicity of phenylmercuric acetate (Delcourt and Mestre, 1978). The binding of phenylmercuric acetate to a definite number of cellular sites was suggested as the possible reason for reduced toxicity of this compound as a larger number of cells meant the availability of a greater number of cellular-binding sites, resulting in reduced toxicity.

pH

pH is one of the most important factors that affects speciation of heavy metals and thereby toxicity to organisms, including algae. In general, heavy metals are more toxic to algae under acidic conditions with decreasing toxicity in the alkaline range. This may be attributed to the presence of metals as free ions in the acidic range whereas at alkaline pH these metals tend to form insoluble precipitates such as carbonates, sulfides, phosphates, oxides, and hydroxides (Forstner and Prosi, 1979). Bioavailability of metals has been shown to be a function of pH and chelation. The growth rate of two diatoms, *Asterionella ralfsii* and *A. formosa* has been shown to be affected by the interaction of pH, EDTA chelation, and free Al ion concentration (Riseng et al., 1991). Low pH enhanced the sensitivity of *A. nidulans* to Zn whereas Mn showed no effect (Lee et al., 1994). A decrease of pH and EDTA concentration resulted in an increase in free Al ion concentration in the algal medium. The Al-EDTA interactions were both pH and concentration dependent. The concentration of reactive Al remained constant at 50 mM EDTA and a pH of 5 or 6, irrespective of the increase in added Al concentration. Free Al ion concentration was shown to be dependent on pH, with a decrease of pH from 7 to 5 giving an increase of 1-4 orders of magnitude of Al ion concentration (Riseng et al., 1991). In acidic environments (pH < 5.5), metals are usually more soluble. Aluminium toxicity to *Nostoc linckia* is strongly pH-dependent, resulting in an increased toxicity with a decrease in pH (Rai et al., 1996). Thus, the LC_{50} doses at pH 7.5, 6.0, and 4.5 were 600 µM, 55 µM, 18 µM for $AlCl_3$ and 500 µM, 40 µM, and 12 µM for AlF_3, respectively. Toxicity to *C. vulgaris* by Al also yielded a similar result (Rai et al., 1998). The LC_{50} concentrations at pH values 6.8, 6.0, and 4.5 are 4000 µM, 220 µM, and 150 µM for $AlCl_3$ and 1500 µM, 180 µM, and 100 µM for AlF_3, respectively.

Poleo et al. (1994) attributed this to the presence of significant amounts of dissolved Al in the solution at the lower pH 4.5. No free Al was found in the solution with pH above 6.0 where reduced toxicity was observed. This suggests that toxicity and free-Al concentration in the solution are related, and pH and the presence of complexing ligands regulate bioavailability and Al toxicity. Further, Rai et al. (1998) demonstrated that the fluoride form of aluminium (AlF_3) was more toxic to the alga than was the chloride form ($AlCl_3$).

The toxicity of Pb to *Selenastrum capricornutum* decreased with an increase in pH, with approximately six times the concentration of Pb (0.5 vs 3.0 mg l^{-1} concentration) required to exert toxicity at pH 8 than at pH 6.2 (Monohan, 1976). Altered availability of Pb to the alga has been suggested as the reason for decreased toxicity at alkaline pH, Pb being more soluble at acidic pH.

Toxicity of inorganic Sn towards the cyanobacterium *Synechococcus aquatilis* was shown to be regulated by pH, with increased toxicity under alkaline conditions compared with that at neutral pH because of the increased bioavailability of the metal under alkaline conditions (Pawlik-Skowronska et al., 1997). Also, it was suggested that toxicity of inorganic Sn was strongly related to its speciation and dependent more on its chemical form than the total metal concentration. Thus, at pH 7 and below, Sn is mainly present as a cationic or neutral form ($Sn(OH)^+$, $Sn(OH)_2^{2+}$, $Sn(OH)_2$, SnO) whereas at high pH it can exist as dissolved anionic forms (SnO_3H^-, SnO_2^{2-}, $Sn(OH)_6^{2-}$), and these anionic forms may enter the cells easily and thus exert increased toxicity.

The effect of Cd under various pHs towards *Anabaena flosaquae* (a cyanobacterium) was examined by using morphometry and X-ray energy-dispersive methodology (Rai et al., 1990). A decrease in the cell size, thylakoid surface area, number and size of polyphosphate bodies, membrane-limited crystalline inclusions, and size and number of cell layers was observed. All these effects were more pronounced at both acidic and alkaline pH than at neutral pH.

Temperature

Information on the effect of temperature on metal toxicity to algae is limited. The toxicity of Zn to *Nitzschia linearis* increased with increasing temperatures between 22 and 32 °C (Patrick, 1971). Huisman (1980) also reported a similar result wherein toxicity of mercury to *Scenedesmus acutus* increased with increase in temperature. In contrast to these results, Cairns et al. (1978) reported a decrease in Zn toxicity with increased temperature in *Chlamydomonas* sp. and *Scenedesmus quadricauda*, the reasons for which were not clearly known. However, same workers reported an increase in

Zn toxicity with increasing temperature for *Cyclotella meneghiniana*, which is in agreement with other reports (Patrick, 1971; Huisman et al., 1980). Enhanced respiratory activity has been suggested as the possible reason for increased toxicity to organisms at higher temperatures (Forstner and Wittman, 1979). Adsorption and desorption of metals and their uptake by algae are also influenced by temperature. For example, accumulation of copper and cobalt in cyanobacteria *A. nidulans* and *Spirulina platensis* was found to be temperature dependent, with an increase in accumulation at higher (25–45 °C) temperatures (Sharma and Azeez, 1998). These workers observed a negative correlation between metal accumulated and survival ratio of algae with increased temperature, further, they have suggested the possible inactivation of a metabolism-dependent exclusion mechanism operating in living organisms as a possible mechanism.

Salinity

Generally, the toxicity of most heavy metals increases with decreasing salinity. A direct relationship between increase in salinity and decrease in Cu -toxicity to *Cladophora* has been reported (Betzer and Kott, 1969). The decreased toxicity of metals at high salinity may be due to greater complexation of the free-metal ions (e.g. free Cd ion) with Cl ion, the inorganic ligand. The effect of salinity on reduction of metal toxicity in sea water (20% salinity) containing 30 µg Cu^{2+} ml^{-1} has been demonstrated by Eisler and Gardner (1973). These investigators observed a 17% loss in Cu^{2+} within 1 h followed by 95% loss in 8 h and 99.8% loss in 24 h; all this loss was ascribed to metal precipitation as Cl.

Hardness

Water hardness is usually caused by bicarbonates, carbonates, or hydroxides of calcium, magnesium, etc. Addition of Ca and Mg reduced the toxicity of metals (Cu, Zn, Cd, Hg, etc.) to various algae. Further, Ca salts were found to be more effective than Mg salts in reducing metal toxicity (Rai et al., 1981 a,b). The formation of complexes of carbonate and hydroxy compounds or precipitation and co-precipitation with Ca and Mg (Forstner and Prosi, 1979) probably reduces the metal bioavailability and toxicity to the biota as it is hard for the complexed form of metals to cross biological membranes.

Presence of Chelators

Algal exudates have been shown to complex toxic metals thereby reducing metal toxicity and accumulation by cells. For example complexation of Cu, Zn and Cd by algal exudates has been reported by many authors (Davies, 1976; McKnight and Morel, 1979; van den Berg et al., 1979; Brown et al., 1988). DOC in water affects the toxicity and availability of metals to algae. For example, Fisher and Frood (1980) found that in DOC-rich water, metal had a lower impact on phytoplankton as compared with that in low-DOC water. Cadmium has been shown to be strongly concentrated by cells of *Prorocentrum micans* in water containing low concentrations of Cd and low DOC (Rabsch et al., 1984). Harter and Naidu (1995) have recently reviewed the role of DOC on metal mobility in soils. They report that the bioavailability of certain metals may be enhanced by DOC through the formation of stable metal-DOC complexes.

Copper toxicity to algae has been shown to be controlled by free cupric ion and chelators, whether natural or artificial, affected bioavailability of metal and hence its toxicity to the organism. Copper concentration greater than 2 μM was toxic to the diatom *Phaeodactylum tricornutum* in the absence of chelator, but was not toxic even at 5 mM concentration in the presence of EDTA (Spencer, 1957). Further, Sunda et al. (1976) demonstrated that Cu toxicity to algae and Cu content in the algal cells were dependent on free cupric ion activity. The uptake of Cu by free and immobilized cells of *N. calcicola* was reduced by soil extract (1% organic content), EDTA, dilute spent medium, and pond water, with soil extract being most effective (Singh et al., 1992). Uptake of uranium by microalgae was inhibited by the presence of carbonate due to formation of stable complex ions ($UO_2(CO_3)_2^{2-}$, and $UO_2(CO_3)_3^{4-}$) (Sakaguchi et al., 1978). Toxicity of Zn to *A. nidulans* was reduced in the presence of EDTA (Lee et al., 1994).

Often, heavy metals do not react in natural waters as predicted by their chemistry. The interaction of metals with both organic and inorganic matrices in natural waters influences the bioavailability and toxicity to algae. Dissolved organic matter (DOM) such as humic and fulvic acids and inorganic materials in waters can influence metal speciation and hence bioavailability and associated toxicity. Presence of humic acid in the medium reduced the uptake of toxic metals chromate and Ni^{2+} by immobilized *Anabaena doliolum* and *Chlorella vulgaris* (Mallick and Rai, 1993). Toxicity of tin towards the cyanobacterium *Synechococcus aquatilis* was found to be reduced by humic acids (Pawlik-Skowronska et al., 1997).

Redox Transformations

Speciation and bioavailability of several metals (eg. Fe, Cu, Co, Mn, Cr, Hg, Ag) are influenced by their redox transformations as many metals exist in more than one oxidation state in natural environments. Thus, different oxidation states have different chemical properties in terms of their solubility, ionic charge, complexation capability with ligands, etc. For instance, Fe occurs in surface waters in the mainly thermodynamically stable Fe^{3+} state, which is only slightly soluble in the absence of organic chelation. Biological and photochemical reduction of Fe(III) to Fe(II) may result in more soluble and bioavailable Fe, which forms weaker complexes than Fe(III) does. Ferrous(II) iron is unstable and readily undergoes oxidation to Fe(III), especially at high pH. This redox transformation of Fe plays an important role in algal uptake of this compound (Miller et al., 1995; Hudson and Morel, 1990). Likewise, the redox stage of the toxic metals Cr, Ag, and Hg greatly influences bioavailability and toxicity to organisms. Photochemical and biological reduction of Cr(VI), which is the more soluble, bioavailable, and toxic form, to Cr(III) results in a less soluble, less toxic, and biologically less available form. Similarly, Sn(II) was found to be more toxic than Sn(IV) toward the growth of planktonic cyanobacterium (Pawlik-Skowronska et al., 1997). Also, biological and photochemical reduction of stable Hg(II) and Ag(I) results in substantial decreases in bioavailability and toxicity. Moreover, Hg(II) and Ag(I), which have a strong binding capacity with organic ligands, are transformed to elemental forms which are non-reactive to complex formation in addition to elemental Hg being volatile and elemental Ag insoluble.

METAL-METAL AND METAL-XENOBIOTIC INTERACTIONS

Toxicant interactions are important in order to obtain valid information relating to their environmental impact. The study of the effects of metal combinations on algae is gaining importance. Heavy metals and other xenobiotic substances are rarely found alone in the environment, but are in combination with other toxicants including various synthetic and naturally occurring organic and inorganic substances. For example, As and DDT (dichlorodiphenyl trichloroethane) are present in combination in cattle dip sites of Australia. It has been well recognized that the response of an organism to a combination of metals can be different than to a single metal. Thus two or more metals can interact additively, synergistically or antagonistically, toward the organism. Information on the interaction of metals and metal-other xenobiotic interactions is limited.

A combination of Hg and pesticides (either atrazine or permethrin) added simultaneously at Day 0 interacted antagonistically toward the cyanobacterium *Anabaena inaequalis* (Stratton, 1985). When Hg was added at Day 0 and the pesticides at Day 1, Hg and permethrin interacted synergistically whereas Hg and atrazine interacted additively. Interestingly, in the same study (Stratton, 1985), when pesticides were added at Day 0 and Hg at Day 1, the reverse occurred. The toxic effect of Cu and Cd on *Anacystis nidulans* was shown to increase in the presence of Fe and Cd respectively, whereas the presence of Ca, Fe, or Zn caused a decrease in toxicity of Cd to this alga (Whitton and Shehata, 1982). The observed lack of toxicity of Cu on this alga was attributed to the production of strong Cu-complexing agent by the alga. Rai et al. (1998) have reported severe inhibition of growth and metabolism in a green alga (*C. vulgaris*) by a combination of $AlCl_3^+$ and NaF when compared with the metals present alone. This interaction caused an additive effect at pH 6.8, and a synergistic effect at pH 6.0 and 4.5. Also, these investigators observed the formation of AlF_4^- at pH 6.0 and 4.5 in $AlCl_3$- and NaF-supplemented medium, suggesting that AlF_4^- formation occurs under acidic conditions. AlF_4^- is known as a structural analog of PO_3^{4-}, and hence its toxicity to the alga was influenced by the presence of PO_3^{4-} in the growth medium, which competed for binding sites. Interactions were found to be dependent on the nature of the toxicity criterion chosen. For example, Hg and Cd interacted antagonistically when tested for growth in *Anabaena inaequalis*, but synergistically for photosynthesis and nitrogen-fixation (Stratton and Corke, 1979). Combinations of Cd with Cu and Cr(VI) showed a synergistic interaction towards growth of natural algae from water of the Nile (Lasheen et al.,1990), followed by a clear reduction in the size and composition of the algae. *Oscillatoria* sp. had emerged as the only tolerant alga. However, the toxic effect and metal accumulation in algae in the Nile followed the order: Cd–Cu > Cd–Cr > Cd > Cu > Cr. In another study, mixed algal flora exposed to 1.0–4.0 mg ml^{-1} Cr(VI) resulted in an almost unialgal culture of *Oscillatoria* sp. (Filip et al., 1979). Similarly, the growth of a cyanobacterium *Chroococcus paris* has been shown to be adversely affected by a combination of Cd, Cu, and Zn at concentrations greater than 1.0 µg ml^{-1} (Les and Walker, 1984).

Interactions among toxic metals (Cu and Zn) and nutrient metals (Mn and Zn) seem to be important in controlling metal toxicity and nutrient limitation to the algae. Toxic metals are often taken up by the organisms via nutrient metal transport systems, resulting in nutrient deficiencies among the organisms. Thus, these toxic metals often compete with nutrient metals for binding sites, thereby increasing nutrient metal deficiency among algae. However, certain algal species can counteract these problems by other protective mechanisms (such as binding the accumulated toxic metals to phytochelatins, intracellular ligands, or by excretion of toxic metals, mentioned elsewhere in this chapter. Copper inhibition of Zn uptake at low concentrations of Zn^{2+} ion resulting in Zn deficiency and inhibition of Mn uptake by high Cu and Zn concentrations have been demonstrated in *Chlamydomonas* and *T. pseudonana* (Sunda and Huntsman, 1983, 1996, 1998). Also, inhibition of Fe uptake by elevated levels of Cd^{2+} causing Fe deficiency in *Thalassiosira weissflogii* (Harrison and Morel, 1983), and inhibition of cobalt uptake by high Zn^{2+} levels causing cobalt deficiency in *Emiliania huxleyi* (Sunda and Huntsman, 1995) have been reported. These results indicate that interactions among metals are quite complex and these greatly influence the cellular metal accumulation, toxicity, and nutrition among different algal flora.

Interaction of Bivalent Cations

There is substantial evidence to suggest that heavy-metal toxicity is regulated by the interactive effect of bivalent metal cations such as Ca^{2+}, Mg^{2+}, Mn^{2+}, Ni^{2+}, Co^{2+}, and Zn^{2+} present in the algal growth medium or aquatic environment. These bivalent metals play a significant role in physiological and biochemical processes of algae. Nontoxic bivalent metals such as Ca^{2+}, Mg^{2+}, and Mn^{2+} have been shown to antagonise the toxic effects of Cr and Sn on growth, nutrient uptake, and certain enzyme activities of *A. doliolum* whereas toxic cations such as Co, Zn, and Ni increased the toxicity of test metals (Rai and Dubey, 1989). In another study, all the tested bivalent cations (Ca^{2+}, Mg^{2+}, Cu^{2+}, and Ni^{2+}) antagonized the Hg^{2+} and CH_3Hg^+ toxicity to growth and nutrient uptake of *N. calcicola*; however, Cu^{2+} and Ni^{2+} were less effective compared to Ca^{2+} and Mg^{2+} ions (Singh and Singh, 1992). Available evidence suggests that the antagonistic effect of Ca^{2+} ions to heavy-metal toxicity toward growth and nutrient uptake in these organisms is due to complexation and competitive inhibition of metal uptake. Thus, it is clear that antagonistic and synergistic interaction effects of metals may influence metal toxicity to organisms; thus, it is important to consider the composition of aquatic bodies before waste metals are discharged into it. A similar effect of bivalent cations such as Ca and Mg on heavy-metal bioavailability has been reported for plants (see Chapter 3).

Monovalent Cations

Toxicity of Cs^+ to cyanobacteria is primarily due to the intracellular replacement of K^+ by Cs^+ (Avery et al., 1991, 1992). In *Nostoc muscorum*, Cs^+ is more effective in inhibiting Rb^+ uptake and accumulation than Rb^+ is in inhibiting Cs^+ uptake. It is also known that higher concentrations of Na^+ or K^+ inhibits Cs^+ uptake and accumulation. This suggests that Na^+, K^+ and Rb^+ alone or in combination regulate the intracellular concentration of Cs^+ in cyanonobacterial populations thriving in Cs-polluted environments (Singh et al., 1997). Interestingly, the uptake and toxicity of Cs^+ in *N. muscorum* was found to be dependent on diazotrophy and repressible by NH_4^+ (Singh et al., 1997).

METAL BIOAVAILABILITY AND UPTAKE BY ALGAE: IMPLICATIONS TO BIOREMEDIATION

The ability of algae to adsorb metals has been recognized for many years. In natural environments, algae play a major role in controlling metal concentration in lakes and oceans (Sigg, 1985, 1987). Mixed cultures of algae and cyanobacteria isolated from a wetland constructed to treat uranium-contaminated waste water were able to concentrate U between

8,000 and 24,000 times above solution concentration (S. Ragusa, M. Megharaj, and R. Naidu, unpublished data).

Algal biomass (live or dead) can serve as a base for biosorbents that can be regenerated for multiple reuse. They can be highly selective, efficient, and cheap, competing with commercial ion-exchange resins and activated carbon (Gadd, 1990). Process arrangement when using biosorbents is almost identical to that used for conventional materials (i.e. resins and activated carbon). The use of algal biomass as a biosorbent has potential application in both environment control and metal-recovery operations.

Algal biomass may accumulate metals due to metabolic activity, termed bioaccumulation, or passively bind metals when biomass is metabolically inactive or dead, termed biosorption. Conventional chemical methods (e.g. precipitation, oxidation, or reduction) are either ineffective or very expensive when large volumes of waste solutions with relatively low metal concentrations (between 1 and 100 μg ml^{-1}) are being cleaned up (Gadd, 1990). Similarly, the use of ion-exchange resins is inappropriate in some cases due to the high price of materials, although in some instances the cost may be partially offset by the value of the metal recovered.

The use of biosorbents has gained favor as the ability of biosorbents to selectively adsorb metals has been developed (Volesky, 1987). Metal adsorption is usually carried out using a stirred tank or packed column arrangement. In a stirred tank contactor, the biosorbent is suspended in the metal-laden liquid and run as a batch or continuous flow system. In this case, the metal-laden biosorbent must be recovered from suspension and the metal eluted before the biosorbent is reused. A fixed packed bed contactor has the biosorbent packed into a column as beads, the metal-laden solution is passed over the beads, and the metals adsorbed on the biomass. In this case, the biosorbent becomes saturated with the metal at the inlet end of the column first and the metal saturation front gradually progresses toward the outlet of the column. When the biomass is fully saturated (breakthrough point), the column must be regenerated by elution of metal from biosorbent. A third configuration is the fluidized bed reactor. These are run at high flow rates in order to keep the biosorbent particles in suspension and well mixed. Again, when the biosorbent is saturated, the metals are eluted (i.e. biosorbent regenerated) and the column is ready to be reused.

The influence of algae is limited to the upper few centimetres of moist soil. Given the fact that most of the pollutants tend to occur in the surface layers of the soil, these organisms could play a significant role during the initial stages. But not much work has been done in this important area. As discussed previously, metal uptake by algae will be controlled by factors that influence metal bioavailability.

CONCLUSION

All heavy metals, including essential micronutrients, are toxic to algae at higher concentrations. The magnitude of the toxicity varies depending on the nature of metal and algal species. In general, toxicity of metals to algae is manifested via their inhibitory effects on growth, pigments, macromolecules, ATP generation, photosynthesis, nitrogen fixation, nutrient uptake, glutamine synthetase, nitrate reductase, cell volume reduction, membrane permeability, and cell morphology. The sensitivity or tolerance towards heavy metals varies among different algal species and further physiological and genetic basis of their tolerance is well recognized, but needs more research for a better understanding. Heavy-metal pollution affects the abundance and species composition of algae, resulting in the survival of a few tolerant algal species. Occurrence of a few tolerant forms of algae as dominant species in metal-polluted environments has been noticed. Several factors, including environmental factors and speciation, govern heavy-metal toxicity to algae. Although the effect of speciation and the role of environmental factors (pH, temperature, chelators, etc.) on heavy-metal toxicity towards algae has been recognized, stil much more research needs to be done in this area. The presence of a combination of different metals and metal and other pollutant mixtures results in additive, synergistic, or antagonistic effects on algae. There is a paucity of information on metal-metal interactions and metal-nonmetal pollutant interactions towards algae, which warrants further studies in this direction. Greater understanding of physiological and physicochemical processes in metal toxicity is necessary to fully understand metal bioavailability.

On the other hand, field survey of aquatic and terrestrial contaminated habitats provide useful information about the extent and pattern of contamination and complement the toxicity data. Also, field surveys aid in the identification of sensitive species that may be affected by the contaminant. A field survey of contaminated sites in relation to carefully chosen uncontaminated reference sites provides information on the effects of contaminants on structure and functioning of the communities of populations.

Most of the information available on heavy metal toxicity to algae is based on aquatic organisms. However, given the similarity of algal growth medium to soil pore water in terms of the range of ionic concentrations of major anions and cations (G.S.R. Krishnamurti, M. Megharaj, and R. Naidu, unpublished data), algal studies conducted using a defined growth medium such as the Bold's basal medium (Bold and Wynne, 1978) may be useful to soil as well. Even though algae form an important component of the soil ecosystem, studies involving soil algae in relation to heavy-metal toxicity are scarce. The reasons for this may be partly because of the ease of handling aquatic algae compared with soil algae and negligence (Megharaj et al., 2000) or limited expertise of soil algae among researchers. The few studies available on the effects of heavy-metal pollution on soil algae demonstrated that heavy metals affect the abundance and composition of these populations and their activities.

The ability of algae to sequester heavy metals may provide a means to strip heavy metals from large volumes of wastewater containing low concentrations of metals. They offer advantages when large volumes of water contaminated with low levels of metal need to be cleaned up.

REFERENCES

Ahluwalia, A.S. and Kaur, M (1989) Nickel toxicity on growth and heterocyst formation in a nitrogen-fixing blue-green alga. *Phykos* **28**: 196–200.

American Society for Testing and Materials (1994) Standard guide for conducting static 96h toxicity tests with microalgae. E1218-90. *In Annual Book of ASTM Standards*. ASTM, Philadelphia, PA.

Asthana, R.K., Pandey, P.K. and Singh, S.P. (1990) Nickel regulation of photoautotrophy in a cyanobacterium. *Air Water Soil Pollut.* **52**: 263–276.

Avery, S.V., Codd, G.A. and Gadd, G.M. (1991) Caesium accumulation and interaction with other monovalent cations in the cyanobacterium *Synechocystis* PCC 6803. *J. Gen. Microbiol.* **137**: 405–413

Avery, S.V., Codd, G.A. and Gadd, G.M. (1992) Replacement of cellular potassium by caesium in *Chlorella emersonii*: Differential sensitivity of photoautotrophic and chemoheterotrophic growth. *J. Gen. Microbiol.* **138**: 69–76.

Bariaud, A and Mestre, J.-C. (1984) Heavy metal tolerance in a cadmium-resistant population of *Euglena gracilis*. *Bull. Environ. Contam. Toxicol.* **32**: 597–601.

Bartless, P.D., Craig, P.J. and Mortan, S.F. (1977) Behaviour of mercury species in isolated estuarine sediments. *Nature* **267**: 606–608.

Bates, S.S., Tessier, A., Campbell, P.G.C. and Letourneau, M. (1985) Zinc-phosphorus interaction and variation in zinc accumulation during growth of *Chlamydomonas variabilis* (Chlorophyceae) in batch culture. *Can. J. Fish Aquat. Sci.* **42**: 86–94.

Becker, E.W. (1983) Limitations of heavy metal removal from waste water by means of algae. *Water Res.* **17**: 459–466.

Betzer, N. and Kott, Y. (1969) Effect of halogens on algae. II Cladophora sp. *Water Res.* **3**: 257–264.

Bold, H.C., and Wynne, M.J. (1978) *Introduction to the Algae: Structure and Reproduction*. Prentice Hall, Englewood Cliffs, NJ.

Brown, L.N., Robinson, M.G. and Hall, B.D. (1988) Mechanisms for copper tolerance in *Amphora coffeaeformis*- internal and external binding. *Marine Biology* **97**: 581–586.

Brooks, P.C., McGrath, S.P. and Heijnen, C. (1986) Metal residues in soils previously treated with sewage-sludge and their effects on growth and nitrogen fixation by blue-green algae. *Soil Biol. Biochem.* **18**: 345–353.

Burns, R.C., andHardy, R.W.F. (1975) *Nitrogen Fixation in Bacteria and Higher Plants*. Springer, Berlin.

Butler, M., Haskew A.E.J. and Young, M.M. (1980) Copper tolerance in the green alga. *Chlorella vulgaris*. *Plant Cell Environ.* **3**: 119–126.

Canterford, G.S. and Canterford, D.R. (1980) Toxicity of heavy metals to the marine diatom *Dithylum brightwellii* (West) Grunow: correlation between toxicity and metal speciation. *J. . Biol. Ass. U.K.* **60**: 227–242.

Cairns, J. Jr., Buikema, A.L. Jr., Heath, A.G. and Parker, B.C. (1978) *Effects of Temperature on Aquatic Organisms Sensitive to Selected Chemicals*. Virginia Water Resouces Research Centre, Virginia Polytechnic Institute and State University, Bulletin 106.

Carr, H.P., Carino, F.A., Yang, M.S. and Wong, M.H. (1998) Characterization of the cadmium-binding capacity of *Chlorella vulgaris*. *Bull. Environ. Contam. Toxicol.* 60: 433–440.

Cedeno-Maldonado, A. and Swader, J.A. (1974) Studies on the mechanism of copper toxicity in *Chlorella*. *Weed Sci.* 22: 443–449.

Chen, C.-Y and Lin, K.-C. (1997) Optimisation and performance evaluation of the continuous algal toxicity test. *Environ. Toxicol. Chem.* 16: 1337–1344.

Clarke, S.E., Stuart, J., and Sander-Loehr, J. (1987) Induction of siderophore activity in *Anabaena* spp. and its moderation of copper toxicity. *Appl. Environ. Microbiol.* 53: 917–922.

Corradi, M.G, and Gorbi, A.G. (1993) Chromium toxicity on two linked trophic levels II. Morphophysiological effects on *Scenedesmus acutus*. *Ecotoxicol. Environ. Safety* 25: 72–78.

Couture, P., Blaise, C., Cluis, D. and Bastien, C. (1989) Zirconium toxicity assessment using bacteria, algae and fish assays. *Water Air Soil Pollut.* 47: 87–100.

Filip, D.S., Peters, T., Adams, V. D. E., and Middlebrooks, J. (1979) Residual heavy metal removal by an algae-intermittent sand filtration system. *Water Res.* 13: 305–313.

Daniel, G.F., and Chamberlain, A.H.L. (1981) Copper mobilisation in fouling diatoms. *Bot. Mar.* 24: 229–243.

Davies, A.G. (1976). An assessment of the basis of mercury tolerance in *Dunaliella tertiolecta*. *J. Mar. Biol. Assoc. U.K.* 56: 39–57.

Davies, A.G. (1978) Pollution studies with marine plankton. Part II Heavy metals. *Adv. Mar. Biol.* 15: 381–478.

Davies, A.G., and Sleep, J.A. (1980) Copper inhibition of carbon fixation in coastal phytoplankton assemblages. *J. Marine. Biol. Assoc. U.K.* 60: 841–850.

De Filippis, L.F., and Pallaghy, C.K. (1976 a) The effect of sublethal concentrations of mercury and zinc on *Chlorella*: I. Growth characteristics and uptake of metals. *Z. Pflanzenphysiol.* 78: 197–207.

De Filippis, L.F., and Pallaghy, C.K. (1976 b) The effect of sub-lethal concentrations of mercury and zinc on *Chlorella*. III. Development of possible resistance to metals. *Z. Pflanzenphysiol. Bd.* 79: 323–335.

Devi Prasad, P.V., and Devi Prasad, P.S. (1982) Effect of cadmium, lead and nickel on three freshwater green algae. *Water Air Soil Pollut.* 17: 263–268.

Delcourt, A., and Mestre, J.C. (1978) The effects of phenylmercuric acetate on the growth of *Chlamydomonas variabilis* Dang. *Bull. Environ. Contam. Toxicol.* 20: 145–148.

Dohler, G. (1986). Impact of UV-B radiation on [^{15}N] ammonia and [^{15}N] nitrate uptake of *Dictylum brightweilli*. *Photobiochem. Photobiophys* 11: 115–121.

Edding, M. and Tala, F. (1996) Copper transfer and influence on a marine food chain. *Bull. Environ. Contam. Toxicol.* 57: 617–624.

Eisler, R., and Gardner, G.R. (1973) Acute toxicology to an estuarine teleost of mixture of cadmium, copper and zinc salts. *J. Fish Biol.* 5: 131–142.

Errecalde, O., Seidl, M., and Campbell, P.G.C. (1998) Influence of low molecular weight metabolite (citrate) on the toxicity of cadmium and zinc to the unicellular green alga *Selenastrum capricornutum*: An exception to the free-ion model. *Water Res.* 32: 419–429.

Fisher, N.S. and Frood, D. (1980) Heavy metals and marine diatoms: influence of dissolved organic compound on toxicity and selection for metal tolerance among four species. *Mar. Biol.* 59: 85–93.

Fisher, N.S., Jones, J., and Nelson, D.M. (1981) Effects of Cu and Zn on growth, morphology and metabolism of *Asterionella japonica* Cleve. *J. Exp. Mar. Biol. Ecol.* 51: 37–56.

Forstner, U., and Prosi, F. (1979) Heavy metal pollution in freshwater ecosystems. In *Biological Aspects of Water Pollution*. Ravera, O. (Ed.). Pergamon Press, N.Y., pp. 129–161.

Forstner, U., and Wittman, G.T.W. (1979) *Metal Pollution in the Aquatic Environment*. Springer-Verlag, Berlin.

Foster, P.L. (1977) Copper exclusion as a mechanism of heavy metal tolerance in a green alga. *Nature (London)* **269**: 322–323.

Gadd G.M. (1990) Biosorption. *Chem and Ind*. **July**: pp. 421–426.

Gimmler, H., Treffny, B., Kowalski, M. and Zimmermann, U. (1991) The resistance of *Dunaliella acidophila* against heavy metals: the importance of the zeta potential. *J. Plant Physiol*. **138**: 708–716.

Gowrinathan, K.P., and Rao, V.N.R. (1991) Uptake and accumulation of copper by two microalgae. *Phykos* **30**: 13–22.

Greene B., and Darnall, D.W. (1990) Microbial oxygenic phototrophs (cyanobacteria and algae) for metal ion binding. In *Microbial Mineral Recovery*, Ehrlich, H.L. and Brierley, C.L. (Eds.). McGraw-Hill Publishing Company, NY, 1990.

Hall, J., Healey, F.P., and Robison, G.G.C. (1989) The interaction of chronic copper toxicity with nutrient limitation in chemostat cultures of *Chlorella*. *Aquat Toxicol*. **14**: 15–26.

Harrison, G.I., and Morel, F.M.M. (1983) Antagonism between cadmium and iron in the marine diatom *Thalassiosira weissflogii*. *J. Phycol*. **19**: 495–507.

Harding , J.P.C., and Whitton, B.A. B.A. (1976) Resistance to zinc of *Stigeoclonium tenue* in the field and laboratory. *Br. Phycol. J*. **11**: 417–426.

Harding, J.P.C., and Whitton, B.A. (197) Environmental factors reducing the toxicity of zinc to *Stigeoclonium tenue*. *Brit. Phycol .J*. **12**: 17–21.

Hart, B.A., and Scaife, B.D. (1977) Toxicity and bioaccumulation of cadmium in *Chlorella pyrenoidosa*. *Environ. Res*. **14**: 401–413.

Hall, J., Healey, F.P., and Robison, C.G.E. (1989) The interaction of chronic copper toxicity with nutrient limitation in two chlorophytes in batch culture. *Aquat. Toxicol*. **14**: 1–13.

Harter, R.D.R., and Naidu, R. (1995) Role of metal-organic complexation in metal sorption by soils. *Adv. Agron*. **55**: 219–264.

Henriksson, L.E., and DaSilva, E.J. (1978) Effects of some inorganic elements on nitrogen-fixation in blue-green algae and some ecological aspect of pollution. *Zeitschrift fur Allgemeine Mikrobiologie* **18**: 487–494.

Hollibaugh, J.T., Seibert, D.L.R., and Thomas, W.H. (1980) A comparison of the acute toxicities of ten heavy metals to phytoplankton from Saanich Inlet, B.C., Canada. *Est. Coastal Mar. Sci*. **10**: 93–105.

Hudson, R.J.M., and Morel, F.M.M. (1990) Iron transport in marine phytoplankton: kinetics of cellular and medium coordination reactions. *Limnol. Oceanogr*. **35**: 1002–1020.

Huisman, J., Ten Hoopen, H.J.G., and Fuchs, A. (1980) The effect of temperature upon the toxicity of mercuric chloride to *Scenedesmus acutus*. *Environ. Pollut*. **A22**: 133–148.

Hutchinson, T.C. (1973) Comparative studies on the phototoxicity of heavy metals to phytoplankton and their synergistic interactions. *Water Pollut. Res. Can*. **8**: 68–89.

International Standards Organization (1987) Water quality-algal growth inhibition test. Draft International Standard ISO/DIS 8692, Geneva, Switzerland

Jennings, J.R. (1979) The effect of cadmium and lead on the growth of two species of phytoplankton with particular reference to the development of tolerance. *J. Plankton Res*. **1**: 121–136.

Jensen, T.E., Rachlin, J.W., Jani V., and Warkentine, B. (1982) An X-ray energy dispersive study of cellular compartmentalisation of lead ans zinc in *Chlorella saccharophila* (Chlorophyta), *Navicula closterium* (Bacillariophyta). *Environ. Expl. Bot*. **22**: 319–328.

Kaplan, D., Heimer, Y.M., Abeliovich, A. and Goldsbrough, P.B. (1995) Cadmium toxicity and resistance in *Chlorella* sp. *Plant Sci.* **109**: 129–137.

Kayser, H (1976) Waste-water assay with continuous algal cultures: The effect of mercuric acetate on the growth of some marine dinoflagellates. *Mar. Biol.* **36**: 61–72.

Khalil, Z (1997) Toxicological response of a cyanobacterium, *Phormidium fragile*, to mercury. *Water Air Soil Pollut.* **98**: 179–185.

Knauer, K., Ahner, B., Xue, H.B., and Sigg, L. (1998) Metal and phytochelatin content in phytoplankton from freshwater lakes with different metal concentrations. *Environ. Toxicol. Chem.* **17**: 2444–2452.

Kong, F.X., Sang, W.L., Hu, W., and Li, J.J. (1999) Physiological and biochemical response of *Scenedesmus obliquus* to combined effects of Al, Ca, and low pH. *Bull. Environ. Contam. Toxicol.* **62**: 179–186.

Kumar, H.D. and Prakash, G. (1971) Toxicity of selenium to the blue-green algae, *Anacystis nidulans* and *Anabaena variabilis*. *Ann. Bot.* **35**: 697–705.

Kumar, D., Jha, M., and Kumar, H.D. (1985) Heavy-metal toxicity in the cyanobacterium *Nostoc linckia*. *Aquat. Bot.* **22**: 101–105.

Kumar, D., Jha, M., and Kumar, H.D. (1985) Copper toxicity in the fresh water cyanobacterium *Nostoc linckia*. *J. Gen. Appl. Microbiol.* **31**: 165–169.

Kumari, J., Venkateswarlu, V., and Rajkumar, B. (1991) Heavy metal pollution and phytoplankton in the river Moosi (Hyderabad), India. *Int. J. Environ. Stud.* **38**: 157–164.

Kuwabara, J.S., and Leland, H.V. (1986) Adaptation of *Selenastrum capricornutum* (Chlorophyceae) to copper. *Environ. Toxicol. Chem.* **5**: 197–203.

Lasheen, M.R., Shehata, S.A., and Ali, G.H. (1990) Effect of cadmium, copper and chromium (VI) on the growth of Nile water algae. *Water Air Soil Pollut.* **50**: 19–30.

Lee, H.L., Lustigman, B., Schwinge, V., Chu, I.Y., and Hsu, S. (1992) Effect of mercury and cadmium on the growth of *Anacystis nidulans*. *Bull. Environ. Contam. Toxicol.* **49**: 272–278.

Lee, L.H., Lustigman, B., and Dandorf, D. D. (1994) Effect of manganese and zinc on the growth of *Anacystis nidulans*. *Bull. Environ. Contam. Toxicol.* **53**: 158–165.

Lehman, J.L., and Vas Cancelos, A.C. (1979) Physiology of copper and mercury stress in the marine diatom *Cylindrotheca closterium*. *J. Phycol.* **15** (Suppl.): 18.

Les, A. and Walker, R.W. (1984) Toxicity and binding of copper, zinc, and cadmium by the blue-green alga, *Chroococcus paris*. *Water Air Soil Pollut.* **23**: 129–139.

Li, W.K.W. (1978) Kinetic analysis of interactive effects of cadmium and nitrate on growth of *Thalassiosira fluviatilis* (Bacillariophyceae). *J. Phycol.* **14**: 454–460.

Mallick, N., and Rai, L.C. (1993) Influence of culture density, pH, organic acids and divalent cations on the removal of nutrients and metals by immobilised *Anabaena doliolum* and *Chlorella vulgaris*. *World J. Microbiol. Biotechnol.* **9**: 196–201.

Mason, R.P., Reinfelder, J.R., and Morel, F.M.M. (1995) Bioaccumulation of mercury and methyl mercury. *Water Air Soil Pollut.* **80**: 915–921.

McCann, A.E., and Cullimore, D.R. (1979) Influence of pesticides on soil algal flora. *Residue Rev.* **72**: 1–31.

McKnight, D.M., and Morel, F.M.M. (1979) Release of weak and strong copper-complexing agents by algae. *Limnol. Oceanogr.* **24**: 823–837.

Megharaj, M., Kantachote, D., Singleton, I., and Naidu, R. (2000) Effects of long-term contamination of DDT on soil microflora with special reference to soil algae and algal transformation of DDT. *Environ Pollut.* **109**: 35–42.

Meisch, H-U., and Schmitt-Beckmann, I. (1979) Influence of tri-and hexavalent chromium on two Chlorella strains. *Z. Pflanzenphysiol. Bd.* **94**: 231–239.

Metting, B. (1981) The systematics and ecology of soil algae. *Bot. Rev.* **47**: 195–312.

Millington, LA., Goulding, K.H., and Adams, N. (1988) The influence of growth medium composition on the toxicity of chemicals to algae. *Water Res.* **22**: 1593–1597.

Miller, W.L., Lin, K., King, D.W., and Kester, D.R. (1995) Photochemical redox cycling of iron in coastal seawater. *Mar. Chem.* 50: 63–78.

Monahan,T.J. (1976) Lead inhibition of Chlorophyceaen microalgae. *J. Phycol.* 12: 358-362.

Monteiro, M.T., Oliveira, R., and Vale, C. (1995) Metal stress on the plankton communities of Sado River (Portugal). *Water Res.* 29: 695–702.

Nagano T, Miwa, M., Suketa,Y., and Okada, S. (1984) Isolation, physicochemical properties and amino acid composition of a cadmium-binding protein from cadmium treated *Chlorella ellipsoidea*. *J. Inorg. Chem.* 21: 1–71.

Nagel, K., Adelmeier, U., and Voigt, J. (1996) Subcellular distribution of cadmium in the unicellular alga *Chlamydomonas reinhardtii*. *J. Plant Physiol.* 149: 86–90.

Nakajima, A., Horikoshi, T., and Sakaguchi, T. (1979) Ion effects on the uptake of uranium by *Chlorella regularis*. *Agric. Biol. Chem.* 43: 625–629.

Nyholm,N., and Kallqvist, T. (1989) Methods of growth inhibition toxicity tests with fresh water algae. *Environ. Toxicol. Chem.* 8: 689–703.

Olafson, R.W. (1986) Physiological and chemical characterization of cyanobacterial metallothioneins. *Environ. Health Perspect.* 65: 71–75.

Organization for Economic Cooperation and Development (1984) *Guideline for testing chemicals.* 201. Algal growth inhibition test. Paris, France.

Parent, L., Twiss, M.R., and Campbell, P.G.C. (1996). Influence of natural dissolved organic matter on the interaction of aluminium with the microalga *Chlorella*: A test of the free-ion model of trace metal toxicity. *Environ. Sci. Technol.* 30: 1713–1720.

Patrick, R. (1971) *Report in Water Quality Criteria.* National Academy of Sciences, National Academy of Engineering, Washington DC.

Pawlik-Skowronska, B., Kaczorowska,R., and Skowronski, T. (1997) The impact of inorganic tin on the planktonic cyanobacterium *Synechococcus aquatilis*: the effect of pH and humic acid. *Environ. Pollut.* 97: 65–69.

Pearson R. (1973) *Hard and soft acids and bases.* Dowder, Hutchinson and Ross, Philadelphia.

Pistocchi, R., Guerrini, F., Balboni, V., and Boni, L. (1997) Copper toxicity and carbohydrate production in the microalgae *Cylindrotheca fuciformis* and *Gymnodinium* sp. *Eur. J. Phycol.* 32: 125–132.

Planas, D., and Healey, E.P. (1978) Effects of arsenate on growth and phosphorus metabolism of phytoplankton. *J. Phycol.* 14: 337–341.

Poleo, A.B.S., Lyderson, E., Kroglud, B.O., Salbu, B., Vogt, R.D., and Kvellestad, A. (1994) Increased mortality of fish due to changing Al-chemistry of mixing zones between limed streams and acidic tributaries. *Water Air Soil Pollut.* 75: 339–351.

Rabsch, U., Wolter, K., and Krischker, P. (1984) Influence of low cadmium and zinc concentrations on batch culture *Prorocentrum micans* (Dinophyta) containing low levels of dissolved organic carbon. *Mar. Ecol. Prog. Ser.* 14: 275–285.

Rana, B and Kumar, H.D. (1974) Effects of toxic waste and wastewater components on algae. *Phykos* 13: 67–83.

Rai, L.C. and Kumar, H.D. (1980) Effects of certain environmental factors on the toxicity of zinc to *Chlorella variabilis*. *Microbias Lett.* 13: 79–84.

Rai, L.C., Gaur, J.P., and Kumar, H.D. (1981a) Phycology and heavy metal pollution. *Biol. Rev.* 56: 99–151.

Rai, L.C., Gaur, J.P., and Kumar, H.D. (1981b) Protective effects of certain environmental factors on the toxicity of zinc, mercury and methyl mercury to *Chlorella vulgaris*. *Environ. Res.* 25: 250–259.

Rai, L.C. and Raizada, M. (1988) Impact of chromium and lead on *Nostoc muscorum*: Regulation of toxicity by ascorbic acid, glutathione, and sulfur-containing amino acids. *Ecotoxicol. Environ. Safety* **15**: 195–205.

Rai, L.C. and Dubey, S.K. (1989) Impact of chromium and tin on a nitrogen-fixing cyanobacterium *Anabaena doliolum*: Interaction with bivalent cations. *Ecotoxicol. Environ. Safety* **17**: 94–104.

Rai, L.C., Singh, A.K., and Mallick, N. (1990) Employment of CEPEX enclosures for monitoring toxicity of Hg and Zn on in situ structural and functional characteristics of algal communities of river Ganga in Varanasi. *Ecotoxicol. Environ. Safety* **20**: 211–221.

Rai, L.C., Singh, A.K. and Mallick, N. (1991) Studies on photosynthesis, the associated electron transport system and some physiological variables of *Chlorella vulgaris* under heavy metal stress. *Plant Physiol.* **137**: 419–424.

Rai, L.C., Husaini, Y., and Mallick, N. (1996) Physiological and biochemical responses of *Nostoc linckia* to combined effects of Aluminium, fluoride and acidification. *Environ. Exp. Bot.* **36**: 1–12.

Rai, L.C., Husaini, Y. and Mallick, N. (1998). PH-altered interaction of aluminium and fluoride on nutrient uptake, photosynthesis and other variables of *Chlorella vulgaris*. *Aquat. Toxicol.* **42**: 67–84.

Riseng, C.M, Gensmer, R.W., and Kilham, S.S. (1991) The effect of pH, aluminium, and chelator manipulations on the growth of acidic and circumneutral species of *Asterionella*. *Water Air Soil Pollut.* **60**: 249–261.

Rivkin, R.B. (1979) Effects of lead on growth of the marine diatom *Skeletonema costatum*. *Mar. Biol.* **50**: 239–247.

Sakaguchi, T., Horikoshi, T., and Nakajima, A. (1978) Studies on the accumulation of heavy metal elements in biological systems VI. Uptake of uranium from sea water by microalgae. *J. Ferment. Technol.* **56**: 561–565.

Say, P.J., Diaz, B.M., and Whitton, B.A. (1977) Influence of zinc on lotic plants. I. Tolerance of *Hormidium* species to zinc. *Freshwater boil.* **7**: 357–376.

Say, P.J., and Whitton, B.A. (1977) Influence of zinc on lotic plants. II. Environmental effects on toxicity of zinc to *Hormidium rivulare*. *Freshwater Biol.* **7**, 377–384.

Sharma, R.M., and Azeez, P.A. (1998) Accumulation of copper and cobalt by blue-green algae at different temperatures. *Int. J. Environ. Anal. Chem.* **32**: 87–95.

Shehata, F.H.A., and Whitton, B.A. (1982) Zinc tolerance in strains of the blue-green alga *Anacystis nidulans*. *Br. Phycol. J.* **17**: 5–12.

Shehata, S.A., Lasheen, M.R., Kobbia, I.A., and Ali, G.H. (1999) Toxic effect of certain metals mixture on some physiological and morphological characteristics of freshwater algae. *Air Water Soil Pollut.* **110**: 119-135.

Sigg L. (1985) Metal transfer mechanisms in lakes; Role of settling particles. In *Chemical Processes in Lakes*, Stumm, W. (Ed.) Wiley Interscience, NY, pp: 283–310.

Sigg L. (1987) Surface chemical aspects of the distribution and fate of metal ions in lakes. In *Aquatic surface chemistry*, Stumm, W. (Ed.), Wiley Interscience, NY, pp:319–348.

Singh, D.P. (1985) Cu2+ transport in the unicellular cyanobacterium *Anacystis nidulans*. *J. Gen. Appl. Microbiol.* **31**: 277–284.

Singh, C.B., and Singh, S.P. (1992) Prottive effects of Ca^{2+}, Mg^{2+}, Cu^{2+}, and Ni^{2+} on mercury and methyl mercury toxicity to a cyanobacterium. *Ecotoxicol. Environ. Safety* **23**: 1–10.

Singh, S.P., Singh, R.K., Pandey, P.K. and Pant, A. (1992) Factors regulating copper uptake in free and immobilised cyanobacterium. *Folia Microbiol.* **37**: 315–320.

Singh, S.P., and Yadav, V. (1983) Cadmium induced inhibition of nitrate uptake in *Anacystis nidulans*: interaction with other divalent cations. *J. Gen. Appl. Microbiol.* **29**: 297–304.

Singh, S., Singh, A.K., Chakravarthy, D., Singh, T.P.K., and Singh, H.N. (1997) Characteristics of a caesium-resistant (Cs+-r) mutant of the N2-fixing cyanobacterium *Nostoc muscorum*: dependence on Cs+ or Rb+ for normal diazotrophy and osmotolerance. *New Phytol.* **136**: 223–229.

Skaar, H., Rystad, B., and Jensen, A. (1974) The uptake of ^{63}Ni by diatom *Phaeodactylum tricornutum*. *Physiol. Plant.* **37**: 353–358.

Silverberg, B.A., Stokes, P.M. and Ferstenberg, L.B. (1976) Intranuclear complexes in a copper-tolerant green alga. *J. Cell Biol.* **69**: 210–214.

Silverberg, B.A., Wong, P.T.S., and Chau, Y.K. (1977) Effect of tetramethyl lead on freshwater green algae. *Arch. Environ. Contam. Toxicol.* **5**: 305–313.

Sorentino, C. (1985) Copper resistance in *Hormidium fluitans* (Gay) Heering (Ulotrichaceae, Chlorophyceae). *Phycologia* **24**: 366–368.

Spencer, C.P. (1957) Utilisation of trace elements by marine unicellular algae. *J. Gen. Microbiol.* **16**: 282–285

Spencer, D.F., and Greene, R.W. (1981) Effects of nickel on several species of freshwater algae. *Environ. Pollut.* **A25**: 241–247.

Stauber, J.L. and Florence, T.M. (1989) The effect of culture medium on metal toxicity to the marine diatom *Nitzschia closterium* and the fresh water green alga *Chlorella pyrenoidosa*. *Water Res.* **23**: 907–911.

Stratton, G.W. and Corke, C.T. (1979) The effect of mercuric, cadmium, and nickel ion combinations on a blue-green alga. *Chemosphere* **8**: 731–740.

Stratton, G.W., Huber, A.L. and Corke, C.T. (1979) Effect of mercuric ion on the growth, photosynthesis, and nitrogenase activity of *Anabaena inaeequalis*. *Appl. Environ. Microbiol.* **38**: 537–543.

Stratton, G.W. (1985) Interaction effects of mercury-pesticide combinations towards a cyanobacterium. *Bull. Environ. Contam. Toxicol.* **34**: 676–683.

Stokes, P.M. (1983) Response of fresh water algae to metals. In: *Progress in Phycological Research*, vol 2, Round, F.E. and Chapman, V.J.(Eds.), vol. 2. pp: 87–111.

Stokes, P.M., Hutchinson, T.C., and Krauter, K. (1973) Heavy metal tolerance in algae isolated from contaminated lakes near Sudbury, Ontario. *Can. J. Bot.* **51**: 2155–2168

Sunda, W. and Guillard, R.R.L. (1976) The relationship between cupric ion activity and the toxicity of copper to phytoplankton. *J. Mar. Res.* **34**: 511–529.

Sunda, W.G. and Huntsman, S.A. (1983) Effect of competitive interactions between manganese and copper on cellular manganese and growth in estuarine and oceanic species of the diatom *Thalassiosira*. *Limnol. Oceanogr.* **28**: 924–934.

Sunda, W.G. and Huntsman, S.A. (1995) Cobalt and zinc interreplacement in marine phytoplankton: Biological and geochemical implications. *Limnol. Oceanogr.* **40**: 1404–1417.

Sunda, W.G. and Huntsman, S.A. (1996) Antagonism between cadmium and zinc toxicity and manganese limitation in a coastal diatom. *Limnol. Oceanogr.* **41**: 373–387.

Sunda, W.G. and Huntsman, S.A. (1998) Interactions among Cu2+, Zn2+, and Mn2+ in controlling cellular Mn, Zn, and growth rate in the coastal alga *Chlamydomonas*. *Limnol. Oceanogr.* **43**: 1055–1064.

Takamura, N., Kasai, F., and Watanabe, M.M. (1990) Unique response of cyanophyceae to copper. *J. Appl. Phycol.* **2**: 293–296.

Twiss, M.R., Welbourn, P.M., and Schwartzel, E. (1993) Laboratory selection for copper tolerance in *Scenedesmus acutus* (Chlorophyceae). *Can. J. Bot.* **71**: 333–338.

USEPA (US Environmental Protection Agency) (1989) Ecological Assessment of Hazardous Waste Sites: A Field and Laboratory Reference Document. Report No. EPA/600-/ 3–89/013. US Environmental Protection Agency, Office of Research and Development, Corvallis, Oregon.

Vasseur, P and Pandard, P. (1988) Influence of some experimental factors on metal toxicity to *Selenastrum capricornutum*. *Tox. Assess.* **3**: 331–343.

Van Den Hoek, C.H. (1978) *Algen- einfuhrung in die phykologie*. Thieme verlag, Stuttgart.

Van den Berg, C.M.G., Wong, P.T.S., and Chau, Y.K. (1979) Measuring of complexing materials excreted from algae and their ability to ameliorate copper toxicity. *J. Fish Res. Bd. Can.* **36**: 901–905.

Voigt, J., Nagel, K., and Wrann, D. (1998) A cadmium-tolerant *Chlamydomonas* mutant strain impaired in photosystem II activity. *J. Plant Physiol.* **153**: 566–573.

Volesky B. (1987) Biosorbents for metal recovery. *TIBTECH* **5**:96-101.

Weiss-Magasic, C., Lustigman, B., and Lee, L.H. (1997) Effect of mercury on the growth of *Chlamydomonas reinhardtii*. *Bull. Environ. Contam. Toxicol.* **59**: 828–833.

Whitton, B.A. (1970) Toxicity of heavy metals to freshwater algae: a review. *Phykos* **9**: 116–125.

Whitton, B.A., and Shehata, H.A. (1982) Influence of cobalt, nickel, copper and cadmium on the blue-green alga *Anacystis nidulans*. *Environ. Pollut.* **A27**: 275–281.

Whitton, B.A. (1984). Algae as monitors of heavy metals in freshwaters. In: *Algae as Ecological Indicators*, Shubert, L.E. (Ed.), Academic Press, London, pp. 257–280

Xue, H-B., Stumm, W., and Sigg, L. (1988) The binding of heavy metals to algal surfaces. *Water Res.* **22**: 917–926.

Zarnowski, J. (1972) The effect of ethylenediamine tetraacetic acid on growth of *Chlorella pyrenoidosa* and its role in dynamics of metabolism and accessiability of iron and calcium. *Acta Hydrobiol.* **14**: 353–373.

6

Absorption and Translocation of Chromium by Plants: Plant Physiological and Soil Factors

R.H. Loeppert, J.A. Howe, H. Shahandeh, L.C. Wei and L.R. Hossner

INTRODUCTION

There is substantial literature on the uptake of essential elements by plants, but there has been considerably less attention given to the absorption of toxic metals. Lately, there has been much more emphasis on this latter topic because of environmental concerns of heavy metal toxicity, carcinogenicity, and mutagenicity. Also, there is increasing interest in the potential use of plants for the reclamation of contaminated soil and water by the processes of phytoremediation, phytoextraction, and phytostabilization.

An understanding of the conditions under which toxic metals are absorbed by plants requires knowledge of the characteristics of heavy-metal tolerance, uptake, translocation, and storage by plants, and the complex soil-plant interactions that influence metal mobilization, bioavailability, and absorption.

In this chapter, we give major emphasis to chromium, which is required by animals in small concentrations, but is toxic to plants. One of the predominant oxidation states of Cr, Cr(VI), is especially toxic and mutagenic to plants and animals. Following a review of the literature, we will relate our recent research experiences with Cr, especially soil and plant factors that influence its availability to plants.

CHEMISTRY OF SOIL CHROMIUM

Predominant Oxidation States

Chromium exists predominantly in the 3+ and 6+ oxidation states (Rai et al., 1987; Palmer and Wittbrodt, 1991). The intermediate states of 4+ and 5+ are metastable and are rarely encountered in soils.

Soil & Crop Sciences Dept., Texas A&M University, College Station, TX 77843, USA

Speciation and Solubility of Cr(III)

The solubility of Cr(III) in soil is dependent on pH (Palmer and Wittbrodt, 1991) and decreases markedly at pH > 4.5. Chromium (III) in most soils exists predominantly as poorly soluble $Cr(OH)_3$ or $(Fe,Cr)(OH)_3$. At pH < 3, Cr^{3+} is an important species of dissolved Cr(III) (Weckhuysen et al., 1996). With increasing pH, Cr^{3+} is hydrolyzed, and between pH 2.6 and 14, $CrOH^{2+}$ and possibly $Cr(OH)_3^0$ and $Cr(OH)_4^-$ are the dominant mononuclear species of Cr(III) in solution (Rai et al., 1987). $Cr(OH)^{2+}$, $Cr_2(OH)_2^{4+}$, $Cr_3(OH)_4^{5+}$ (Baes and Mesmer, 1976), and $Cr_4(OH)_4^{6+}$ (Stunzi and Marty, 1983) also contribute to total dissolved Cr(III). Chromium(III) can form stable complexes with organic ligands, including oxalate, citrate, malate, EDTA, and DTPA; at pH values as high as 7.5 (James and Bartlett, 1983), and with natural organic polymers (humates and fulvates). These complexes form slowly, but once formed, are difficult to break. Also, the kinetics of exchange of Cr^{3+} between ligands can be slow (Cotton and Wilkinson, 1980). Stable Cr^{3+}-siderophore and -phytosiderophore (mugineic acid) complexes can be produced in the laboratory, but there is as yet no evidence that they exist in appreciable quantities under natural conditions. The presence of any of the organic complexes can significantly influence the concentration of total dissolved Cr(III) in the soil solution.

Speciation, Solubility and Adsorption of Cr(VI)

Chromium(VI) species are considerably more soluble than Cr(III) species. Chromate, CrO_4^{2-}, which is the predominant dissolved Cr(VI) form at pH > 6, exists in pH-dependent equilibrium with other species of Cr(VI) such as $HCrO_4^-$ and dichromate ($Cr_2O_7^{2-}$). The anionic Cr(VI) species can be adsorbed by positively charged soil colloids such as Al and Fe oxides at pH values below their zero point of charge (zpc) (Ainsworth et al., 1989) as well as at positive charge sites at the surfaces of plant roots. Adsorption of Cr(VI) species is considerably less at neutral to alkaline pH than at acidic pH values due to the greater negative charge character of mineral and organic surfaces at higher pH values. Adsorption will depend on the type and quantity of soil components as well as pH and the presence of competing ligands such as phosphate.

Redox Transformations Involving Cr

Chromium(III) is the dominant oxidation state and the thermodynamically stable form in most soils (Bartlett and James, 1979). Chromium(VI) is only thermodynamically stable in highly alkaline, highly oxidizing environments, and is readily reduced to Cr(III) (Cary et al., 1977a). The transformation of Cr(VI) to Cr(III) in soils can occur as a result of reduction by Fe^{2+} in solution and at mineral surfaces, reduced sulfur compounds, or soil organic matter (Bartlett, 1991; Charlet and Manceau, 1992; Losi et al., 1994; Fendorf, 1995). The reduction of Cr(VI) to Cr(III) by organic matter is more rapid in acid than in alkaline soils (Cary et al., 1977a). Although oxidation of Cr(III) to Cr(VI) is possible, especially in the presence of manganese oxides (Fendorf, 1995), oxidation usually only occurs under moist conditions and not appreciably in dry soils (Bartlett and James, 1979). Soil Cr is sometimes mobilized when soils are flooded and then drained or incubated with organic matter, by the oxidation to Cr(VI) or association of Cr(III) with soluble organic complexing agents.

ROLE OF CHROMIUM IN PLANTS AND ANIMALS

Chromium is essential to humans and animals for glucose metabolism; however, there is no evidence that Cr has any physiological function in plants (Huffman and Allaway, 1973). Chromium is toxic to agronomic plants, but the concentration at which the Cr is toxic is dependent on plant species. Numerous studies have reported that Cr(VI) is more toxic than Cr(III) (e.g., Mortvedt and Giordano, 1975; Skeffington et al., 1976). The symptoms of Cr toxicity, which vary among plant species, include chlorosis, purpling (especially along the veins, of the stems, or at the leaf edges), wilting, blanching of leaves, and necrosis of the plant tissue (especially of the older leaves).

Absorption of Toxic Metals by Plants

The uptake of beneficial metals occurs predominantly by way of channels and transporters in the root plasma membrane. Plants characteristically exhibit a remarkable capacity to absorb what they need and exclude what they do not. But most vascular plants absorb toxic and heavy metals through their roots to some extent, though to varying degrees, from negligible to substantial. The mechanisms of absorption of toxic metals, including Cr, by plants can be divided into three general categories:

- *specific absorption* of a toxic metal by specific uptake mechanisms for the metal or class of metals. This general mechanism is prevalent for some hyperaccumulating plant species that are endemic to soils that are especially high in that metal;
- *"tag-along" absorption* in which the toxic metal is similar in chemistry to a beneficial nutrient and is absorbed by a channel or transporter intended for the beneficial nutrient;
- *non-specific absorption* through root plasma membrane channels, pores, or defects. This is the most general of the various categories of uptake of toxic metals, and is especially important for the absorption of Cr, Pb, and the actinides.

Absorption of Cr by Plants

Skeffington et al. (1976) concluded, from the use of metabolic inhibitors, that plant uptake of Cr(VI) is an active process, whereas uptake of Cr(III) is a passive process, i.e. no energy expenditure by the plant was required. They also observed that chromate absorption follows Michaelis Menton kinetics at low concentration and that absorption is completely inhibited in the presence of excess sulfate. This result indicates that chromate absorption might occur through the sulfate channel. If the absorption of Cr(III) is a passive process, then it is logical to conclude that plants should be able to continuously absorb Cr(III) if it is soluble in the medium, due to the negative membrane potentials that are likely to exist across the root plasma membrane. However, the absorption of Cr(III) is usually limited due to its low concentration in the soil solution.

Many studies have reported that plants absorb Cr(VI) more readily than Cr(III), based solely on plant concentration data and the observation that Cr(VI) is more toxic to plants than Cr(III) (Hara and Sonoda, 1979; Lee et al., 1981; Peterson, and Girling, 1981; Barcelo et al., 1986). A difficulty of any study of this type is that the equilibrium solubilities of Cr(VI) and Cr(III) are not comparable under the same conditions (McGrath, 1982). In a study of Cr uptake by oat, McGrath (1982) used a flowing culture technique in which equal concentrations of soluble Cr(VI) and Cr(III) were maintained, and found that the plants absorbed Cr(VI) and Cr(III) equally well.

Plant species differ significantly in Cr uptake capacity and distribution within the plant (Soane and Saunder, 1959; Grubinger et al., 1994). Cary et al. (1977a,b) compared Cr uptake by various dicotyledonous and monocotyledonous food crops and found that, in general, dicots, especially green leafy vegetables (Zayed et al., 1998), absorbed more Cr and transported more Cr to shoots than did monocots such as corn and barley.

Translocation of Cr in Plants

Plants must have the ability to translocate Cr from the root to the shoot, or to compartmentalize it, for the plant to continue absorption of a toxic metal from the medium. There are at least two physiologic advantages to the plant for translocation: (1) it will reduce metal concentration and thus reduce the toxicity potential to the root, and (2) translocation to the shoot is one of the mechanisms of resistance to high Cr because for some plants the high concentration of Cr will be lost when leaves fall. Baker et al. (1994) observed high metal uptake by the roots of Thalaspi species. Zinc, Cd, Co, Mn, and Ni were readily transported to the shoot whereas Al, Cr, Cu , Fe, and Pb were predominantly immobilized in the roots.

Most research using non-hyperaccumulator plants has shown that Cr is mainly accumulated in the roots, and much less of the total Cr in a plant is in the leaves (Shewry and Peterson, 1974; Cary et al., 1977b). The distribution of Cr between the root and shoot in hyperaccumulator plants has indicated that the leaves sometimes contain a much higher Cr concentration than that of non-hyperaccumulator plants, suggesting better translocation of Cr from root to shoot in hyperaccumulator plants. The percentage of Cr found in the shoot tended to increase with increasing concentration of CrO_4^{2-} in solution, but remained approximately the same for Cr(III) (Skeffington et al., 1976). Although the detailed mechanisms of Cr translocation are not thoroughly understood, there are reports that Fe-deficient and P-deficient plants can better translocate Cr from roots to shoots (Cary et al., 1977b; Bonet et al., 1991). These and similar observations have lead to the hypothesis that Fe- and P-deficiency induced accumulation of organic acids, e.g., citric acid (Landsberg, 1981; Ric De Vos et al., 1986; Johnson et al., 1994), may play an important role in Cr translocation. Using a hyperaccumulator plant, *Leptospermum scoparium*, Lyon et al. (1969a,b) found that soluble Cr in leaf tissue was present as the Cr(III) trioxalate ion, $[Cr(C_2O_4)_3]^{3-}$. It is known that even when taken up as Cr(VI), Cr will soon be reduced in the plant to Cr(V), Cr(IV) and finally Cr(III) (Micera and Dessi, 1988; Liu et al,. 1995; Hunter et al., 1997; Lytle et al., 1998; Zayed et al., 1998). Inorganic chromium(III) is only slightly soluble at the physiological pH; therefore, organic complexing agents must be required to increase the solubility of Cr(III) within the plant, and thus to enhance Cr translocation. These data suggest that hyperaccumulator plants, which translocate more Cr than non-hyperaccumulator plants, probably have a different metabolism from that of non-hyperaccumulator plants. More organic complexing agents might be produced to increase Cr translocation from root to shoot.

Iron is translocated in plants predominantly as the Fe^{3+}-citrate complex (Tiffin, 1966), and Ni is readily complexed by organic acids and is very mobile in the plant. The majority of Ni in the plant is present in the leaves rather than in the roots (Baker and Brooks, 1989). Soluble Cr concentration is positively correlated with soluble Fe (Cary et al., 1977b) and Ni (Shewry and Peterson, 1974) in plant leaves and roots. The positive correlation indicates that some Fe or Ni hyperaccumulator plants might also be Cr hyperaccumulators, and implies that Cr might have similar translocation and compartmentation mechanisms to those of Fe and Ni.

There is little information about Cr compartmentation in plants. The vacuole is considered to be the major storage site for most heavy metals (e.g., Cd, Zn, Mn, and Ni) (Wagner et al., 1995). Little is known about the form and fate of Cr in plants and how plants detoxify Cr; hence, it is very important to determine where Cr is compartmentalized.

Plant Detoxification of Absorbed Cr

It is widely accepted that detoxification of metal ions within plant tissue involves chelation by appropriate ligands and transport of the complex to cellular compartments (e.g. vacuoles) or to regions of the plant where the metal has less influence on plant metabolic activity (e.g. to old leaves). Some organic acids such as citrate, malate, and malonate are good metal binding agents and are often found in high concentrations in leaves of plants. Reeves (1992) has pointed out that these anions are present constitutively in *Alyssum* in substantial amounts and cannot account for the metal specificity or species variability of Ni hyperaccumulators. Andrew et al. (1995) suggested that the Ni hyperaccumulation trait of *Alyssum* is associated with the ability of the root systems to produce substantial amounts of histidine. An exogenous supply of histidine to *Allysum montanum* increased tolerance to Ni and Ni uptake (Kramer et al., 1996). Recent experiments with other plants, including *T. caerulescens, Steptanthus polygaloides* (Brassicaceae), and *Berkheya coddii* (Asteraceae), tend to support the role of histidine in tolerance and accumulation of Ni, but histidine is apparently not important in the case of Zn. In addition to the organic acids and amino acids, plants can also contain metal-binding phytochelatins and metallothioneins. Phytochelatins are a family of peptides with the general structure $[-GluCys]_n-Gly$, where $n > 1$ (Rauser, 1990; Rauser, 1995; Rauser and Meuwly, 1995). Metallothioneins are Cys-rich metal complexing proteins, which are synthesized by mRNA translation.

Intercellular reducing agents, e.g. glutathione (GSH) (Bose et al., 1992; Liu et al., 1997), ascorbic acid (Zhang and Lay, 1996; O'Brien and Woodbridge, 1997) and cysteine (Lay and Levina, 1996), can be involved in the reduction of Cr(VI) to Cr(III) or possibly to the intermediate oxidation states, Cr(IV) or Cr(V) (Cieslak-Golonka, 1996). These reactions have been more thoroughly characterized in animal than in plant systems because of the interest in carcinogenic and mutagenic effects of Cr(VI) in the former. As Cr(VI) is considered to be a hard acid and sulfur a soft base, the formation of the Cr-O bond rather than the Cr-S bond is expected to be favored; however, Cr is often found to be coordinated to the S atom (Brauer et al., 1996; Cieslak-Golonka, 1996).

Hyperaccumulation of Metals by Plants

The environmental and economic success of phytoremediation is dependent on the existence of plant genotypes that hyperaccumulate metals (Baker and Brooks, 1989; Baker et al., 1991). Hyperaccumulation of metals, especially Zn and Ni, by certain plant species has been thoroughly documented. For example, from results of a field study conducted by the Rothamsted Experiment Station in England, it was estimated that *Thalaspi careulescens* has the potential to remove more than 40 kg Zn ha^{-1} yr^{-1} (McGrath et al., 1993). *Alyssum betolonii*, which is endemic to serpentine soils, is known for its high concentration of Ni (> 10,000 mg kg^{-1} in leaves). Indian mustard (*Brassica juncea*) has an ability to accumulate Cr, Cd, and Ni in its shoots (Kumar et al., 1995).

In comparative adsorption experiments of Ni, Co, and Zn, excised roots of *Alyssum bertolonii* did not show preferential uptake of a specific metal (Gabbrielli et al., 1991). The plant roots tended to accumulate Ni, Co, and Zn, without discriminating between them. Clones of two Zn hyperaccumulators accumulated Pb mainly in their roots, compared to a pea mutant (E107), which accumulated Pb in its shoots (Huang et al., 1997). Hydroponically grown Indian mustard (*Brassica juncea* (L.) Czem) and sunflower (*Helianthus annuus* L.) effectively removed Cu, Cd, Cr, Ni, Pb, and Zn from aqueous solutions (Dushenkov et al., 1995). Baker et al. (1994) suggested similar mechanisms of absorption and transport of several metals by *Thalaspi* species. They observed high uptake by the roots for all metals studied; however, Zn, Cd, Co, Mn, and Ni were readily transported to the shoot, whereas, Al, Cr, Cu, Fe, and Pb were predominantly immobilized in the roots.

Only a few plant species, when grown on high Cr soils, have been identified as Cr hyperaccumulators. Several of the previously identified Cr hyperaccumulator plants are dicots. Wild (1974) has reported high concentrations of Cr in leaves of *Dicoma niccolifera* (1,500 µg g^{-1} dry weight) and *Sutera fodina* (2,400 µg g^{-1}) (Baker and Brooks, 1989) (both species were located in Zimbabwe); *Lepertospermum scoparium* from an abandoned chromite mine in New Zealand contained up to 1% Cr (Lyon et al., 1971). Another plant species from Zimbabwe, *Pearsonia metallifera*, was also reported to contain high concentrations of Cr (Wild, 1974). The plant species *Leptospermum scoparium* 'J. R. et G. Forst' (Myrtaceae) is an accumulator of Cr, and there was a significant correlation between plant and soil Cr concentration (Lyon et al., 1969a,b). This species can accumulate up to 20,000 µg Cr g^{-1} in the foliage ash when grown on serpentine soils. Peterson and Girling (1981) reported 48,000 and 30,000 µg Cr g^{-1}in the ash of *Sutera fodina* and *Dicoma niccolofera*, respectively. The greater uptake of Cr by dicots suggests that they might be reasonable target species for Cr hyperaccumulation.

SOIL FACTORS INFLUENCING THE ENHANCED ABSORPTION OF TOXIC METALS

Numerous studies have demonstrated that the absorption of toxic metals by plants is strongly influenced by the concentration of dissolved metal in the soil solution. Solubility in solution is strongly influenced by soil pH and the presence of organic complexes of the metal in solution. For plants to be able to hyperaccumulate Cr from soil, where most Cr exists as insoluble Cr(III), the plants have to be efficient in a series of processes including solubilization of Cr in soil, absorption of soluble Cr, and translocation, compartmentation, and detoxification of absorbed Cr within the plant. The failure of any of these processes will prevent the plant from hyperaccumulating Cr from soils. Given the chemistry of Cr in the soil, solubilization of Cr can be the limiting process. Several studies have reported that plant uptake of Cr increased with increased soluble Cr in the media (Cary et al., 1977b; McGrath, 1982). Yet enhanced solubility does not ensure absorption, because the complexed metal must also be able to cross the root plasma membrane as an intact complex or via a ligand-exchange process.

RESULTS AND DISCUSSION OF CURRENT RESEARCH

Uptake of Cr(III) and Cr(VI) from Hydroponic Culture

The following is a summary of our recent studies on the absorption and translocation of Cr by plants grown in hydroponic culture. Emphasis in this discussion is on subterranean clover. Subterranean clover was selected because of the relatively high proportion of Cr that is translocated from the roots to the leaves.

Toxicity of Cr Sources

The relative toxicities of the various Cr sources generally decreases in the following order: Cr(VI) >> inorganic Cr(III) > Cr(III)-organic complexes. For example, yields of biomass of subterranean clover in hydroponic culture are shown in Table 6.1. Other studies have shown this same general trend (e.g., Skeffington et al., 1976, with barley). With subterranean clover toxicity (Table 6.1) symptoms included reduced biomass, purpling of the plant tissue (especially along the stems, veins and leaf margins), wilted and dried leaves, and eventual death of the plant in the case of the highest concentration chromate treatments. Cr(VI) was especially toxic to the plant, and even with the intermediate Cr treatment, the plants were nearly dead after the 2 week exposure to Cr. Plants exposed to Cr DTPA exhibited some slight reductions in biomass, but visual symptoms of Cr toxicity were not evident except at the highest Cr DTPA concentration.

Absorption of Cr by Plants as Affected by Cr Source and Concentration

The shoot Cr concentrations with the three Cr sources decreased as follows: Cr(VI) >> $CrCl_3$ > Cr(III)DTPA (Table 6.2). Treatment with 1 mM Cr(VI)

Table 6.1: Effect of Cr source and concentration on growth of Koala subclover.[a]

Treatment (mM)		Shoot/Dry/Weight g plant^{-1}	Root Dry Weight g plant^{-1}
Cr(VI)$_2$0..	0.04	2.14	0.29
	0.20	1.59	0.23
	1.00	1.28	0.22
Cr(III)Cl$_3$	0.04	7.43	2.11
	0.20	3.70	1.42
	1.00	2.21	0.33
Cr(III)DTPA	0.04	7.36	1.90
	0.20	7.60	1.98
	1.00	6.88	2.17

[a] The plants were precultured in a balanced Hoagland nutrient solution for two weeks. The Cr treatment (2 weeks) was in a balanced Hoagland solution initially buffered to pH 5.0 with MES buffer.

resulted in a Cr concentration of >10,000 µg Cr g^{-1} dry weight. The roots of these plants were severely damaged, and their selectivity for nutrients was also likely severely impaired. The high Cr contents of the leaves of these plants could have been influenced by continued plant transpiration and nonspecific absorption of Cr(VI), even as the function of the roots was severely impaired. This uptake pattern is frequently observed with plants with chemically damaged roots systems.

Table 6.2: Effect of Cr source and concentration on Cr concentration in roots and shoots of Koala subterranean clover grown at -Fe and +Fe conditions.[a]

	Treatment mM	Shoot µg g^{-1}		Root µg g^{-1}		Shoot Cr[b]	
		-Fe	+Fe	-Fe	+Fe	-Fe	+Fe
Cr(VI)O$_4^{2-}$	0.04	227	237	1080	993	61.9	63.8
	0.2	2560	2650	1780	1780	91.2	91.1
	1	8570	8030	3360	3350	94.1	93.3
Cr(III)Cl$_3$	0.04	55.8	5.3	1870	2320	12.9	0.8
	0.2	200	111	13900	13500	5.8	2.1
	1	1290	748	38000	38600	18.2	11.5
Cr(III)DTPA	0.04	7.8	2.7	77.1	82.7	32.0	11.3
	0.2	41.1	18.9	142.8	95.3	60.0	43.2
	1.0	166	167	173	105	82.3	83.4

[a] The plants were precultured in a balanced Hoagland nutrient solution for two weeks. The Cr treatment (2 weeks) was in a balanced Hoagland solution (with or without Fe) initially buffered to pH 5.0 with MES buffer.

[b] $\dfrac{\text{shoot Cr}}{\text{total plant Cr}} \times 100\%$

The CrCl$_3$ treatment resulted in the highest root Cr concentration. The roots of these plants had a slight blue-green tint, similar in color to freshly precipitated Cr(OH)$_3$. An examination of the CrCl$_3$. treated roots by scanning electron microscopy indicated the presence of very tiny irregular particles and aggregates on the root surface (Howe, 1999). Microprobe analysis corroborated that these particles were high in Cr. Scanning electron microscopy of longitudinal root sections provided no evidence of precipitated $Cr(OH)_3$ or $CrPO_4$ inside the root. Therefore, the high concentration of Cr associated with the roots of $CrCl_3$-treated plants was attributable to the precipitation of $Cr(OH)_3$ and possibly $CrPO_4$ on the root surface (Howe, 1999). In the present case, the negatively charged root surface provided a surface for adsorption, nucleation, and crystal growth of the $Cr(OH)_3$ phase. Similar hydrolysis and precipitation processes are expected to occur in soils. The absorption of Cr from $Cr(OH)_3$ is strongly influenced by pH because of the influence of pH on solubility of $Cr(OH)_3$. In the current experiment, the pH was initially buffered at pH 5.0 with 1 mmol MES L^{-1}, but even in the presence of this buffer, the pH decreased to < 4 due to acidity released during the hydrolysis of Cr^{3+} (Howe, 1999):

$$Cr^{3+} + 3H_2O \Leftrightarrow Cr(OH)_3 + 3H^+$$

At pH 4, the solubility of $Cr(OH)_3$ is appreciable; therefore, in this experimental system, the absorption of Cr^{3+} and translocation to the shoots (~1000 µg Cr g^{-1}) was appreciable. In similar experiments that were more highly buffered, the absorption of Cr^{3+} was considerably less. In soils, the absorption of Cr from $Cr(OH)_3$ is usually small and dependent on pH and pH-buffering capacity of the soil.

The absorption and translocation from Cr(III)-DTPA was considerably less than with the inorganic Cr(III) and Cr(VI) sources and was generally greater with the -Fe than with the +Fe hydroponic culture. The absorption of Cr(III) from the Cr(III)-DTPA complex could have occurred by either (A) the direct absorption of the intact DTPA complex or (B) a ligand exchange process involving the exchange of Cr^{3+} between DTPA and a plant metabolic ligand, e.g. citrate (Fig. 6.1). The overall low absorption of Cr from the DTPA complex indicates that both the intact absorption of Cr(III) DTPA and the ligand exchange of Cr must be relatively slow processes.

Translocation of Cr in plants as affected by Cr source and concentration

With subterranean clover under each Fe condition, shoot Cr percentage, shoot Cr/total plant Cr x 100%, followed the trend of Cr(VI) > Cr(III) DTPA >> $CrCl_3$ and increased with increasing Cr treatment concentration for each Cr source (Table 6.2). The percentage of shoot Cr compared to total plant-retained Cr ranged from 0.8% to 11.5% for the +Fe $CrCl_3$ treatments, but as discussed previously, the Cr associated with the root

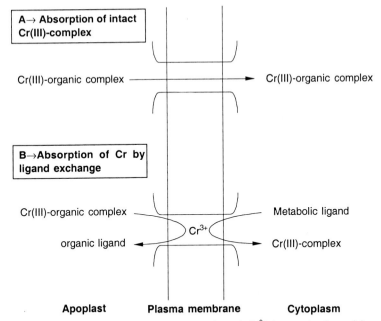

Fig. 6.1: The absorption of organically complexed Cr^{3+} by a plant must involve either (A) absorption of the intact Cr(III)-organic complex, or (B) the ligand exchange of Cr^{3+} between the organic ligand and a plant metabolic ligand.

occurred principally as $Cr(OH)_3$ precipitated at the root surface. The highest shoot Cr percentage (11.5%) occurred with the highest Cr treatment level. Among the $CrCl_3$ treatments, the pH reduction was greatest at the highest $CrCl_3$ treatment concentration, resulting in the highest solubilization, absorption, and translocation of Cr^{3+}.

With subterranean clover, the shoot Cr percentage ranged from 63.8 to 93.3 for the +Fe Na_2CrO_4 treatments and 11.3 to 83.4 for the +Fe Cr(III) DTPA treatments. These results with subclover with Na_2CrO_4 as the Cr source differ considerably from results with several other plant species in hydroponic culture (Table 6.3). The differences between species reflect different predominant uptake and transport mechanisms.

Effect of Nutrient Deficiency

Plants respond to some nutrient-deficiency stresses by means of physiological responses that increase mobilization and absorption of the target nutrient. For example, some plant species respond to P-deficiency stress by citrate accumulation and the excretion of citric acid from roots. Manganese deficiency can cause citric acid accumulation in microorganisms and can possibly have the same effect on plants. Iron deficiency has been shown to result in the exudation of Fe^{3+}-chelating agents (phytosiderophores) by grasses, and exudation of H^+, increased root

Table 6.3: Plant growth and Cr uptake of selected plant species.[a]

Cr Source	Cr source	Cr concentration $\mu g\ g^{-1\ (\%)}$		Shoot Cr[b]
		Shoot	Root	
Koala subclover	Na_2CrO_4	941	976	77.4
	$CrCl_3$	18.8	14800	0.45
	Cr oxalate	36.4	167	45.7
Indian mustard	Na_2CrO_4	67.2	1040	17.7
(*Brassica juncea*)	$CrCl_3$	8.5	13400	0.16
	Cr oxalate	35.6	2920	3.52
Sparkle pea	Na_2CrO_4	23.4	1410	5.43
(*Pisum sativum*)	$CrCl_3$	6.0	8860	0.22
	Cr oxalate	7.0	1490	1.90
E107 pea	Na_2CrO_4	20.6	1294	5.48
(*Pisum sativum*)	$CrCl_3$	9.5	12600	0.31
	Cr oxalate	11.3	49.6	44.1
Canola	Na_2CrO_4	11.6	343	13.7
(*Brassica napa*)	$CrCl_3$	1.0	14500	0.03
	Cr oxalate	0.4	1050	0.24

[a] The plants were precultured in a balanced Hoagland nutrient solution for 2 weeks. The Cr treatment (2 week) was in a balanced Hoagland solution initially buffered to pH 5.0 with MES buffer.

[b] $\dfrac{\text{shoot Cr}}{\text{total plant Cr}} \times 100\%$

plasma-membrane reductase activity, and increased production of metabolic acids by dicots. Experiments have been conducted to evaluate the influence of Fe, P, and Mn deficiencies on Cr absorption by plants.

For $CrCl_3$ and Cr(III)-organic complex treatments (e.g. the Cr(III) DTPA treatments in Table 6.2), total shoot Cr and shoot Cr percentage were usually higher with Fe-deficient plants than with Fe-sufficient plants. Results have been inconsistent with Cr(VI) treatments. In the case of the $CrCl_3$ treatment, there are several possible contributing factors. The Fe-deficiency stress response of subterranean clover involves an accelerated release of H^+ from the root plasma membrane (Wei et al., 1997; Wei et al., 1998). This process would result in decreased apoplast pH and increased solubility of the precipitated $Cr(OH)_3$. Also, the Fe-deficiency stress response involves the increased production of metabolic complexing agents, e.g. citrate, that are involved in the transport of Fe^{3+} and the possible transport of Cr^{3+} within the plant. The fact that the -Fe treatment does not consistently result in a significant change in Cr uptake and translocation with CrO_4^{2-} treatments, indicates that the Fe-deficiency induced Fe^{3+} reductase might not influence chromate reduction and absorption, or that CrO_4^{2-} might retard Fe^{3+}-reductase activity and the effectiveness of the Fe-deficiency stress response.

As previously discussed, the absorption and translocation from Cr(III) DTPA was generally greater with -Fe than +Fe plants. As the Fe-deficiency stress response of clover involves the increased production of plant metabolic ligands, the difference in uptake and translocation between -Fe and +Fe plants for Cr(III)DTPA supports the hypothesis that Cr absorption might involve a ligand-exchange process (Fig. 6.1, mechanism B). The enhanced production of metabolic acids by the Fe-deficient plant might be expected to influence the rate of mechanism B (ligand exchange) but not mechanism A.

Phosphate-deficient treatments, while not significantly influencing shoot P concentration, seem to have a consistent positive influence on total root P and negative influence on shoot P percentage (Table 6.4). This phenomenon is attributable to the increase in total root growth under P deficiency conditions and decreased shoot/root ratio of the -P treated plants.

Table 6.4: Effect of plant phosphate status on Cr uptake from 0.6 mM Cr-trioxalate by Koala subclover.[a]

Treatment	Dry weight g plant^{-1}		Cr Concentration $\mu g\ g^{-1}$		Total Cr mg		Shoot Cr[b] %
	Shoot	Root	Shoot	Root	Shoot	Root	
Complete + Cr	0.54	0.14	192	142	103	20.2	83.6
–P + Cr	0.38	0.25	269	128	104	61.6	76.6
LSD (p = 0.05)	0.10	0.03	38.3	10.3	15.4	3.3	4.2

[a] The plants were precultured in a balanced Hoagland nutrient solution for 1 week. The Cr(III) oxalate treatment (2 wk) was in a balanced Hoagland solution initially buffered to pH 5.0 with MES buffer.

[b] $\dfrac{\text{shoot Cr}}{\text{total plant Cr}} \times 100\%$

EFFECT OF ORGANIC ACID ON Cr UPTAKE BY PLANTS

As mentioned previously, the absorption of a Cr(III) organic complex by the plant must involve either the direct absorption of the Cr complex or the ligand exchange of Cr between the source ligand and a plant metabolic ligand. To obtain clues concerning the predominant mechanism, experiments have been conducted to evaluate the effect of free organic acid (either applied to the nutrient solution or injected through the stem into the plants) on Cr absorption and translocation by plants which were initially supplied with a Cr(III) organic complex. Several experiments with subterranean clover are summarized here.

The Cr(III) oxalate treatment resulted in greater Cr uptake than the other Cr complex treatments (Table 6.5). It is not known whether this result was attributable to the greater uptake of the intact Cr(III) oxalate complex or the greater rate of exchange of Cr^{3+} across the root plasma membrane (Fig. 6.1). In this experiment, the percent shoot Cr for the Cr complex treatments ranged from 33% to 46%, indicating that adsorbed Cr was readily translocated within the plant.

Table 6.5: Effect of organic acid complexed Cr on the growth and Cr uptake by Koala subclover.[a]

Cr Source	Cr Concentration $\mu g\ g^{-1}$		Shoot Cr^b %
	Shoot	Root	
Cr oxalate	36.4	167	45.9
Cr malate	15.6	104	33.5
Cr citrate	13.4	51.1	45.4
CrCl$_3$	18.8	14800	0.46

[a] The plants were precultured in a balanced Hoagland nutrient solution for 1 week. The Cr(III) treatments (2 weeks) were in a balanced Hoagland solution initially buffered to pH 5.0 with MES buffer.

[b] $\dfrac{\text{shoot Cr}}{\text{total plant Cr}} \times 100\%$

The comparison of the CrCl$_3$ treatment with the Cr complex treatments is complicated by the large proportion of added Cr precipitated either in the nutrient solution or at the root surface following the addition of CrCl$_3$. The Cr oxalate, Cr citrate, and Cr(III) malate complexes were relatively stable under the conditions of this experiment; therefore, the actual concentrations of Cr in solution from these sources were considerably higher than in the case of the CrCl$_3$.

Free citric, malic, or oxalic acid applied to the nutrient solution did not affect either shoot or root growth compared to the control (Table 6.6). However, each of these treatments resulted in decreased shoot- and root-Cr concentration and total quantity of absorbed Cr. Shoot Cr percentage was not significantly influenced by any of the organic acid treatments compared to the control. The decrease in Cr uptake in the presence of citrate, malate, and oxalate, as compared with the control, could possibly be due to a retardation of ligand exchange of Cr^{3+} at the root plasma-membrane surface due to the presence of free ligand in the nutrient solution. This result also seems to support the hypothesis that absorption of Cr^{3+} from the Cr^{3+}-oxalate complex might involve ligand exchange rather than direct absorption of the intact Cr^{3+}-oxalate complex.

Table 6.6: Effect of organic acid applied in nutrient solution on Cr uptake by subterranean clover.[a]

Treatment	Dry weight g plant^{-1}		Cr Concentration $\mu g\ g^{-1}$		Shoot Cr[b] %
	Shoot	Root	Shoot	Root	
1 mM Cr(III) oxalate	0.54	0.14	192	142	83.6
1 mM Cr(III) trioxalate +0.6 mM oxalic acid	0.54	0.14	129	91.4	84.2
1 mM Cr(III) trioxalate +0.6 mM citric acid	0.54	0.14	129	91.4	84.2
1 mM Cr(III) trioxalate +0.6 mM malic acid	0.53	0.14	157	98.6	85.6
LSD (p=0.05)	0.10	0.03	38.3	10.3	4.2
1 mM Cr(III) EDTA	0.58	0.17	204	157	81.5
1 mM Cr(III) EDTA +1.2 mM EDTA	0.37	0.13	762	942	69.7
LSD (p = 0.05)	0.10	0.03	38.3	10.3	4.2

[a] The plants were precultured in a balanced Hoagland nutrient solution for 1 week. The Cr(III) trioxalate or Cr(III) EDTA treatment (2 weeks) plus excess free oxalate, citrate, malate, or EDTA was in a balanced Hoagland solution initially buffered to pH 5.0 with MES buffer.

B $\dfrac{\text{shoot Cr}}{\text{total plant Cr}} \times 100\%$

EDTA had the opposite effect of the metabolic organic acids and resulted in an increase in Cr content and concentration of the plant roots and shoots. Plant biomass was substantially reduced and the visual symptom of Cr toxicity (purple coloration of the stems and leaves) was increased by the free-EDTA treatment. These results agree with results of soil studies in which EDTA also had a significant influence on absorption of Cr from Cr(III)-organic complexes or CrO_4^{2-}. It is probable that the free EDTA had a disruptive effect on the root plasma membrane and resulted in an increase in the nonspecific absorption of Cr^{3+} oxalate.

Stem injection of oxalic acid into the plants significantly increased shoot Cr concentration and total quantity of shoot Cr for both Koala and Karridale subclover (Table 6.7), but without significant effect on the shoot and root growth for the experimental period. The root-Cr concentration of Karridale but not of Koala was increased by the stem injection treatment. The shoot Cr percentage was increased by injection of oxalic acid, particularly for Karridale subclover. Injection of oxalic acid contributes to the pool of Cr^{3+}-complexing agent. The effect of oxalic acid was greater for Karridale than for Koala, probably due to its poorer Fe-deficiency stress response and the lower tendency for Fe-deficiency induced accumulation of organic acid in the former. The results from this experiment indicate that the enhanced availability of a Cr^{3+}-complexing ligand inside the plant might contribute to Cr absorption and translocation, even in the present case in which the source of Cr was Cr^{3+}-trioxalate. This observation also supports the hypothesis that Cr from Cr^{3+}-trioxalate might involve ligand exchange rather the direct absorption of the Cr^{3+}-trioxalate.

Table 6.7: Effect of stem injection of oxalic acid on Cr uptake by Koala and Karridale subclover.[a]

Subclover Cultivar	Treatment	Dry weight g plant^{-1}		Cr Concentration $\mu g\ g^{-1}$		Shoot Cr[b] %
		Shoot	Root	Shoot	Root	
Koala	Control	0.54	0.14	192	142	83.6
	+ injection	0.48	0.13	270	141	87.5
Karridale	Control	0.45	0.12	159	138	80.9
	+ injection	0.39	0.11	370	170	88.3
LSD (p=0.05)		0.12	0.04	42.8	11.5	4.7

[a] The plants were precultured in a balanced Hoagland nutrient solution for 1 week. The Cr(III) trioxalate treatment (2 weeks) was in a balanced Hoagland solution initially buffered to pH 5.0 with MES buffer. Oxalic acid (10 µl, 0.02 µmol mL^{-1}) was injected into the base of the stem of the plant on Days 2, 4, and 6 of the Cr treatment.

[b] $\dfrac{\text{shoot Cr}}{\text{total plant Cr}} \times 100\%$

Reduction of Cr(VI)

Spectroscopic techniques are useful for the localization and determination of oxidation states and bonding of Cr within the plant. Electron paramagnetic resonance (EPR) spectroscopy, X-ray absorption spectroscopy (XAS), and synchrotron X-ray fluorescence (SXRF) spectroscopy are especially useful procedures, since they can be used as non-destructive techniques that do not require sample homogenization or extraction. Examples from a study of the absorption of CrO_4^{2-} by subclover (Howe, 1999) are presented next.

EPR spectroscopy is useful for the study of transition metals, e.g. Cr, with unpaired electrons. The EPR spectra of both leaves and roots of plants grown in different concentrations of CrO_4^{2-} varied considerably (Figs. 6.2 and 6.3). The spectra of leaves of plants grown in 0.04 mmol CrO_4^{2-} L^{-1} produced a Mn^{2+} signal at $g = 2$ (3100 gauss; attributable to Mn absorbed from the nutrient solution) and a very broad faint peak in the $g = 4$-5 region 1100 gauss; (Fig. 6.2). The leaves of plants grown in 2 mmol CrO_4^{2-}L^{-1} produced a similar signal in the $g = 4$-5 region; however, the signal was much stronger (Fig. 6.3). Chromium(VI) is non-paramagnetic (does not have unpaired electrons) and does not produce an EPR signal (Fig. 6.2). The peak at $g = 4$-5 is attributable to a Cr(III) organic complex . Therefore, at least a portion of the Cr(VI) had reduced to Cr(III) and was present in the plant as an undetermined Cr(III) organic complex (Howe, 1999). The leaves of plants grown in 2 mmol CrO_4^{2-}L^{-1} produced a very strong signal at $g = 2$ (Fig. 6.3), which was absent from the plants grown

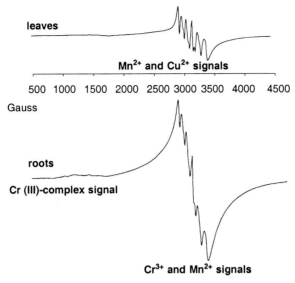

Fig. 6.2: The electron paramagnetic resonance (EPR) spectra of a Cr(VI) standard and leaves and roots of plants grown in a Hoagland nutrient solution with 0.04 mmol CrO_4^{2-} L^{-1}.

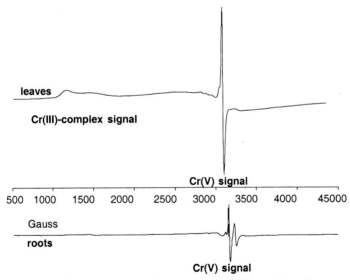

Fig. 6.3: The electron paramagnetic resonance (EPR) spectra of leaves and roots of plants grown in a Hoagland nutrient solution with 2 mmol CrO_4^{2-} L^{-1}

at the lower concentration of CrO_4^{2-} (Fig. 6.2). This peak is attributable to a metastable Cr(V) species. Cr(V) has been previously identified in plant systems (Micera and Dessi, 1988; Liu et al., 1995). In synthetic systems, it has been shown that Cr(V) can result from the reaction of Cr(VI) with reducing agents such as glutathione (GSH) (Bose et al., 1992; Liu et al., 1997), ascorbic acid (Zhang and Lay, 1996; O'Brien and Woodbridge, 1997), and cysteine (Lay and Levina, 1996).

The EPR spectra of roots of plants grown in 0.04 and 2 mmol CrO_4^{2-} L^{-1} also differ considerably (Figs. 6.2 and 6.3). The low concentration of CrO_4^{2-} produced a signal in the $g = 2$ region, which resembles a mix between $Cr(OH)_3$ and Mn^{2+} signals (Fig. 6.2), but no Cr(V) signal was observed (Howe, 1999). This spectrum indicates the reduction of Cr(VI) to Cr(III) occurred either in or on the plant root. The spectra of the roots of the plant exposed to high CrO_4^{2-} concentration produced a Cr(V) signal at $g = 2$ (Fig. 6.3). The probable presence of Cr(V) in the leaves and roots of plants grown at the high CrO_4^{2-} concentration but its absence at the low concentration suggests that in the latter situation the plant had sufficient reductive capacity to totally reduce the absorbed Cr(VI) to Cr(III) (Howe, 1999). But at the high Cr(VI) concentration, the plant did not have sufficient reductive capacity to totally reduce the adsorbed Cr(VI) and the Cr(V) intermediate to Cr(III). The plants grown at the low Cr concentration were still viable after the 3-week exposure period, compared with the plants grown at the high Cr(VI) concentration that were dead after a 2-week exposure. Plants differ significantly in their reductive capacity. Chromium(V) is a reactive intermediate that is mutagenic and carcinogenic in animal systems and hence of considerable environmental interest.

Localization of Cr(III) and Cr(VI) in the Plant

Comparison of Cr Ka X-ray absorption (XAS) spectra of Cr(VI) and Cr(III) standards (Fig. 6.4) illustrates the usefulness of X-ray absorption near edge spectroscopy (XANES) to determine oxidation state of Cr in the plant. Cr(VI) has a pre-edge peak that is absent in the Cr(III) spectrum, thus allowing identification and quantification of Cr(VI) and Cr(III) (Bajt et al., 1993).

Chromium Ka XANES spectra of roots and leaves of subterranean clover plants were obtained (Howe, 1999) and data were normalized and quantified to determine Cr(VI) concentration. The Cr Ka XANES spectra of root material from 1.6 mmol CrO_4^{2-} L^{-1} indicated that Cr(VI) accounted for less than 25% of the total Cr in the roots. Chromium(VI) was not detected in the roots of plants grown in 0.04 mmol $CrO_4^{2-}L^{-1}$. The Cr Ka XANES spectra of leaf material from plants treated with 1.6 and 0.04 mmol $CrO_4^{2-}L^{-1}$ indicated the complete loss of the Cr(VI) pre-edge feature, indicating undetectable Cr(VI) concentrations in the leaves. The presence of Cr(VI) in the nutrient solution and the predominant Cr(III) found in the root supports the conclusion of Cr(VI) reduction by the plant. The

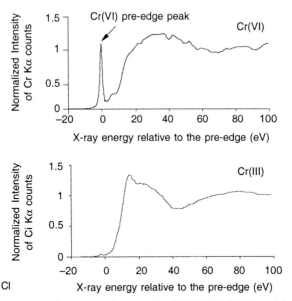

Fig. 6.4: X-ray absorption near edge spectra (XANES) of Cr(VI) and Cr(III) standards.

reduction mechanism must be located primarily in the root since Cr(VI) was not detected in the leaf. The apparent absence of Cr(VI) in the roots of the 0.04 mmol CrO_4^{2-} L^{-1} treated plants and the appreciable Cr(VI) in the roots of the 1.6 mmol $CrO_4^{2-}L^{-1}$ treated plants indicates that the reductive capacity of the roots at the higher concentration was not sufficient to neutralize all the absorbed Cr(VI).

The synchrotron X-ray fluorescence (SXRF) microprobe scans of leaves of plants grown in 0.04 mmol $CrO_4^{2-}L^{-1}$ for 21 days showed a predominant accumulation of Cr at the margin of the leaf and a secondary accumulation in the veins (Howe, 1999). SXRF microprobe scans of transects of leaves grown with 1.6 mmol CrO_4^{2-} L^{-1} for 3 days showed highest levels of Cr in the veins with a slight accumulation at the leaf margin and areas around the midvein. Thus, there were significant differences in leaf Cr accumulation pattern between the low and high Cr(VI) treatments. Plants grown with 1.6 mmol CrO_4^{2-} L^{-1} developed purple coloration along the veins, which gradually became blanched and necrotic. The plants died within 10 days of treatment initiation. Plants grown in 0.04 mmol CrO_4^{2-} L^{-1} exhibited increasing red pigmentation of older leaves but were viable even after 21 days of Cr treatment. At the high CrO_4^{2-} concentration, the presence of Cr(V) and Cr(VI) severely affected plant metabolism; however, at the low CrO_4^{2-} concentration, the subterranean clover plants had sufficient reductive and complexing capacity to mitigate the toxic effects of Cr(VI) and the Cr(III) product.

Absorption of Cr(III) and Cr(VI) from Soils

Thirty-six plant species and cultivars have been tested under soil conditions for their tolerance to and absorption and translocation of Cr(III) and Cr(IV) (Table 6.8). The species and cultivars utilized in these studies were selected based on their agronomic importance, dry matter production, and general tolerance to heavy metals.

Table 6.8: Plant species used for soil screening studies.

Common name	Scientific name	Cultivar/source
Dicotyledonous crops		
Indian mustard	*Brassica juncea*	426308
Sunflower	*Helianthus annus*	Hybrid 571
Canola	*Brassica napa*	Cathy
Ragweed	*Ambrosia artemisiifolia*	Texas
Saltbush	*Atriplex canescens*	Texas
Monocotyledonous crops (Field crops)		
Oats	*Avena sativa*	Texas
Corn	*Zea mays*	4673B
Barley	*Hordum vulgaris*	Texas
Sorghum	*Sorghum bicolor*	Agr1ppo
Wheat	*Triticum aestivum*	Tam 200
Monocotyledonous crops (Cool and warm season grasses)		
Alkali Sacaton	*Sporobolus airoides*	Texas
C. bermudagrass	Cynadon dactylon	Texas
Annual ryegrass	*Colium multiflorum*	Oregon
Wild ryegrass	*Elymus sp.*	Texas
Perennial ryegrass	*Eolium perenne*	Texas
Switchgrass	*Panicum virgatum*	Alamo
Big bluestem	*Andropogan gerardi*	Texas
Little bluestem	*Schizachyrium scoparium*	Texas
Vetiver grass	*Vetiveria zizanioides*	Texas
Seashore paspalum	*Paspalum vaginatum*	Texas
Brassica family		
Broccoli	*Brassica oleracea*	Waltham 29
Brown mustard	*Brassica juncea*	"Serub"
Brussels sprouts	*Brassica oleracea*	Long Island
Cauliflower	*Brassica oleracea*	Snowball
Chinese cabbage	*Brassica perkinensis]*	Repollo chino
Collards	*Brassica oleracea*	Vates
Kale	*Brassica oleracea*	Col crespa
Kohlrabi	*Brassica oleracea*	White Vienna
Mustard green	*Brassica juncea*	Tender green
Rutabaga	*Brassica napus*	—
Turnip	*Brassica rapa*	Purple top
Root crops		
Beets	*Beta vulgaris*	Detroit red
Sugarbeet	*Beta vulgaris*	Tonnage
Swiss chard	*Beta vulgaris*	Lucullus

Chromium Absorption and Translocation by Plant Species

Chromium concentrations in roots and shoots of plant species grown in soils contaminated with 100 mg Cr(III) or Cr(VI) kg^{-1} soil are presented in (Figs. 6.5 and 6.6). In all cases, considerably more Cr was associated with the roots and translocated to the shoots (approximately a 10-fold increase) with the Cr(VI) source than with the Cr(III) source. This difference can be attributed principally to the greater solubility of Cr(VI) compared with that of Cr(III). Inorganic Cr(III) is readily hydrolyzed and precipitated, and largely unavailable to the plant.

In all cases, a higher concentration of Cr was associated with the roots than with the shoots. Root-to-shoot Cr-concentration ratios were generally in the range of 10–50 (Figs. 6.5 and 6.6), and were generally higher with the Cr(III) than the Cr(VI) treatments. Plant roots were thoroughly washed to remove soil particles, but it is not possible to totally remove polymerized Cr(III) from association with the roots. Therefore, it is not possible to determine whether the Cr from the Cr(III) source resides in or on the roots.

The Cr concentration in the shoots of all plant species grown in soil contaminated with 100 mg Cr(III) kg^{-1} was < 10 μg g^{-1}, and in many cases < 3 μg g^{-1}. The dicots generally translocated more Cr than monocots did, and the root crops and *Brassica* species generally had slightly higher shoot Cr concentrations. With the Cr(VI) source, the dicots including the Brassica and root crop species also generally had higher shoot Cr concentrations than the monocots did, though there were exceptions. For example, vetiver grass and oats had high shoot Cr concentrations.

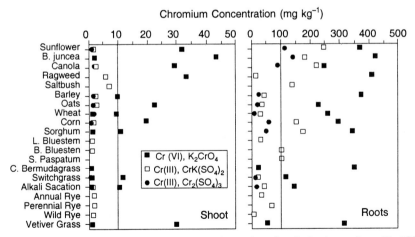

Fig. 6.5: Chromium concentration in shoots and roots of plant species with 100 mg kg^{-1} of Cr(III) or Cr(VI) applied to the soil.

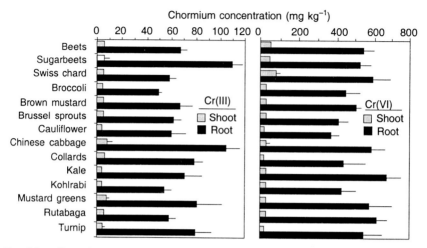

Fig. 6.6: Chromium concentration of root crops and *Brassica* species in shoots compared to roots at 100 mg kg⁻¹ Cr applied as Cr(lll) or Cr(Vl) to soil.

The Cr(III) sources did not result in significant reductions in biomass whereas with the Cr(VI) source biomass was significantly reduced. Toxic effects of Cr(VI) were even observed in soils contaminated with only 25 mg Cr(VI) kg⁻¹. The toxic effects were attributable to the relatively high concentration of Cr in the plant tissue. Other researchers have similarly observed that Cr(VI) is more readily accumulated and more toxic to plants than Cr(III) (Mortvedt and Giordano, 1975; Bartlett and James, 1979).

Effect of EDTA on Cr Accumulation

The effect of supplemental EDTA on the accumulation of Cr from an inorganic Cr(III) source is presented in Fig. 6.7. Plant species responded differently to the presence of the chelating agents. EDTA was effective in increasing the shoot Cr concentration of some plants, e.g. Indian mustard and sunflower Fig. 6.7. Addition of EDTA resulted in an increase in Cr concentration in sunflower shoots of more than ten fold. Wallace et al. (1976) also observed that EDTA applied with $Cr_2(SO_4)_3$ in the ratio 2:1 resulted in decreased bush bean yields and increased Cr concentration; however, $Cr_2(SO_4)_3$ applied without EDTA did not affect yield. In the current study, addition of EDTA resulted in reduced root Cr concentration in barley, corn, oats, wheat, and vetiver grass. Addition of EDTA significantly increased Cr concentration in canola, Indian mustard, sunflower, sorghum, and alkali sacaton roots. EDTA addition had no effect on Cr concentration of other plant species. However, EDTA addition to soil affected plant performance as evidenced by increased chlorosis of leaves and lower plant biomass.

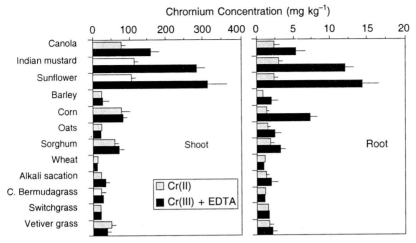

Fig. 6.7: Effect of chelating agent (EDTA) on Cr concentration in shoots and roots of plant species with 100 mg kg^{-1} Cr(III) (1:1) applied to the soil.

Cunningham and Ow (1996) have stated that the most successful amendments to enhance metal uptake are the addition of EDTA to soil. EDTA addition to soil increases the soil solution concentration of the metal and its shoot/root partitioning in the plant. Skeffington et al. (1976) explained that the greater uptake and translocation of Cr(III) EDTA compared to Cr^{3+} was because Cr(III) EDTA is not retarded by ion exchange in plant tissues. EDTA might also have a disruptive effect on the root plasma membrane and result in an increase in the nonspecific absorption of the target metal. The reduced yields in the current study in the presence of EDTA could be attributed to this disruptive (toxic) effect of the EDTA and the increased toxicity resulting from the increased absorption, concentration, and translocation of Cr in the plant.

CONCLUSIONS

The major conclusions from these studies are as follows:

Hydroponic Studies

- The relative toxicity to Cr was generally observed to decrease in the following order: Cr(VI) >> Cr^{3+} > Cr(III) organic complexes. In soil studies, it is common to see the relative toxicity of Cr^{3+} and Cr(III) organic complexes in the reverse order of that given above, due to the low solubility of Cr^{3+} in highly buffered soil systems and the considerably higher solubility of the Cr(III) organic complexes. The visible Cr toxicity symptoms included purple coloration along the stems, leaf veins, and leaf edges; reduced biomass production;

severe necrosis and blanching, principally along the veins and edges; and plant death.

- The relative rate of uptake of Cr by plants generally decreased in the same relative order as (1) above. The slow rate of uptake of Cr^{3+} is attributed to its rapid hydrolysis and precipitation. Precipitation can occur at the root surface, especially in hydroponic systems. The slow rate of absorption of the Cr(III) organic complexes is due to the absence of specific transport channels in the root plasma membrane and the kinetic stability of these complexes.
- The plant tolerance to Cr(VI) was highly species dependent. Also, the translocation of Cr to the leaves was highly dependent on plant species. For example, in hydroponic studies, the Cr contents of the shoots were 45%-90% (Koala clover), 4.2% (*Brassica juncea*), 0.3-2% (pea), and 0.3% (canola) of the total plant Cr. At the highest Cr(VI) treatment, the Cr concentration on a dry matter basis approached 1%.
- In hydroponic studies, Cr from the $Cr(III)Cl_3$ treatment remained largely associated with the roots, due to hydrolysis and precipitation as $Cr(OH)_3$ at the root surface. There was no evidence of $Cr(OH)_3$ precipitation inside the root. In most cases, <1% of the total plant Cr was transported to the shoot.
- The absorption and translocation of Cr from Cr^{3+} and Cr(III) organic complex sources was increased when subterranean clover was grown under Fe-deficiency conditions, but the absorption and translocation of Cr from Cr(VI) was not influenced by the Fe treatment. The translocation of Cr from root to shoot with Cr(III)-organic complex sources was generally increased when subterranean clover was grown under P-deficiency conditions.
- The rate of Cr absorption by the roots from the Cr-organic complexes was in all cases very low. There was generally a greater rate of absorption from Cr(III) oxalate than from Cr(III)-citrate, -malate, -EDTA, and -DTPA.
- The addition of excess free metabolic ligand, e.g. oxalate, citrate and malate, to the growth medium generally resulted in decreased absorption of Cr from a Cr-trioxalate source. These results provide indirect evidence that absorption of Cr from the Cr(III) trioxalate complex might involve ligand exchange at the root plasma membrane. The addition of free EDTA to the growth medium had the opposite effect, i.e. increased (sometimes substantially increased) absorption of Cr from Cr(III)-trioxalate. EDTA might have had a disruptive effect on the root plasma membrane, which resulted in an increased nonspecific absorption of Cr(III)-trioxalate. Similar effects of EDTA were observed in soil studies.

- With the Cr(VI) treated plants, only a trace of Cr(VI) was identified in the leaves and only at the highest Cr treatment. This result indicates that Cr(VI) reduction occurred predominantly in the roots, and the Cr was translocated to the shoots predominantly as a Cr(III) organic complex. At the high Cr(VI) treatment levels, Cr(V) was identified in the roots and shoots, but at low Cr treatment levels no Cr(V) was identified in roots or shoots. Cr(V) is both toxic and mutagenic, hence its presence can have implications to both agriculture and environment. At low Cr(VI) concentrations, the subterranean clover had sufficient reductive capacity to totally reduce the absorbed Cr(VI) to Cr(III). Under this condition, the clover plants remained viable for the 4 week treatment period. At the highest Cr(VI) level, the subterranean clover did not have sufficient reductive capacity, and the plants died within approximately 1 week of ·initiation of the Cr treatment.
- At low Cr(VI) treatment levels, the Cr was translocated predominately to the leaf margins of subterranean clover. This is apparently an adaptive mechanism of the plant, i.e. to translocate the Cr to a region of the plant where it would have a minimal detrimental influence on plant metabolic processes.
- At this time, we do not have conclusive evidence for the predominant mechanism of absorption of Cr(III) from the Cr(III) organic complexes, though current data principally supports hypothesis B (see Fig. 6.1).

  Hypothesis A: non-specific transport of the intact Cr(III) organic complex across the root plasma membrane.

  Hypothesis B: ligand exchange of Cr^{3+} at the root plasma membrane from the source ligand to a plant metabolic ligand.
- Absorption of Cr(VI) and Cr(III) organic complexes is enhanced under conditions where the root plasma membrane has been damaged, e.g. by treatment with free EDTA or CrO_4^{2-}. Under these conditions, root plasma membrane selectivity might be impaired and mass transport would contribute a greater proportion to the total metal uptake.
- Synchrotron XAS procedures and EPR spectroscopy were used successfully to nondestructively evaluate, in situ, the reduction uf Cr(VI) and localization of absorbed Cr(VI) and its reduction products.

Soil Studies

- Plant species differed significantly in Cr(III) and Cr(VI) uptake and translocation. In general, dicots exhibited greater absorption and translocation of Cr. Sunflower and Indian mustard accumulated more Cr than the other plant species tested. High biomass producers such as mustard and sunflower are potential candidates for phytoremediation of Cr. These results agree with results of previous researchers, e.g. Huang et al. (1997).

- Cr(VI) was toxic to most plants at a soil concentration greater than 100 mg kg^{-1} and was severely damaging to plants at concentrations greater than 25 mg kg^{-1}.
- With all Cr sources and concentrations, Cr accumulated predominantly in the roots. In all cases, the absorption and translocation of Cr was greater with Cr(VI) than with Cr(III). This relationship is largely attributable to the greater solubility of Cr(VI) and the greater ease of hydrolysis and precipitation of Cr(III).
- The addition of free EDTA to a soil contaminated with Cr(III) variably influenced absorption and tranlocation of Cr. With sunflower, Indian mustard, and corn, the absorption and translocation of Cr was substantially increased. This phenomenon is attributable to both the increased solubility of the Cr(III) EDTA complex and the disruptive influence of EDTA on the root plasma membrane. This latter factor resulted in increased non-specific absorption of the Cr(III) EDTA complex.

Practical Implications

This series of studies and similar studies have helped to provide insight to the physiologic, soil, and environmental conditions associated with (1) enhanced absorption of Cr from Cr(III)-organic complexes and Cr(VI), and (2) enhanced reduction and plant tolerance to absorbed Cr(VI). These types of studies have important implications to the management of Cr(VI) contaminated sites and the selection and bioengineering of plants for the phytoremediation of Cr-contaminated sites.

References

Ainsworth, C.C., Girvin, D.C., Zachara, J. M., and Smith, S.C. (1989) Chromate adsorption on goethite: Effects of aluminum substitution. *Soil Sci. Soc. Am. J.* **53**: 411–418.

Andrew, J., Smith, C., Kramer, U., and Baker, J. M. (1995) Role of metal transport and chelation in nickel hyperaccumulation in the genus *Alyssum*. In *Fourteenth Annual Symposium, Current Topics in Plant Biochemistry, Physiology and Molecular Biology*, University of Missouri-Columbia, pp. 31-33.

Baes, C. F., and Mesmer, R. E. (1976) *The Hydrolysis of Cations*. Wiley, New York.

Bajt, S., Clakr, S.B., Sutton, S.R., Rivers, M.L., and Smith, J.V. (1993) Synchrotron x-ray microprobe determination of chromate content using x-ray absorption near-edge structure. *Anal. Chem.* **65**: 1800–1804.

Baker, A.J.M., and Brooks, R.R. (1989) Terrestrial higher plants which hyperaccumulate metal elements - A review of their distribution, ecology, and phytochemistry. *Biorecovery* **1**: 81–126.

Baker, A. J. M., Reeves, R.D., and Hajar, A.S.M. (1994) Heavy metal accumulation and tolerance in British populations of the metallophyte Thalaspi. *New Phytol* **127**: 61–68.

Baker, A.J.M., Reeves, R.D., and McGrath, S.P. (1991) In situ decontamination of heavy metal polluted soils using crops of metal accumulating plants? A feasibility study. In In situ *Bioreclamation*, Hinchee, R.L., Olfenbuttel, R.F. (Eds), Butterworth-Heinemann, Boston.

Barcelo, J., Poschenrieder, C., and Gunse, B. (1986) Water relations of chromium VI. treated bush bean plants (*Phaseolus vulgaris* L. cv. Contender) under both normal and water stress conditions. *J. Expt. Bot.* **37**: 178–187.

Bartlett, R.J. (1991) Chromium cycling in soil and water: links, gaps and methods. *Environ. Health Perspective* **92**: 17–24.

Bartlett, R.J. and James, B. (1979) Behavior of chromium in soils; III. Oxidation. *J. Environ. Qual.* **5**: 379–386.

Bonet, A., Pocshenrieder, C., and Barcelo, J. (1991) Chromium interaction in iron sufficient and iron deficient bean plants. *J. Plant Nutr.* **14**: 415–428.

Bose, R.N., Moghaddas, S., and Gelinter, E. (1992) Long-lived chromium(IV) and chromium(V) metabolites in the chromium(VI)-glutathione reaction: NMR, ESR, HPLC, and kinetic characterization. *Inorg. Chem.* **31**: 1987–1994.

Brauer, S.L., Hneihen, A.S., McBridge, J.S., and Wetterhahn, K.E. (1996) Chromium(VI) forms thiolate complexes with gamma-glutamylcysteine, N-acetylcysteine, cysteine, and the methyl ester of N-acetylcysteine. *Inorganic Chemistry* **35**: 373–381.

Cary, E.E., Allaway, W.H., and Olson, O.E. (1977a) Control of chromium concentrations in food plants. 2. Chemistry of chromium in soils and its availability to plants. *J. Agric. Food Chem.* **25**: 305–309.

Cary, E.E., Allaway, W.H, and Olson, O.E. (1977b) Control of chromium concentrations in food plants. 1. Absorption and translocation of chromium by plants. *J. Agric. Food Chem.* **25**: 300–304.

Charlet, L., and Manceau, A. (1992) In situ characterization of heavy metal surface reactions: the chromium case. *Intern. J. Environ. Anal. Chem.* **46**: 97–108.

Cieslak-Golonka, M. (1996) Toxic and mutagenic effects of Cr(VI): A review. *Polyhedron* **15**:3667–3689.

Cotton, F.A., and Wilkinson, G. (1980) *Advanced Inorganic Chemistry*. 4th Edn., Wiley, New York.

Cunningham, S.D., and Ow, D.W. (1996) Promises and prospects of phytoremediation. *Plant Physiol.* **110**: 715–719.

Dushenkov, V., Nanda Kumar, P.B.A., Motto, H., and Raskin, I. (1995) Rhizofiltration-the use of plants to remove heavy metals from aqueous streams. *Environ. Sci. Technol.* **29**:1239–1245.

Fendorf, S.E. (1995) Surface reactions of chromium in soils and waters. *Geoderma* **67**:55–71.

Gabbrielli, R., Mattioni, C., and Vergnano, O. (1991) Accumulation mechanisms and heavy metal tolerance of a nickel hyperaccumulator. *J. Plant Nutr.* **14**: 1067–1080.

Grubinger, V. P., Gutenmann, W. H., Doss, G.J.M., Rutzke and Lisk, D.J.C. (1994) Chromium in Swiss chard grown on soil amended with tannery meal fertilizer. *Chemosphere* **4**: 717–720.

Hara, T., and Sonoda, Y. (1979) Comparison of the toxicity of heavy metals to cabbage growth. *Plant Soil* **51**: 127–133.

Howe, J.A. (1999) Absorption of chromium and reduction of chromate by subterranean clover. M.S. Thesis, Texas A&M University, College Station.

Huang, J., Chen, W.J., Berti, W.R., and Cunningham, S.D. (1997) Phytoremediation of lead contaminated soils: Role of synthetic chelates in lead phytoextraction. *Environ. Sci. Technol.* **31**: 800–805.

Huffman, E.W.D., and Allaway, W.H. (1973) Growth of plants in nutrient culture containing low levels of chromium. *Plant Physiol.* **52**: 72–75.

Hunter, D.B., Bertsch, P.M.K., Kemner, M., Clark, S.B. (1997) Distribution and chemical speciation of metals and metalloids in biota collected from contaminated environments by spatially resolved XRF, XANES and EXAFS. *J. Phys.* IV 7(C2): 767–771.

James, B.R., and Bartlett, R.J. (1983) Behavior of chromium in soils: V. Fate of organically complexed Cr(III) added to soil. *J. Environ. Qual.* **12**: 169–172.

Johnson, J.F., Allan, D.L., and Vance, C.P. (1994) Phosphorus stress-induced proteoid roots show altered metabolism in *Lupinus albus*. *Plant Physiol.* **104**: 657–665.

Kramer, U., Cotter-Howells, J.D., Charnock, J.M., Baker, A.J.M., and Smith, J.A.C. (1996) Free histidine as a metal chelator in plants that accumulate nickel. *Nature* **379**: 635–638.

Kumar, P.B.A.N, Dushenkov, V., Motto, H., and Raskin, I. (1995) Phytoextraction - The use of plants to remove heavy metals from soils. *Environ. Sci. Technol.* **29**: 1232–1238.

Landsberg, E.C. (1981) Organic acid synthesis and release of hydrogen ions in response to Fe deficiency of mono- and dicotyledonous plant species. *J. Plant Nutr.* **3**: 579–591.

Lay, P.A. and Levina, A. (1996) Kinetics and mechanism of chromium(VI) reduction to chromium(III) by l-cysteine in neutral aqueous solution. *Inorganic Chemistry* **35**: 7709–7717.

Lee, C.R., Sturgis, T.C., and Landin, M.C. (1981) Heavy metal uptake by marsh plants in hydroponic solution cultures. *J. Plant Nutr.* **3**: 139–151.

Liu, K. J., J. Jiang, X. Shi, H. Gabrys, T. Walczak and H. M. Swartz. 1995. Low frequency EPR study of chromium(V) formation from chromium(VI) in living plants. *Biochem. Biophys. Res. Commun.* **206**: 829–834.

Liu, K.J., Shi, X.L., and Dalal, N.S. (1997) Synthesis of Cr(IV)-GSH, its identification and its free g-hydroxyl radical generation: A model compound for Cr(VI) carcinogenicity. *Biochem. and Biophys. Res. Commun.* **235**: 54–58.

Losi, M.E., Amrhein, C., and Frankenberger, W. T. (1994) Factors affecting chemical and biological reduction of hexavalent chromium in soils. *Environ. Toxicol. Chem.* **13**: 1727–1735.

Lyon, G.L., Peterson, P.J., and Brooks, R.R. (1969a) Chromium-51 distribution in tissues and extracts of *Leptospermum scoparium*. *Planta* **88**: 282–287.

Lyon, G.L., Peterson, P.J., and Brooks, R.R. (1969b) Chromium-51 transport in the xylem sap of *Leptospermum scoparium* (Manuka). *N. Z. J. Sci.* **12**: 541–545.

Lyon, G.L., Peterson, P.J., Brooks, R.R., and Butler, G.W. (1971) Calcium, magnesium and trace elements in a New Zealand serpentine flora. *J. Ecol.* **59**: 421–429.

Lytle, C.M., Lytle, F.W., Yang, N., Qian, J. H., Hansen, D., Zayed, A., and Terry, N. (1998) Reduction of Cr(VI) to Cr(III) by wetland plants: Potential for in situ heavy metal detoxification. *Environ. Sci. Technol.* **32**: 3087–3093.

McGrath, P.S., Sidoli, C.M.D., Baker, A.J.M., and Reeves, R.D. (1993) The potential for the use of metal-accumulating plants for the in situ decontamination of metal-polluted soils. In *Integrated Soil and Sediment Research: A Basis for Proper Protection*, Eijsackers, H.J.P. and Hamers, T. (Eds.), Kluwer Academic Publishers, Dordrecht, pp. 673–676.

McGrath, S.P. (1982) The uptake and translocation of tri- and hexa-valent chromium and effects on the growth of oat in flowing nutrient solution and in soil. *New Phytol.* **92**: 381–390.

Micera, G., and Dessi, A. (1988) Chromium adsorption by plant roots and formation of long-lived Cr(V) species: An ecological hazard? *J. Inorg. Biochem.* **34**: 157–166.

Mortvedt, J.J., and Giordano, P.M. (1975) Response of corn to zinc and chromium in municipal wastes applied to soil. *J. Environ. Qual.* **4**: 170–174.

O'Brien, P., and Woodbridge, N. (1997) A study of the kinetics of the reduction of chromate by ascorbate under aerobic and anaerobic conditions. *Polyhedron* **16**: 2081–2086.

Palmer, C.D. and Wittbrodt, P.R. (1991) Processes affecting the remediation of chromium-contaminated sites. *Environ. Health Perspective* **92**: 25–40.

Peterson, P.J,. and Girling, C.A. (1981) Other trace metals: Chromium. In. Effect of Heavy Metal Pollution on Plants, *1*. *Applied Science*, Lepp, N.W (Ed.), Applied Science London, pp. 222-229.

Rai, D., Sass, B.M., and Moore D.A. (1987) Chromium (III) hydrolysis constants and solubility of chromium(III) hydroxide. *Inorg. Chem.* **26**: 345–349.

Rauser, W.E. (1990) Phytochelatins. *Annu. Rev. Biochem.* **59**: 61–86.

Rauser, W.E. (1995) Phytochelatins and related peptides: structure, biosynthesis and function. *Plant Physiol* **109**: 1141–1149.

Rauser, W.E., and Meuwly, P. (1995) Retention of cadmium in roots of maize seedlings. Role of complexation by phytochelatins and related thiol peptides. *Plant Physiol.* **109**: 195–202.

Reeves, R.D. (1992) The hyperaccumulation of nickel by serpentine plants. In *The Ecology of Ultramafic (Serpentine) Soils*, Baker, A.J.M., Proctor, J., and Reeves, R.D. (Eds.). Intercept, Andover.

Ric De Vos, C., Lubberding, H.J., and Bienfait, H. F (1986) Rhizosphere acidification as a response to iron deficiency in bean plants. *Plant Physiol.* **81**: 842–846.

Shewry, P.R., and Peterson, P. J. (1974) The uptake and translocation of chromium by barley seedlings (*Hordeum vulgare* L). *J. Exp. Bot.* **25**: 785–707.

Skeffington, R.A., Shewry, P.R., and Peterson, P. J. (1976). Chromium uptake and transport in barley seedlings. *Planta* **132**: 209–214.

Soane, B.D., and Saunder, D.H. (1959) Nickel and chromium toxicity of serpentine soils in southern Rhodesia. *Soil Sci.* **94**: 322–329.

Stünzi, H., and Marty, W. (1983) Early stages of the hydrolysis of chromium(III) in aqueous solution. 1. Characterization of tetrameric species. *Inorg. Chem.* **22**: 2145–2150.

Tiffin, L.O. (1966) Iron translocation I. Plant culture, exudate sampling, iron-citrate analysis. *Plant Physiol.* **41**: 510–514.

Wagner, G. J., Salt, D., Gries, G., Donachie, K., Wang, R., and Yan, X. (1995) Biochemical studies of heavy metal transport in plants. In *Proceedings/Abstracts of the Fourteenth Annual Symposium: Current Topics in Plant Biochemistry, Physiology and Molecular Biology.* University of Missouri, Columbia, pp. 21–22.

Wallace, A., Soufi, S. M., Cha, J.W., and Romney, E. M. (1976) Some effects of chromium toxicity on bush bean plants grown in soil. *Plant Soil* **44**: 471–473.

Weckhuysen, B.M., Wachs, I.E., and Schoon, R.A. (1996) Surface chemistry and spectroscopy of chromium in inorganic oxides. *Chem. Rev.* **96**: 3327–3349.

Wei, L.C., Loeppert, R. H., and Ocumpaugh, W. R. (1997) Fe-deficiency stress response in Fe-deficiency resistant and susceptible subclover. *J. Expt. Bot.* **48**: 239–246.

Wei, L.C., Loeppert, R.H., and Ocumpaugh, W. R. (1998) Factors affecting Fe-deficiency stress response of subclover. *Physiologia Plantarum* **103**: 443–450.

Wild, H. (1974) Indigenous plants and chromium in Rhodesia. *Kiekia* **9**: 233-241.

Zayed, A., Lytle, C.M., Qian, J.H., and Terry, N. (1998) Chromium accumulation, translocation and chemical speciation in vegetable crops. *Planta* **206**: 293–299.

Zhang, L.B., and Lay, P.A. (1996) EPR spectroscopic studies of the reactions of Cr(VI) with L-ascorbic acid, L-dehydroascorbic acid, and 5,6-0-isopropylidene-L-ascorbic acid in water. Implications for chromium(VI) genotoxicity. *J. Am. Chem. Soc.* **118**: 12624–12637.

Plant Soil Metal Relationships from Micro to Macro Scale

K. Bujtas[1], A.S. Knox[2], I. Kadar[1] and D.C. Adriano[3]

INTRODUCTION

Experimental data on fate, bioavailability, and effects of metals in the soil plant system are often derived from various culture techniques:

- pot experiments (microcosm),
- larger soil columns or lysimeters (mesocosm), or
- field experiments (macrocosm).

In some studies, solution culture is used instead of soil medium. As the different experimental techniques have various advantages and may yield information on several aspects of the same issue, selection of the most suitable experimental technique is important. However, there are inherent difficulties in interpreting data on plant growth, on bioaccumulation, and phytotoxicity of metals established using different techniques.

Problems associated with assessment of metal phytoavailability (defined as metals taken up by plant via root uptake) and long-term risks associated with accumulation of high levels of these elements in the soils may be better interpreted if the limitations of the various experimental techniques are known.

This chapter summarizes laboratory and field investigations aimed at studying the advantages and disadvantages of various culture techniques and their applicability as a measure of metal bioavailability (see Chapter 3 for other techniques of bioavailability assessments) and as a tool in the risk assessment of heavily contaminated soils. The experimental studies include

1) greenhouse pot experiments in which various forms of metals originating from flue dust applications to the soil were measured using

[1] Research Institute for Soil Science and Agricultural Chemistry of the Hungarian Academy of Sciences, H-1022 Budapest, Herman Otto ut 15, Hungary
[2] Westing House Savannah River Company, Building 773-63A, Room 6, Aiken SC 29808, USA
[3] University of Georgia, Savannah River Ecology Lab, Aiken, SC 29802, USA

the sequential extraction technique and the effects of metals on the growth and their uptake into experimental plants were assessed, both in the presence and absence of ameliorants;

2) use of large soil columns, either under field conditions as lysimeters to assess movement of the metals in the soil profile or in the laboratory for studying plant availability of the metals after addition of metal-enriched sewage sludge to cropped, large, undisturbed soil monoliths; and

3) long-term small-plot field experiments aimed to study the effects of 13 potentially toxic elements on various crops and their potential harm in further steps of the food chain.

All these experiments involve application of several metals, separately or in combination with the growth medium. This chapter will focus on forms, phytoavailability, movement, and plant uptake of zinc (Zn) and chromium (Cr) as examples of an essential, relatively mobile, and a nonessential, more tightly bound element, respectively.

MICROCOSMS: POT EXPERIMENTS

Advantages and Limitations

Pot experiments are preferred over columns and small field plot trials since they are more cost-, time-, and labor-effective, but the results may vary greatly according to experimental conditions (size of pots, set-up, etc.), and their applicability to real field situations is questionable.

Pot experiments were introduced in fertilizer and plant nutrition research at the turn of the century (Wagner, 1883; 'Sigmond, 1904). In the beginning, the pots were larger than what they are in today's general practice. This helped to simulate field conditions better than what the smaller (1–2 kg) pots would, and the plants were raised till maturity. However, the interpretational problems were discovered early on. As Sigmond noted in 1904, "It is a general observation that the effects of fertilisation in the fields and in the pots are only exceptionally identical, at best the field and pot results may show some definite proportionality. This discrepancy may be fully explained by the fact that the circumstances of growth in the pots are only randomly and rarely similar to the natural soil conditions. The soil in the pots is not in its natural physical state, and this situation in itself may cause great differences among the experimental pots. Moreover, while the soil moisture conditions may be very favourable in the pots, the ventilation of the soil air phase is worse than in the field soil; and the development of roots and the extent of the surface cover by the plants is not such as usual (normal)".

This more than 90-year-old citation identifies most of the problems associated today with pot experiments. Climatic factors such as temperature, humidity, intensity and quality of light have been suspected as the primary reason for discrepancies between pot and field experiments. Plants growing in the shadow, i.e. under smaller light intensities in the greenhouse, may need less nutrient, as indicated by the experiments of Pusztai and Kadar (1980).

With the limited soil volume in pot experiments, rhizosphere effects and fertilizer distribution may be markedly different in the pots and in the field plots. The utilization of soil nutrients may be greater because of the low soil/root ratio; the solubility of the nutrients is greater because of the optimal soil moisture conditions, which are generally higher than in the field; and the fertilizer effects appear more substantial because other factors are generally kept optimum and the volume of the soil available for the roots is much smaller (Atanasiu, 1966).

On the other hand, pot experiments have some advantages over field studies. In addition to being simpler, cheaper and quicker, their management is much easier and their reproducibility is generally better. Pot experiments may be designed to assess the exhaustivity of the soil nutrient supply, which is difficult to achieve in the field. A further advantage is that pot experiments need less sophisticated statistical treatment, with each pot representing not only the average sample but also the whole sampled area (Kadar, 1992).

The problems and advantages of pot experiments originally described for nutrient studies are similar to those occurring in studies concerning the fate of various soil-polluting trace elements. Pot experiments generally overestimate phytoavailability of the metals found under field conditions. Page and Chang (1978) and De Vries and Tiller (1978) demonstrated that plants absorbed more Cd from Cd-enriched soils when grown in large pots in the greenhouse than under field conditions on the same soils enriched identically with Cd. In both situations, the uptake of Cd by the Swiss chard plants increased linearly with increasing amounts of Cd applied in the form of sewage sludge, but the slope of the line was higher (about 1.5-fold) for the greenhouse pots. Also, the higher correlation coefficient of the linear relationship in the greenhouse experiment indicated the more marked and reliable response of the plants under the controlled conditions. Page et al. (1981) suggested that the different root development (in the field plots the roots may grow into the uncontaminated soil below the polluted zone) may be the reason for the different uptake characteristics. McBride (1995) also stressed that the rooting pattern seems to be most critical when comparing the results of several sewage-sludge application experiments in the field. Well-developed root systems of deep-rooting crops (cereals, corn) may reach soil depths that contain none of the applied sludge, and this may explain the smaller metal uptake in some field experiments. In addition, the rooting pattern characteristics and distribution of applied sludge in the profile as well as other important soil and climatic characteristics may vary significantly among experimental sites, and this may be a cause of the high degree of unexplained variability in the results from field experiments (McBride, 1995).

However, in greenhouse pot experiments, where the treatments are well defined, and the growth of plants is more uniform because of the controlled environmental factors (moisture, humidity, temperature) it is possible to assess the effect of one or only a few factors at the same time:

1) The presence of trace elements may be selected as single elements or in any desired combination, using a wide concentration range;

2) There may be planned, experimental variations in the growth medium (e.g. when comparing different soils in the same experiment under the same condition, or examining the impacts of microbial activities, such as mycorrhizal infection of the roots by using native and sterilized soil samples, etc.).

Assessment of phytotoxicity thresholds may be done reliably, provided a suitable indicator is chosen for this purpose. This can be difficult even in relatively simple systems when the trace elements are applied in their inorganic forms, without the presence of additional organic matter from the sludge. Among the characteristics most often used to assess phytotoxicity, i.e. dry matter accumulation, elemental concentrations in soil (in soil solution or mostly in soil extracts) and plant tissues, the last one may be regarded to be the best for estimating phytotoxicity thresholds (Schmidt, 1997). Beckett and Davis (1977) indicated that plant trace element concentrations produced a fairly consistent phytotoxicity threshold under various circumstances while critical elemental concentrations in the liquid phase corresponding to yield decrease were different at various yield levels.

Case Studies: Application of Lime as an Ameliorant to Zn Contaminated Soil

Materials and methods

A Ultisol Appling silt loam soil was used in pots that contained 7 kg of surface soil. The main soil characteristics were as follows: pH(H_2O), 5.4; particle size <0.02 mm, 21.5 %; organic matter (OM), 2.5 %; and cation exchange capacity (CEC), 3.50 cmol kg^{-1}.

In the first experiment flue dust (FD) from the Owen Steel Recycling Plant, Columbia, SC, was added to the soil. The flue dust had a pH of 8.0, a calcium carbonate equivalence (CCE) of 29.1 %, and contained 0.12% Cd, 0.23% Cu, 0.45% Na, 0.92% K, 1.26% Mn, 1.63% Mg, 3.12% Pb, 10.5% Ca, 14.3% Fe, and 28.0% Zn as major components. Flue dust was added in doses corresponding to 150, 300, and 600 mg Zn kg^{-1} dry weight of soil. Dolomitic lime was added as an ameliorant to the soil at variable rates to adjust the pH to about 6.5. For all treatments, there were four replicates, and the randomly distributed pots were equilibrated at soil moisture holding capacity for 4 weeks prior to planting.

Three crops were grown consecutively on the potted soil: 12 maize (*Zea mays* L. cv. Pioneer 3165), 12 spring barley (*Hordeum vulgare* L., cv. Boone), or 8 radish (*Raphanus sativus* L. cv. Red Devil) plants. Maize was harvested at mature age (after 3 months), spring barley before earing at the boot stage, radish at mature age, i.e. 25 days after germination. Mature maize and radish plants were separated into various parts: maize to roots, old leaves, young leaves and grain, radish to tubers and leaves. Aboveground parts of barley were taken for further analysis. Plant samples were washed with deionized water and dried to constant weight at 60°C. The dried plant tissues were finely ground, and 5 g of the dry sample was ashed at 450°C for 16 h. Plant ash was dissolved with 10 ml of HCl (1:1, v/v) on a hot plate and then filtered through a soft filter into a 50-ml flask.

In the second experiment, metals (Cd, or Cu, or Ni, or Pb, or Zn) were added to the soil as a defined mixture of various metal sources (40% sulfate, 25% carbonate, 20% oxide, 15% chloride). Zn was added at rates of 1000 and 2000 mg kg^{-1} soil. After the soils were equilibrated for 4 weeks, lime was added as an ameliorant to adjust the soil pH to a value of 6.5. First rye (*Secale cereale* L.) and then maize were grown on the pots, each plant for 6 weeks. Aboveground parts of both plant species were analyzed for metal content. Plant samples were washed with deionized water and dried to constant weight at 60 °C. Elemental content of the plant samples was determined after wet digestion of 0.5 g of dried and finely ground tissues in 10 ml of concentrated HNO_3 on a hot plate.

Soil samples taken from the pots in both experiments were dried at ambient temperature (about 21 °C), ground, and passed through a 2-mm sieve, which was used to determine soil properties such as: pH, OM, CEC, etc. Samples were further homogenized (by grinding in an agate mortar to break down into particles of <0.01 mm diameter) to analyze the nature and total metal content in the soil. A sequential extraction procedure (Tessier et al., 1979) was used to assess the amount of Zn in the following operationally defined fractions: exchangeable (1.0 M $CaCl_2$, pH 7), carbonate (1.0 M NaOAc, pH 5), iron-manganese oxides (0.04 M NH_4OH • HCl), organic (0.02 M HNO_3 + H_2O_2; 3.2 M NH_4OAc), and residual (aqua regia and HF). The total Zn contents in the control and treated soil samples, and in the flue dust were determined by digesting 0.5 g sample in a mixture of 1.5 ml of aqua regia and 5 ml of concentrated HF.

Zn contents of all soil and plant samples were analyzed by flame atomic absorption spectroscopy. Data quality was controlled by measurements of blank digests, duplicate samples, and internal plant and soil standards. The data were subjected to analysis of variance, and Tukey's multiple range test was used to determine whether the treatment means were significantly different.

Plant yield

Strong yield reductions of maize, barley, and radish were observed in the first experiment in the presence of FD, with great differences among the plant species in their susceptibility to excess amounts of metals (Fig. 7.1). Lime added as an ameliorant practically eliminated the yield reduction in barley and maize, and was beneficial also for radish, greatly reducing phytotoxicity at the 150 and 300 mg kg^{-1} FD rates. However, radish plants were still unable to grow at the highest (600 mg kg^{-1}) FD rate, even when lime was added.

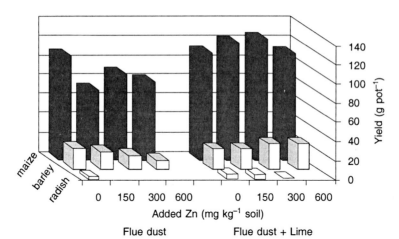

Fig. 7.1: Total yield of maize, barley, and radish, in a pot experiment with Zn-containing flue dust and with lime as ameliorant.

In the second experiment, where Zn was applied as a defined mixture of various inorganic Zn salts at provocatively high rates (1000 and 2000 mg kg^{-1} soil), both rye and maize showed poor germination, and died one week after germination when Zn was added to the soil without any ameliorant. In the lime treatment, rye still germinated poorly and plants showed Zn toxicity symptoms during growth. Rye yields were very low (43% and 3.6% of the control plants at 1000 and 2000 mg Zn kg^{-1} soil rates, respectively). Maize was more relieved of Zn toxicity by lime treatment, but still the yield of plants was reduced by 77% at 2000 mg Zn kg^{-1} soil.

Zn uptake by plants

In general, uptake of Zn was influenced by plant species, plant part, and the presence of lime. Zinc content of the tissue increased dramatically with increasing amounts of added FD (Figs. 7.2 and 7.3). Lime considerably decreased Zn uptake into barley leaves, and its effect was more

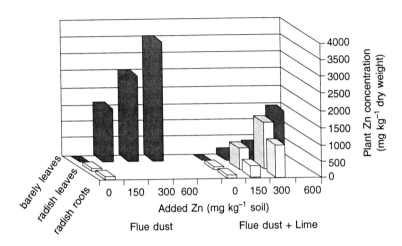

Fig. 7.2: Zinc content in barley leaves, radish leaves, and radish roots, in a pot experiment with Zn-containing flue dust and with lime as ameliorant.

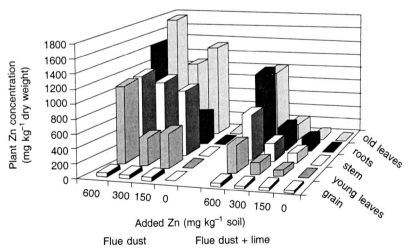

Fig. 7.3: Zn concentration in mature maize tissues, in a pot experiment with Zn-containing flue dust and with lime as ameliorant

pronounced when less FD was applied (Fig. 7.2). At the 150 mg kg^{-1} FD rate, lime additions reduced Zn uptake by 88%, whereas there was 42% reduction at the 600 mg kg^{-1} FD rate.

Radish plants took up more Zn than the other two plant species in those treatments where radish could grow because of addition of lime (Figs. 7.2 and 7.3). Excess Zn was apparently transported effectively to the radish leaves, as indicated by the higher Zn levels in the leaves than in the tubers. In contrast, distribution of Zn in the control radish plants was fairly uniform between the tubers and leaves both in the presence of lime or without lime, in agreement with the physiological role of Zn as a micronutrient.

Data for the mature maize plants (Fig. 7.3) indicate which plant parts that accumulate Zn. With FD alone, the old leaves displayed higher Zn contents than the roots did, indicating the intense translocation of Zn during the early stages of growth to these tissues. Among the shoot tissues, the next higher accumulation of Zn was observed in the stem, followed by the young leaves, then by the grain which had the least Zn content. Grain Zn contents increased about twofold in all FD treatments but showed no increase with increasing FD rates. This distribution pattern indicates the transport route and also suggests that the younger vegetative and especially the reproductive plant parts are protected from accumulating excess amounts of metals by a more intense bioaccumulation in the lower (more mature) plant parts. Addition of lime greatly reduced Zn uptake, especially at the lower FD rates, as indicated by the much lower Zn contents in the older leaves for the two lower FD doses and when compared with the root Zn contents.

In the second experiment, the Zn content in young rye and maize plants grown in the presence of high rates of Zn salts + lime greatly exceeded the Zn concentration of plants from the control soil. For example, Zn content in rye reached 3450 mg kg^{-1} dry weight and in maize 5700 mg kg^{-1} dry weight at the 2000 mg Zn kg^{-1} soil rate. These values mean 140-fold and 310-fold increases, respectively, as compared with plant Zn concentrations in the control + lime treatment.

Exchangeable Zn in soil

The content and chemical forms of Zn in the soil were influenced by the dose of FD, the mixture of Zn salts, and the addition of lime as ameliorant. In the control soil of the first experiment, Zn was primarily associated with the residual fraction. The next abundant form of Zn in the control soil was in the Fe-Mn oxide fraction, and was followed by the exchangeable fraction. The Zn content in the various fractions, of the soil treated with FD generally increased with increasing FD rates. This rise was the smallest in the organic fraction, in agreement with the relatively low OM content of this soil. The distribution pattern of Zn among the various fractions was also altered in the FD treatments as compared with the control, but smaller differences were found in the relative abundance of the Zn forms among the different FD rates. The proportion of exchangeable carbonate and Fe-Mn oxide forms increased, while that of the organic and especially of the residual fractions decreased when FD was added to the soil (Table 7.1). Lime addition somewhat decreased the amount of Zn recovered in the sequential extraction procedure. The decrease was caused mostly by a substantial reduction of the exchangeable fraction.

Table 7.1: Percentage of the soil's Zn content in the various fractions after flue dust (FD) and lime addition

Treatment mg Zn kg^{-1}	Exchangeable %	Carbonate %	Fe-Mn-oxide %	Organic %	Residual %	mg kg^{-1}
FD						
0	9.7	6.7	16.8	8.6	58.2	33.9
150	26.1	12.2	30.4	5.3	26.0	225
300	28.4	11.6	34.7	6.5	18.8	398
600	27.9	13.8	40.1	4.4	13.8	848
FD + lime						
0	9.9	7.8	18.6	8.2	55.5	31.7
150	16.8	15.6	30.8	8.1	28.7	211
300	17.8	12.9	42.1	5.6	21.6	356
600	19.5	16.2	40.2	4.4	19.7	696

Similar changes in the Zn form were observed in the second experiment after addition of Zn salts. The soil's native Zn content (0 treatment) was found mostly in the residual form. The next fractions with high concentrations of Zn were the organic and the Fe-Mn oxide forms, respectively (Table 7.2). In the Zn-contaminated soil, all Zn fractions increased, with the majority of Zn found in the exchangeable form. Addition of lime significantly decreased the exchangeable form of Zn by increasing Zn concentration mostly in the carbonate, Fe-Mn oxide, and residual fractions.

The influence of lime on the exchangeable Zn content was partly due to the increase in soil pH. Chlopecka et al. (1996) observed that Zn, as indicated by the exchangeable fraction, was most immobile when the pH of the polluted soil was over 6.5. Precipitation of metals as insoluble hydroxides, carbonates, and phosphates increases with increasing soil pH (Smith, 1994). Lime immobilizes Zn in the soil by decreasing its solubility in the soil solution. Also, adsorption of heavy metals onto clay minerals and OM is increased by increasing soil pH (Kiekens, 1984). Thus, one of the effects of lime amelioration is a decrease of the availability

Table 7.2: Distribution of soil Zn forms after addition of Zn salts and influenced by lime amendment

Treatment mg Zn kg^{-1}	Exchangeable %	Carbonate %	Fe-Mn-oxide %	Organic %	Residual %	mg kg^{-1}
without lime						
0	3.0	1.6	9.8	14.0	72.1	26.5
1000	37.4	16.4	26.8	2.1	17.3	816
2000	45.7	23.0	20.9	1.2	9.2	1820
with lime						
0	2.6	2.0	11.2	16.5	67.9	24.9
1000	15.2	22.2	29.8	3.3	29.4	965
2000	21.3	34.9	23.6	1.9	18.3	1795

of the metals and a consequent reduction in heavy-metal uptake by plants (Sanders et al., 1986). Researchers have reported that the solubility and bioavailability of Zn are negatively correlated with Ca^{2+} domination of the soil exchange complex (Shuman, 1985; Kabata-Pendias and Pendias, 1992). In the pot experiments presented here, plant concentrations of Zn were much higher when no lime was added and the soil's pH was below 6.0.

Relationship between exchangeable and plant Zn concentrations

There was a significant correlation between plant Zn content and exchangeable Zn in the soil. Although concentrations of the Zn in all aboveground parts of maize plants and in barley leaves correlated with both the exchangeable and the total Zn content of the soil, the correlation was better and the significance higher for exchangeable Zn ($R = 0.56$ and 0.88, all at $P < 0.01$) than for total Zn (R = between 0.35 and 0.45, at $P<0.05$). Root and leaf Zn concentrations of radish plants correlated highly with both the exchangeable and the total Zn contents of the soil (all $R >$ 0.93, at $P < 0.01$). This may be explained by the higher sensitivity of this crop to elevated Zn levels and the very high phytotoxicity found at the higher FD (i.e. Zn) application rates.

Similarly, in the second experiment, when amount of the exchangeable Zn in the soil was very high (Table 7.2) and soil pH was below 6.0, Zn was highly phytotoxic and no plant survived. Lime addition increased soil pH to 6.5, decreased exchangeable Zn, and produced plant yield, although highly reduced.

Mesocosms: Large Soil Columns

Advantages and limitations

Experiments on large soil columns can be viewed as an intermediate situation between the laboratory pot experiments and small-scale field plots, with the advantages and disadvantages of both. This experimental technique may be suitable for quick simulation of field conditions, either for relatively short duration or for longer periods covering the whole lifetime of the selected crop. Water relations and behavior of nutrients and of other elements, with special emphasis on leaching and accumulation processes in the soil profile, are the most frequently studied issues involving the use of soil columns.

The two main types of columns are the disturbed lysimeters and the undisturbed soil monoliths. The former are prepared by collecting soil samples according to the genetic soil layers at the site and filling these into the coat of the lysimeter in the original order and as close to the original volume as possible. Generally, a certain period is necessary until the soil conditions become again more similar to the natural state. One of the basic problems of these lysimeters is the so-called "edge-effect", i.e. the water and the solutes move along the edge (lysimeter wall) generally faster than in the inside because of the poor contact of the disturbed soil structure with the lysimeter wall. To eliminate this problem, lysimeters or soil monoliths of undisturbed structure are used by some researchers. These may be prepared in several ways, e.g. digging around, or hydraulic pressing of tubes into the soil as coating, etc.

Preparation of undisturbed large soil columns according to the description of Homeyer et al. (1981) was modified in Hungary (Nemeth et al., 1991). The modification allowed preparation of these monoliths even from light sandy soils. Such 1.5 m high columns of various Hungarian soil types have been used to follow movement of water and N compounds, with special regard to accumulation and leaching of nitrate and transformation of N compounds (Nemeth et al., 1991b). In recent experiments movement of Cd, Cr, Ni, Pb, and Zn along the soil profile from metal-enriched sewage-sludge was also studied, with emphasis on available amounts of these metals (Bujtas et al., 1995).

The main advantages of using the large lysimeters or soil monoliths are the larger soil volume (as compared with that of pots), the better controlled environmental conditions, and smaller expenses (as compared with those of field plots). Still, many researchers remain skeptical about the possibilities of extrapolating the results to real situations. The soil volume, although greater than in pots, still remains limited and the rooting pattern different from those on the fields. On the other hand, the heterogeneity of the soil at the site may cause large differences between individual undisturbed monoliths, thus decreasing the reproducibility of the results in comparison to pot experiments.

Case study 1: Lysimeter Experiment with Trace Metals

Material and methods

The experiment was set up in 12 lysimeters filled with pseudopodzolic (loamy) sand according to natural genetic horizons to a depth of 80 cm. Size of the lysimeters was 0.5 m × 0.5 m × 0.9 m. An air-dried and homogenized sewage-sludge was mixed carefully after 1 year into the upper 20 cm soil layer, at rates of 5.3 kg per lysimeter. The sludge had a pH of 6.8, its OM content was 34.8% as measured by loss on ignition, and contained 2.8% Ca, 0.54% K, and 0.25% Mg. Its Cr and Zn contents were also high (Cr, 0.20% and Zn, 0.61%); total Ni concentration was 92 mg kg^{-1} dry sludge.

After sludge was added, ryegrass (*Lolium perenne* L.) and white clover (*Trifolium pratense* L.) were planted in four replications in a randomized design. Plants were harvested 4 to 5 times (always at the shooting stage), and lysimetric percolate waters were collected about 6 to 8 times a year.

Metal partitioning was conducted by the sequential extraction procedure of Tessier et al. (1979). Extractions were carried out in triplicates using 1 g samples in 50mL centrifuge tubes. Following each extraction, the mixtures were centrifuged at 2000 g for 30 min at 15 °C. The total trace metal content of the samples was determined after wet digestion with a mixture of 3 ml of concentrated $HClO_4$ and 10 ml of concentrated HF. Metal contents in the plant, percolate water, and soil samples were determined by flame atomic absorption spectroscopy.

Distribution of Zn forms in soil

Evaluation of the environmental effects of trace metals requires a precise knowledge of their speciation and of the plants response to the extracted species. Also, from speciation results, conclusions may be drawn about the mobility of the metals and hence about their potential to move downward in the soil profile after sludge application to land. In the lysimeter experiments cited here, both speciation of Zn and its downward movement in the lysimeter percolate waters were measured.

The Zn content of the applied sewage sludge (6100 mg kg^{-1}) was two orders of magnitude higher than the original soil Zn concentrations, so the addition of sludge significantly increased the amount of Zn in the soil, from 69 to 439 mg kg^{-1}. The distribution patterns of Zn among the separated fractions in the soil and in the sludge itself were markedly different. Although the organic fraction was the most abundant in the soil (42%), followed by the residual (36%) and Fe-Mn oxide fractions (26%), the sludge contained the majority of the metals bound in the oxide fraction (72% of the total Zn concentration), followed by the organic (17%)

and the carbonate fractions (15%). Relative amounts of the residual form of Zn were negligible in the sludge, although the actual amounts exceeded the soil residual Zn levels four fold. The exchangeable form of Zn was relatively low both in the soil and in the sludge (3% of the total Zn content, i.e. 2 and 206 mg Zn kg^{-1} dry matter in the soil and sludge, respectively). Addition of sludge resulted in a 50-fold increase of the exchangeable form of Zn in the soil (from 2 to 113 mg kg^{-1}), so this fraction increased up to 26% of the total Zn content.

Leaching of trace metals in the soil profile

Elemental composition of lysimeter percolate waters showed a two fold increase in Zn concentration (from 515 to 1052 mg kg^{-1} soil), indicating an increased mobility of Zn in the sludge-treated soil. Concentrations of other trace metals that were present in the sludge at elevated rates (Cd, Cr, Cu, Ni, and Pb) increased also about two fold in percolate waters of the sludge-treated lysimeters, but the total amounts of these metals that leached from the soil profile did not exceed 1 mg metal per lysimeter during the three-year-long experiment.

The removal of metals from sludge-treated sites by plant uptake and subsequent crop removal, and by leaching to the subsurface, is normally very small (McGrath et al., 1994; Kabata-Pendias and Adriano, 1995). However, there are contradictory views in the literature regarding the extent to which sludge-borne metals are vulnerable to leaching. For instance, Legret et al. (1988) found appreciable migration of Ni and especially of Cd, but little or no movement of Pb and Cr during a field experiment on a sludge-treated, coarse-textured soil. In contrast, Dowdy and Volk (1983), Chang et al. (1984) and Alloway (1990) reported retention of these elements within the zone of sludge incorporation in both column and field experiments. Differences among experimental conditions may be a reason for the differences observed in the movement of various metals in the soil.

Case Study 2: Application of Metal-enriched Sewage Sludge onto Large Undisturbed, Cropped Monoliths

Description of the method

Experiments on soil columns of a brown forest soil were conducted under controlled environmental conditions to assess the fate and effects of Cd, Cr, Ni, Pb, and Zn in the soil-plant system. Pertinent soil properties in the A and B horizons of the soil profile were as follows: pH-KCl, 5.01–5.29; OM content 10–12 g kg^{-1}, particle size <0.02 mm, 18.9–20.5%; and CEC, 8.5–9.0 cmol kg^{-1}.

The undisturbed, 40-cm diameter, 100-cm long soil monoliths were prepared following the methods originally proposed by Homeyer et al. (1973) and modified by Nemeth et al. (1991). The monoliths were excavated at the selected field sites and their cylindrical surfaces and bottoms were coated with fiber glass cloth impregnated with a synthetic resin. After the coatings solidified, the monoliths were transported to the laboratory. Soil water contents along the soil profile were regulated by saturating the columns from the bottom through a special built-in valve connected to a hanging water column or by sprinkler irrigation at the soil surface. Water was added on the basis of time-domain reflectometry measurements and/or weighing. Supplemental light was provided as 12 hour day/night cycles.

Cadmium, Cr, Ni, Pb, and Zn were applied as communal sewage sludge enriched with nitrate salts of these metals. Dry matter content of the compressed sludge was 20.6% and inorganic matter content was 48.2%. Original total concentrations of the metals in the sludge were as follows: 12.3 mg Cd, 217 mg Cr, 109 mg Ni, 210 mg Pb, and 3026 mg Zn per kg DW. These values are comparable to or less than the limits (15 mg Cd, 1000 mg Cr, 200 mg Ni, 1000 mg Pb and 3000 mg Zn per kg DW) in the Hungarian Technical Directive (1990) for sludge application on agricultural land. Original low-metal sludge was used as the control treatment. Additions of metal nitrates to this sludge were calculated to give metal loadings in the soil equivalent to 10x, 30x, and 100x the permitted loading rates (*L*-values) resulting from an average sludge application practice (500 t/ha sewage sludge with 5% dry matter content is incorporated into a 20-cm surface soil layer). Loadings corresponding to *L*-values were 0.125 mg Cd, 8.33 mg Cr, 1.67 mg Ni, 8.33 mg Pb, and 25 mg Zn per kg soil. Identical amounts of the same sludge were used for each treatment in order to obtain as uniform conditions as possible in terms of such additional factors in the sludge as OM content, nutrient levels, and concentrations of other trace elements. Corn (*Zea mays* L. cv. Favea) and tomato (*Lycopersicon esculentum* Mill. cv. Kecskeméti bőtermő) were grown as test plants until maturity.

At the end of the experiment, plant tops were divided into various parts, and the roots were collected from the four consecutive soil layers from the 0–10, 10–15, 15–20, and 20–30 cm depth intervals. Soil samples were also taken from these layers. Water potentials of the soil samples and relative root distribution were determined. Total potentially available metal concentrations (after 2 mol l^{-1} HNO_3 extraction as described by Andersson, 1976) and plant-available amounts (after acetic ammonium acetate + EDTA-extraction as proposed by Lakanen and Erviö, 1971) were measured in air-dried soil samples; directly plant-available concentrations of the metals were assessed in triplicates in soil solution samples obtained by centrifugation of the initially moist soil samples (Csillag et al., 1995). Metal concentrations in the various plant parts were determined after wet digestion by $HNO_3 + H_2O_2$ in a microwave oven. Elemental concentrations of all plant and soil samples were measured by inductively coupled plasma atomic emission spectrometry.

Evaluation of the method

The experimental conditions in the undisturbed soil monoliths were in some respects closer to the real field situation compared with those in pot experiments or lysimeters. The technique applied for the monolith's preparation has an advantage over other methods: it ensured a very close contact between the outer layers of the soil column and the material of the coating. Part of the coating imbibed the outer macropores, creating a continuum between the soil and the coating; thus, the wall effects were reduced. The metal contamination, together with the sludge, was applied only to the top layer of the soil, as in field situations, instead of being mixed into the whole volume as in most pot experiments. However, this caused some problems in the interpretation of the results, because the larger soil volume, especially the greater depth, allowed the roots to grow beyond the original contaminated upper soil layer. At the higher metal loadings, those plants that barely survived in the initial stages of the experiment, partly recovered later. The onset of recovery presumably coincided with the growth stage when at least some of the roots reached an uncontaminated soil volume. Because of such changes of the growth effects in the same soil column, the determination of any values for phytotoxicity was questionable.

Chemical availability of the metals

One objective of the experiment was to study metal bioavailability by comparing various ecologically significant metal fractions with plant uptake. The three extractants selected gave no direct information on the speciation of the metals, but rather their variously interpreted availabilities: the total potentially available, the presumably plant-available, and the directly plant-available fractions.

Concentrations of the total potentially available metal fraction in contaminated soil reflected the differences among the application rates of the metals (corresponding to the rates permitted in the Hungarian Technical Directive for land application of sludge), and were in the order of Zn > Pb > Cr > Ni > Cd (Table 7.3). In contrast, the AAAc + EDTA extracts released much lower concentration of metals and the trends were also different from the total potentially available metal estimated using the Anderson (1976) method. The EDTA component of the extractant

Table 7.3:   Soil availabilities (SA) of metals in the metal-contaminated layer of soil monoliths

	Control			*Metal-contaminated (100 L Loading Rate)*		
	SA(AE) AAAc+EDTA-available (% total pot. Available)	SA(S) in soil solu-tion (% total pot. Available)	Total potentially available mg kg^{-1}	SA(AE) AAAc+EDTA-available (% of total pot. Available)	SA(S) in soil solu-tion (% of total pot. Available)	Total potentially available mg kg^{-1}
Cd	30.8	*	1.30	60.5	0.79	11.4
Cr	5.4	0.014	13.6	11.3	0.001	687
Ni	18.0	0.028	15.5	44.0	1.73	150
Pb	46.0	*	15.9	61.7	0.001	736
Zn	45.4	0.051	100	58.9	1.37	2092

forms a more stable complex with Cr than with the other 4 trace metals. The stability constants for Cd, Ni, Pb, and Zn-EDTA are between 16.4 and 18.5 (Lindsay, 1979). For Cr(III)EDTA, the stability constant is 24.0, but complexes of Cr(III) attain the equilibrium only very slowly (Dwyer and Mellor, 1964). This might explain the relatively smaller extractability of Cr in the AAAc + EDTA extraction.

Soil solution was separated by centrifugation of fresh moist soil samples (the moisture content of the samples was kept between field capacity and maximum water holding capacity) by applying a centrifugal force that corresponded to the suction exerted by the plants at the conventional wilting point (Csillag et al., 1995). Thus, the soil solution obtained may be considered as energetically directly available for plant uptake. As expected, metal concentrations in the soil solution were many fold lower than the total concentrations, even at the very high, provocative overloading of 100 L. Chromium and Pb concentrations were particularly very low, in agreement with the reported low bioavailability of these elements (Kabata-Pendias and Pendias, 1992; Kabata-Pendias and Adriano, 1995).

The AAAc + EDTA-extractable and the soil solution concentrations expressed as percentages of the total potentially available concentrations of the metals (SA$_{AE}$ and SA$_S$, respectively, Table 7.3) correspond to the soil available factor (SA) used in many instances to describe the distribution and transport of radionuclides in terrestrial ecosystems (Coughtrey et al., 1985). The calculated percentage values characterize the proportion of potentially plant-available (SA$_{AE}$) and the directly plant-available (SA$_S$) metal forms as compared with the total potentially available metal pool of the soil. Metal availabilities were much lower when soil solution concentrations were used in the calculations, than the values calculated from the simple extraction methods. Reported SA values (Coughtrey et al., 1985) generally correspond better with our SA$_{AE}$ than with SA_S values, because literature data are based generally on extraction procedures that yield higher elemental concentrations than those occurring in the soil liquid phase under natural field soil moisture contents. In a study involving the use of several extractants on more than 1000 Hungarian soil samples with the aim to compare the availabilities of the elements in the different methods, AAAc + EDTA-extractable concentrations of the metals were between 39% and 91% of the 0.5 M HNO_3-extractable amounts (Marth, 1990).

Relations between estimated available amounts and actual plant uptake of metals

Accumulation of metals in the aboveground plant tissues slowly increased with increasing metal contamination of the soil but exhibited breakthrough-like increase at the highest metal dose, with the exception of the least-mobile Cr. At the highest metal application rate, the elemental concentrations for each metal, except for Cd, were above the critical tissue levels in tomato, but only Zn concentrations exceeded the critical levels in the metal-tolerant corn variety (Table 7.4).

Table 7.4: Comparison of reported critical tissue levels to shoot metal concentrations in plants grown on soil monoliths treated with metal-enriched sludges

	a	*b*	*c*	*Maize at 100* L	*Tomato at 100* L
			mg kg^{-1}		
Cd	5–10	10–70	5–30	0.67	2.56
Cr	1–2	1–10	5–30	0.94	20.8
Ni	20–30	10–150	10–100	11.1	24.1
Pb	10–20	—	30–300	1.69	46.8
Zn	150–200	100–450	100–400	301	305

Notes:
[a]Kloke et al. (1984), threshold values for sensitive species
[b]MacNicol and Beckett (1985)
[c]Kabata-Pendias and Pendias (1992)

The concentration of Zn in maize shoots was directly proportional to the (AAAc + EDTA)-available soil Zn contents at the lower Zn concentrations, but increased sharply at the highest Zn level (Fig. 7.4a). In contrast, plant Zn concentrations were directly proportional to the logarithm of soil solution Zn concentrations in the entire investigated Zn application range (Fig. 7.4b). Similar relationships were found for Cd. It is reasonable to expect plant uptake to be related to the soil solution concentration of the metals, which is regarded also as an indicator of the mobile pool of metals in soils (Kabata-Pendias and Adriano, 1995). Concentrations of the elements in the liquid phase depend not only on

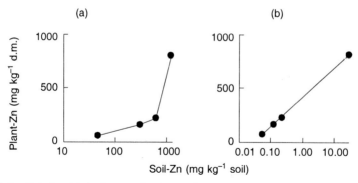

Fig. 7.4: Relationship between plant tissue Zn in maize shoots and (a) (AAAc+EDTA)-extractable soil Zn, or (b) directly plant-available Zn in the soil solution. For the latter, $Zn_{plant} = 268 \times \log Zn_{soil\ solution} + 406$, $R^2 = 0.9995$.

total loading rates and the chemical conditions (pH, redox status, etc.) but also on physical properties such as soil water content and drying-rewetting periods (Csillag et al., 1994). Although the soil moisture conditions were kept fairly constant in the experiment, the soil solution concentrations should be regarded only as "snapshots" at a certain point of time, whereas plant metal concentrations reflect a cumulative availability. Thus, the observed relationship between the plant and soil solution concentrations for the mobile elements is remarkable.

Concentration Factors for Soil-to-plant Transfer of the Metals

Calculation of concentration factors (CFs) involving transfer of radionuclides from the soil to plants and further into the foodchain is universally used in radioecological research and risk assessment (Coughtrey et al., 1985). Risk assessment for sludge application may similarly use concentration factors for the trace elements contained in the sludge. These CFs can be calculated either as overall CFs for larger units of the ecosystems (e.g. soil to plant), or for each consecutive step in the route of the trace elements from the sludge to humans. Experimentally, transfer factors provide a useful measure of the overall efficiency of contaminant movement between the compartments of the ecosystem.

In the present experiments on soil monoliths, simple calculations of the plant:soil metal concentration ratios gave varying results. The slopes calculated from the linear equations should be regarded to correspond better to CFs. The concentration factor of Zn, based on soil solution concentration, was »270 (Fig. 7.4). Reported values are in the range of 0.05-8.0 in natural vegetation, in vegetables, and in cereal grains (Coughtrey et al., 1985). However, the much lower soil solution concentration of Zn explains the difference, as literature data are based mostly on total soil Zn contents. Distribution coefficients of Zn between the soil solids and the soil solution are on average about 20, but may exceed 100 (Coughtrey et al., 1985). In the present experiment SA(S) value for Zn at the 100 L loading rate was 1.37% (Table 7.3). This value fits into the above-mentioned distribution range.

Macrocosms: Field Experiments

Advantages and limitations

Field experiments represent the reality for a given soil-plant system. However, extrapolation of the results to other sites with widely different properties is, in most cases, questionable. The fate of the added trace elements (or any added element) in the soils is determined by adsorption-desorption, precipitation, dissolution, complexation, and decomplexation processes, which involve both inorganic and organic constituents, and which in turn depend on soil pH and environmental conditions (temperature, soil moisture), and exist in a complex system of dynamic equilibria. All of these processes are influenced by the chemical and physical characteristics of the soils, and the variability of these characteristics make the universal applicability of the results obtained at a given site to other different soil types nearly impossible. However, comparison of identically designed experiments conducted over long-term at several different sites may yield better knowledge about the behavior of trace elements.

Long-term experiments involving the application of trace-element-containing sewage-sludge in agriculture or forestry were initiated in the 1970s (Chang et al., 1983, 1997; Juste and Mench, 1992; Csatho, 1994). However, the number of experiments that can be regarded as long-term (i.e. at least 10 years in duration) is limited, and accessibility of the results in international literature is worse. Most of the studies were conducted in USA and in the EU member states. Variations in the mode, pattern and rate of sewage sludge application, in the quality of the sludge, in the soil and climatic factors and in the cultural practices at the various sites often make the comparison among the results of the separate experiments difficult, and may be the reason for conflicting results (Juste and Mench, 1992). Despite heterogeneous results the following, general conclusions may be drawn from the long-term experiments.

Accumulation of metals in the soil from application of metal-containing sludge can be considered permanent, except for quantities taken up by the plants and transported from the site after the harvest, because movement of the trace elements from the site of application was minimal in most cases. However, the possibility of increased mobility of the elements in the future (e.g. as a consequence of acidification of the environment) may not be totally excluded. Slow, constant changes of the environment as a consequence of specific economic practices may have an impact of unknown dimensions on soil processes, and the small quantitative changes may give way to a relatively fast, qualitatively different situation (time-bomb effect). Thus, adsorbed toxic chemicals from heavily polluted soils and sediments may be released if the soil's capacity to bind contaminants is exceeded (Stigliani, 1988).

The main reason for the controversies surrounding land application of sewage-sludge (Chang et al., 1992, 1997; Ryan and Chaney, 1994; McGrath et al., 1994; McBride, 1995; Schmidt, 1997) and are that the USEPA regulations are based partly on extrapolations from short-term, controlled experiments conducted in greenhouses using inorganic forms of the metals rather than studying the long-term behavior of sludge-applied metals under field conditions. The recent debates on the proper risk assessment stresses the importance of conducting comparable long-term field trials. Comparability of the risk assessment in the different countries (i.e. harmonization of their regulations) requires that characteristics of the soil and soil solution equilibria that determine the phytoavailability of the metals should not be encompassed in a single value for all the various soil types and different environmental circumstances, but there should be site-specific determinations. For this reason, long-term trials cannot be avoided (Schmidt, 1997).

Results by Chang et al. (1997), derived from 15 years of field experiments in which the sludge application rate reached 2880 tonne per hectare, show no evidence of the time-bomb effect, but the analysis of these long-term data revealed the existence of the necessary conditions for a time bomb. Although the authors stated that "this experiment has pushed the cumulative loading of sewage sludge beyond what would be expected of normal practice, and it represents a worst case scenario in land application of sewage sludge in terms of pollutant loading", they were not sure how long the field observations must continue for the "time bomb" to appear.

Case Study: Long-term, Small-plot Field Experiment with 13 Trace Elements

Description of the experimental set-up

The experiment has been conducted (two replicates) on a total of 104 plots on a calcareous chernozem soil in Hungary with the aim to study the effects of 13 potentially toxic elements on various crops and their potential harm in domestic animals fed by crop products.

The soil was of a loamy texture, particles of a diameter between 0.02 and 0.05 mm being the most abundant fraction (35%-50%), and clay content (diameter of particles <0.002 mm) about 20%. The main types of clay minerals were illite (47%), chlorite (29%), smectite (16%), and illite-smectite + illite-chlorite (8%) (Fuleky, 1987). Soil pH was between 7.2 and 7.5, $CaCO_3$ content 5% and soil organic matter content 3%. Saturation with the basic cations was significant (40 cmol kg^{-1}), with Ca (80%) and Mg (16%) as the main cations, and 3% K and 1% Na. The soluble salt content was negligible (1 cmol kg^{-1}), with Ca^{2+}, Mg^{2+}, HCO_3^- and SO_4^{2-} as the main ionic forms.

The experimental plots (21 m^2 each) were contaminated in 1991 by $AlCl_3$, As_2O_3, $BaCl_2$, $CdSO_4$, K_2CrO_4, $CuSO_4$, $HgCl_2$, $(NH_4)_6Mo_7O_{24}$, $NiSO_4$, $Pb(NO_3)_2$, Na_2SeO_3, $SrSO_4$, or $ZnSO_4$ at rates of 0 (10 for As, Cd, Hg and Se), 30, 90 and 270 mg element kg^{-1} soil. The experiment was designed in split-plot arrangement, with the 13 elements constituting the main plots and the four doses the subplots. Each treatment was replicated twice. The total net area of the 104 experimental plots was 2184 m^2, with an additional 2008 m^2 serving as buffer strips between the plots. The plots received a uniform basic NPK fertilization each year from 1991, and a different crop was grown on the contaminated site, with usual agronomical practices in Hungary.

The major objectives of the study were to

- study the behavior of the elements in the soil (sorption, leaching, volatilization, etc.);
- determine the effect of metal contamination on the soil biological activity and on the population of the microflora and macrofauna;
- assess the effect of metal loading on the plants (yield, quality, resistance against pests, weediness, interaction with environmental stress factors);
- measure plant uptake into and accumulation of the metals in the various plant parts; and
- follow the effect of the contaminants that entered the plants, on domestic animals in feeding experiments.

Phenological parameters of the crops were constantly monitored, and several growth parameters and metal contents of the various plant parts at various developmental stages (after wet digestion of the dried and ground plant samples with cc. HNO_3 + H_2O_2 in a microwave oven) have been measured. The size of the experiment allows additional studies on the availability of the metals from the contaminated plants to various animal species in feeding experiments (Bokori et al., 1994). Also, composite soil samples (from 20 subsamples) were taken each year from the ploughed layer to assess the availability of the metals, using the AAAc + EDTA extraction method (Lakanen and Erviö, 1971). Every three years, soil samples have been taken from greater depths to follow movement of the metals in the soil profile by a similar extraction procedure. The total element content of the soil samples was also determined in 1992, using a

$HNO_3 + H_2O_2$ digestion method. A detailed description of the first four years of the experiment is given in Kádár (1995).

Effects of trace element contaminants on crop yields

There were significant differences among the behavior and phytotoxicity of the 13 elements. The soil chemical properties influenced greatly the potency of the elements: a universal feature of the experiment was that in the calcareous chernozem those elements that were applied in anionic form were the most toxic and caused significant growth reductions. This may be explained by the influence of the soil pH (7.4).

During the first six years of the experiment, only As, Cd, Cr, Hg, Mo, Pb, and Se showed some phytotoxicity (Mo and Pb only in the first year, Cd only from the fifth year). Effects of the phytotoxic treatments on the yields and metal concentrations causing 20% yield reductions in the edible parts of the crops are summarized in Table 7.5. Maximal As rates (>90 mg kg^{-1} soil) were always toxic, but lower rates caused little yield reductions. A possible explanation for the late appearance of Cd toxicity might be a higher Cd-sensitivity of recently grown crops (beetroot and spinach) than of crops grown earlier. High Cr levels had devastating effects in the first two years, but its phytotoxicity decreased with time, probably because it leached to greater depths. Toxicity of Mo in the first year might be related to the high amount of NH_4^+ ions applied in the Mo salt. This effect later disappeared because excess NH_4^+ was eliminated by its turnover in the N transformation processes of the soil and by subsequent leaching. Selenium was strongly inhibitory in each year and for each plant species. Selenium mobility and toxicity seemed not to decrease but rather to increase with time. This might be explained by transformation of Na-selenite into more toxic Ca-selenate in the oxidative environment of this well-aerated calcareous soil.

Mobility of Cr forms in the soil profile

The decrease in Cr toxicity was found to be related to the leaching of mobile Cr forms in the soil profile. Not surprisingly, three years after contamination, the total Cr concentrations in the most highly contaminated plots (original rate 270 mg Cr kg^{-1} soil) were significantly higher than in the control, in soil samples taken at the 20–40 and 40–60 cm depths (Fig. 7.5). In the ploughed layer, about one-third (100–120 mg kg^{-1}) of the originally applied Cr could be detected in the concentrated $HNO_3 + H_2O_2$ extracts three years after the contamination, and about half of this amount was found in each of the next two 20-cm depth increments.

Table 7.5: Effects of phytotoxic treatments on crop yields in a long-term field experiment

	Elements applied in spring of 1991, mg kg⁻¹				LSD 5%	20 % yield
	0/10*	30	90	270		reduction at
	(t yield ha⁻¹)	(t yield ha⁻¹)	(t yield ha⁻¹)	(t yield ha⁻¹)		mg kg⁻¹ soil
	1991 (maize grain, air-dry)					
As*	7.6	8.6	7.9	6.9		>> 270
Cr	8.1	5.2	1.9	1.6		12
Mo	8.5	8.4	7.4	4.7	1.5	130
Pb	8.9	8.4	7.8	6.4		180
Se*	6.9	7.6	5.7	4.3		110
	1992 (carrot root at harvest)					
As*	17.6	15.1	19.0	13.3		190
Cr	13.0	7.1	-	-		12
Hg	18.6	15.3	13.8	10.8	4.8	40
Se*	12.8	14.4	7.2	-		65
	1993 (potato tuber at harvest)					
As*	12.1	14.4	11.1	10.2		> 270
Cr	12.0	11.3	7.9	4.9	3.5	60
Hg	11.2	9.3	8.0	7.9		35
Se*	12.5	10.5	3.8	1.5		30
	1994 (pea seeds at harvest, air-dry)					
As*	2.4	2.6	2.3	0.4		130
Cr	2.5	2.0	1.9	1.6	0.8	30
Se*	3.4	2.4	-	-		20
	1995 (beetroot at harvest)					
As*	10.9	12.3	12.8	9.4		> 270
Cd*	7.4	9.5	3.7	0.7		61
Cr	14.6	12.8	9.4	2.2	7.2	50
Se*	11.5	8.9	-	-		26
	1996 (spinach at harvest, air-dry shoots)					
As*	5.3	3.8	5.3	3.8		> 270
Cd*	3.6	4.0	3.6	-	1.9	140
Cr	5.5	6.1	6.3	4.7		> 270
Se*	5.4	5.3	-	-		40

Note: * indicate the first contaminant level listed in Page 231

Concentration of the plant-available, i.e. (AAAc + EDTA)-extractable Cr forms, showed an increase with increasing depth. In the upper 20 cm it constituted only 6% of the total Cr, but its relative abundance increased to 25% in the layer between 40 and 60 cm. Even sharper increases were found in the proportion of the water-soluble Cr forms. Although only 1% of the total Cr content was present in the water-soluble form in the ploughed layer, the proportion of water-soluble Cr forms increased to 14% in the next 20 cm, and to 24% in the layer between 40 and 60 cm. Thus, between 40 and 60 cm, practically 100% of the plant-available Cr was in the water-soluble form.

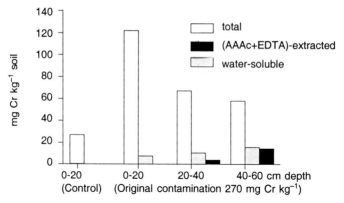

Fig. 7.5: Downward movement of Cr included in a calcareous Chernozem soil 3 years after contamination of the ploughed layer (0–20 cm) by K_2CrO_4 application.

It must be stressed that Cr was originally applied in the hexavalent (Cr(VI)) form, as K_2CrO_4 to the calcareous soil (pH 7.4). Although most of the Cr entering the soil is quickly adsorbed on clay particles or on OM by specific ion effects (Coughtrey and Thorne, 1983) and the Cr(VI) forms are usually easily transformed to Cr(III) (Cary et al., 1977), under certain circumstances Cr(VI) may remain in the soil in appreciable quantities (Bartlett and James 1979). Naidu et al. (1997) indicated that appreciable amounts of Cr contamination of the soil can be recovered as hexavalent Cr in the groundwater when the pH of the water collected from boreholes was slightly alkaline. From the point of environmental risk, these anionic forms of Cr have the greatest significance, because they can move more readily through the vadose zone toward the underlying groundwater systems, are highly toxic for most living organisms, and are also easily transported within the living compartments of the ecosystems. Further, deep core soil samplings are necessary to follow the downward leaching of the water-soluble Cr forms.

Uptake and transport of Zn and Cr in plants

Field experiments provide a better opportunity than pot experiments to follow the accumulation of the elements within the plant tissue. Repeated sampling of the same plant material on the same plot several times during the growth period may provide more reliable data than using plant material from a new pot for each sampling. The elemental composition of plant tissue differs among plant species, among parts of the same plant, and changes during the development. Such factors are standardized when plant analysis data are used to establish the nutrient requirement of the plants. A similar standardization is necessary also to obtain comparable results on the toxic elemental content of the plants for risk assessment.

Accumulation of the essential microelements and the toxic contaminants in the various plant parts may follow different patterns. The plants have various regulatory mechanisms for acquisition and transport of the essential elements (Marschner, 1991), whereas exclusion of the harmful elements from places of intense metabolic activity is one of the defense mechanisms of plants against phytotoxic environmental effects (Peterson, 1983).

Zinc as an essential micronutrient was taken up readily by maize grown in the first year after soil contamination. Zinc concentration in the shoots was higher in the actively growing young plants at the 4–6 leaf stage than in the mature plants at harvest, when Zn was transported to the developing grain (Fig. 7.6). Zinc content increased with increasing Zn application rate not only in the vegetative tissue, but also in the grain, although to a lesser extent. The 40 mg kg^{-1} Zn concentration of the maize seeds at the highest Zn dose was near to the limit value for using the crop as feed or food (permitted Zn content of flour or ground cereals of 30 mg kg^{-1} dry matter).

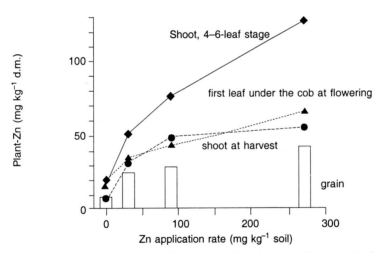

Fig. 7.6: Relationshop between Zn concentration in maize tissue and soil application rate of Zn in a field experiment on a calcareous Chernozem soil. (Zinc was mixed as $ZnSO_4$ into the ploughed layer prior to sowing.)

In contrast, Cr is generally known to be poorly relocated from the roots to the aboveground parts. However, in the present experiment, Cr concentrations of maize shoots, although relatively low in absolute value (max. 5 mg kg^{-1} DW), increased substantially during the development of the plants, being higher at harvest than at the 4–6 leaf stage (Fig. 7.7). In contrast, the reproductive tissues were defended against accumulation of this nonessential element. Chromium was totally excluded from the grain, and even the first leaf under the cob which nourishes the developing seeds had nearly negligible Cr concentration, even at the highest Cr dose which severely impaired the growth and yield of maize.

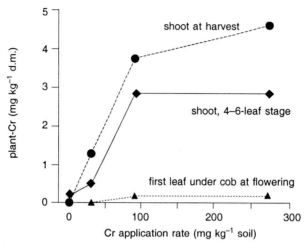

Fig. 7.7:   Relationship between Cr concentration in maize tissue and soil applica-tion rate of Cr in a field experiment on a calcareous Chernozen soil. (Chromium was mixed as K_2CrO_4 into ploughed layer prior to sowing.)

CONCLUSIONS

For assessing risk, it is essential to know the movement of the metals among the various compartments of the environment in order to calculate the predicted environmental concentrations (PECs) for the different compartments. Movement of the metals is determined by soil and plant factors and speciation in soil. The accumulation of these elements by plants depends on their bioavailability in soils, but is largely determined by plant physiological-biochemical processes.

Detailed knowledge of the fundamental processes involved in the above outlined movement of metals in soil-plant systems will be ideal to support risk assessment. A simpler and more practical approach would be to estimate the transport of the metals on the basis of their availability for the various environmental compartments. Ultimately, availability of the metals to the plants can be determined on the basis of actual plant uptake. However, several methods were introduced with the aim to relate the plant metal contents to various fractions of the soil metal contents. These methods, at best, can only characterize the plant availability of the metals.

Nonetheless, a good correlation among the plant and soil metal contents is conducive to effective risk assessment.

The experimental techniques described in this chapter illustrate the various approaches of these investigations to find such correlations:

- Pot experiments are useful to evaluate the effects of excessive pollution rates, which for safety reasons cannot be studied easily in the field, and to compare efficacy of many ameliorants (only lime was selected for discussion here) in a time- and labor-effective manner.
- Experiments on large soil columns (lysimeters or undisturbed) are necessary to follow the movement of the metals among the various available fractions. Also, the larger soil volume in these experiments allowed the separation of the soil solution at natural soil moisture contents.
- The field experiments are indispensable to follow the transport of the metals not only into the various crops but also further into the food chain to follow their downward migration and to assess the long-term fate and effects of the metals under real circumstances. Comparison of plant metal contents in the various organs and at several sampling dates facilitates the selection of a proper sampling procedure for toxicity and risk assessment.

ACKNOWLEDGEMENTS

The undisturbed soil column experiments were carried out as part of PSTC 12.056E project "Fate and Plant-Uptake of Toxic Elements in Cropped Soil Profiles" and were supported under grant No. DHR-5600-G-1056-00, Program in Science and Technology Cooperation, Office of the Science Advisor, U.S. Agency for International Development. The research by Drs. Knox and Adriano was supported by financial assistance through Contact Number DE-FC09-96SR 18546 between the US Department of Energy and the University of Georgia Res. Foundation. Financial help of the Soros Foundation, Hungary, for K.Bujtás under grant 247/2/7001 during evaluation of results and preparation of this paper is gratefully acknowledged.

REFERENCES

Alloway, B.J. (1990) *Heavy Metals in Soils*. Blackie and J. Wiley and Sons, Glasgow, 339 pp.

Andersson, A. (1976) On the determination of ecologially significant fractions of some heavy metals in soils. *Swedish J. Agric. Res.* **6**: 19–25.

Atanasiu, N. (1966) Pflanzenphysiologische Verfahren. In *Handbuch der Pflanzenernährung und Düngung.II.Band*, Scharrer, K., and Linser, H. (Eds). Springer Verlag, Wien-New York, pp. 844–873.

Bartlett, R.J., and James, B. (1979) Behaviour of chromium in soils. III. Oxidation. *J. Environ. Qual.* **8**: 31–35.

Beckett, P.H.T., and Davis, R.D. (1977) Upper critical levels of toxic elements in plants. *New Phytol.* **79**: 95–106.

Bokori, J., Fekete, S., Kádár, I., Albert, M., and Koncz, J. (1994) Effect of Cd load on the Cd content of eggs. In *Proc. 6th Int. Symp. New Perspectives in the Research of Hardly Known Trace Elements*, Pais, I., (Ed), pp. 183–188. Budapest, Hungary.

Bujtás, K., Csillag, J., Pártay, G., and Lukács, A. (1995) Distribution of selected metals in a soil-plant experimental system after application of metal-spiked sewage sludge. In *Proceedings of the XXVth Annual Meeting of ESNA/UIR, Castelnuovo Fogliani (Piacenza), Italy. Working Group 3, Soil-Plant-Relationships*, Gerzabek, M.H., (Ed), Österreichisches Forschungszentrum Seibersdorf Ges.m.b.H., Austria, pp. 98–104.

Cary, E.E., Allaway, W.H., and Olson, O.E. (1977) Control of chromium concentrations in food plants. II. Chemistry of chromium in soils and its availability to plants. *J. Agric. Food Chem.* **25**: 305–309.

Chang, A.C., Page, A.L., Warneke, J.E., Resketo, M.R., and Jones, T.E. (1983) Accumulation of cadmium and zinc in barley grown on sludge-treated soils: A Long-Term Field Study. *J. Environ. Qual.* **12**: 391–397.

Chang, A.C., Warneke, J.E., Page, A.L., and Lund, L.J. (1984) Accumulation of heavy metals in sewage sludge-treated soils. *J. Environ. Qual.* **13**: 87–91.

Chang, A.C., Granato, T.C., and Page, A.L. (1992) A methodology for establishing phytotoxicity criteria for chromium, copper, nickel, and zinc in agricultural land application of municipal sewage sludge. *J. Environ. Qual.* **21**: 521–536.

Chang, A.C., Hyun, H., and Page, A.L. (1997) Cadmium uptake for Swiss chard grown on composted sewage sludge treated field plots: plateau or time bomb? *J. Environ. Qual.* **26**: 11–19.

Chlopecka, A., Bacon, J.R., Wilson, M.J., and Kay, J. (1996) Forms of cadium, lead, and zinc in contaminated soils from southwest Poland. *J. Environ. Qual.* **25**: 69–79.

Coughtrey, P.J., and Thorne, M.C. (1983) Chromium. In *Radionuclide Distribution and Transport in Terrestrial and Aquatic Ecosystems*, Vol.II, Balkema, A.A., Rotterdam/Boston, pp. 1–40.

Coughtrey, P.J., Jackson, D., and Thorne, M.C. (1985) *Radionuclide Distribution and Transport in Terrestrial and Aquatic Ecosystems. A Compendium of Data*, Vol.6, Balkema, A.A., Rotterdam/Boston, 194pp.

Csathó, P. (1994) Contamination of the Environment with Heavy Metals and its Consequences on Agricultural Production. A throughout literature survey, MTA TAKI, Budapest, 176 p. (Hungarian).

Csillag, J., Lukács, A., Molnár, E., Bujtás, K., and Rajkai, K. (1994) Study of heavy metal overloading in soils in a model experiment. *Agrokémia és Talajtan* **43**: 196–210.

Csillag, J., Tóth, T., and Rédly, M. (1995) Relationships between soil solution composition and soil water content of Hungarian salt-affected soils. *Arid Soil Research Rehabilitation* **9**: 245–260.

De Vries, M.P.C., and Tiller, K.G. (1978) Sewage sludge as a soil amendment with special reference to Cd, Cu, Mn, Ni, Pb and Zn - comparisons of results from experiments conducted inside and outside a glasshouse. *Environ. Pollut.* **16**: 231–240.

Dowdy, R.H., and Volk, V.V. (1983) Movement of heavy metals. In *Proc. Symp. on Chemical Mobility and Reactivity in Soil Systems*, Nelson, D.W. *et al.* (Eds). SSSA Spec. Publ. No. 11., Madison, WI, USA, pp 229–240.

Dwyer, F.P., and Mellor, D.P. (Eds.) (1964) *Chelating Agents and Metal Chelates*, Academic Press, New York/London, 530pp.

Füleky, G. (1987) Potassium supply in typical soils of Hungary. *Bull. Univ. Agric. Sci. Gödöllő.* **1**: 113–119.

Homeyer, B., Labenski, K.O., Meyer, B., and Thormann, A. (1981) Herstellung von Lysimetern mit Boden in naturlicher Lagerung (Monolith-Lysimeter) als Durchlauf-, Unterdruck- oder Grundwasserlysimeter. *Z. Pflanzenernähr.Bodenkd.* **136**: 242–245.

Hungarian Technical Directive (1990) *Land- and forest applications of waste waters and sewage sludges.* MI-08-1735–1990. (In Hungarian).

Juste, C., and Mench, M. (1992) Long-term application of sewage sludge and its effects on metal uptake by crops. In *Biogeochemistry of Trace Metals*, Adriano, D.C. (Ed.), CRC Lewis Publishers, Boca Raton, USA, pp. 159–193.

Kabata-Pendias, A., and Adriano, D.C. (1995) Trace Metals. In *Soil Amendments and Environmental Quality*, Rechcigl, J.E., (Ed.), CRC Lewis Publishers, Boca Raton, USA, pp 39–167.

Kabata-Pendias, A., and Pendias, H. (1992) *Trace Metals in Soils and Plants.* CRC Press, Boca Raton, FL, 365 p.

Kádár, I. (1992) *Principles and Methods in Plant Nutrition.* MTA TAKI. 398pp (Hungarian).

Kádár, I. (1995) *Contamination of the Soil-Plant-Animal-Man Food-Chain with Chemical Elements in Hungary.* Ministry for Environmental Protection and Land Development RISSAC, Budapest, 387pp. (Hungarian).

Kiekens, L. (1984) Behavior of heavy metals in soil. In *Utilization of Sewage Sludge on Land: Rates of Application and Long-term Effects of Metals*, Berglund, S., Davis, R.D., and Hermite, P.L., (Eds.) D.Reidel, Dordrecht. pp.126–134.

Kloke, A., Sauerbeck, D.R., and Vetter, H. (1984) The contamination of plants and soils with heavy metals and the transport of metals in terrestrial food chains. In *Changing Metal Cycles and Human Health*, Nriagu, J.O. (Ed.), Dahlem Konferenzen, Springer Verlag, Berlin, pp. 113–141.

Lakanen, E., and Erviö, R. (1971) A comparison of eight extractants for the determination of plant available micronutrients in soils. *Acta Agr. Fenn.* **123**: 223–232.

Legret, M., Divet, L., and Juste, C. (1988) Migration et spéciation des metaux lourd dans un sol soumis à des epandages de boues de station d'epuration à très forte charge en Cd et Ni. *Wat. Res.* 22. 953–959.

Lindsay, W.L. (1979) *Chemical Equilibria in Soils.* John Wiley & Sons, New York.

MacNicol, R.D., and Beckett, P.H.T. (1985) Critical tissue concentrations of potentially toxic elements. *Plant Soil* **85**: 107–129.

Marschner, H. (1991) Plant-soil relationships: acquisition of mineral nutrients by roots from soils. In *Plant Growth: interactions with nutrition and environment.* Porter, J.R., and Lawlor, D.W., (Eds.) Cambridge University Press, UK., pp. 125–155.

Marth, P. (1990) *Comparative study on soil extractants.* Postgraduate Thesis, Agricultural University of Gödöllő, (manuscript in Hungarian).

McBride, M.B. (1995) Toxic metal accumulation from agricultural use of sludge: are USEPA regulations protective? *J. Environ. Qual.* **24**: 5–18.

McGrath, S.P., Chang, A.C., Page, A.L., and Witter, E. (1994) Land application of sewage sludge: scientific perspectives of heavy metal loading limits in Europe and the United States. *Env. Reviews* **2**: 1–11.

Naidu, R., Mahimairaja, S., Mowat, D., Cox, J., Kookana, R.S., McLaughlin, M.J., and Ramasamy, K. (1997) Fate of chromium in contaminated soils: I. Evidence for migration of Cr. In *Proc. of Fourth Internat. Conf. on Biogeochemistry of Trace Elements*, Iskandar, I.K., Hardy, S.E., Chang, A.C., and Pierzynski, G.M., (Eds.) pp. 743–744. Berkley, California, USA.

Németh, T., Pártay, G., Buzás, I., and Mihályné, H.Gy. (1991) Preparation of undisturbed soil monoliths. *Agrokémia és Talajtan* **40**: 236–242. (Hungarian).

Németh, T., Radimszky, L., Fehér, J., and Simonffy, Z. (1991b) Movement and distribution of 15N labelled nitrogen in undisturbed soil columns. In *Proc. Symp. Stable Isotopes in Plant Nutrition, Soil Fertility and Environmental Studies, Vienna, 1990*, IAEA, Vienna, pp. 338–341.

Page, A.L., and Chang, A.C. (1978) Trace elements impact on plants during cropland disposal of sewage sludges. In *Proceedings of the Fifth National Conference on Acceptable Sludge Disposal Techniques*, Information Transfer Inc., Rockville, Maryland, pp. 91–96.

Page, A.L., Bingham, F.T., and Chang, A.C. (1981) Cadmium. In *Effect of Heavy Metal Pollution on Plants. Vol.1. Effects of Trace Metals on Plant Function. Pollution Monitoring Series*, Lepp, N.W. (Ed.) Applied Science Publishers, London and New Jersey, pp. 77–109.

Peterson, P.J. (1983) Adaptation to Toxic Metals. In *Metals and Micronutrients: Uptake and Utilization by Plants*, Robb, D.A., and Pierpoint, W.S., (Eds.). Academic Press Inc. (London) Ltd, pp. 51–69.

Pusztai, A., and Kádár, I. (1980) Investigations on the fate of nitrogen on a chernozem soil in a model experiment. *Agrokémia és Talajtan* **29**: 252–272. (Hungarian).

Ryan, J.A., and Chaney, R.L. (1994) Development of limits for land application of municipal sewage sludge: Risk assessment. In *Transactions of International Congress Soil Science Society., 15th, Acapulco, Mexico, 10-17 July 1994*. International Society for Soil Science, Vienna, Austria. Vol. 3a, pp. 534–553.

Sanders, J.R., McGrath, S.P., and Adams, T.Mc.M. (1986) Zinc, copper and nickel concetnrations in ryegrass grown on sewage sludge-contaminated soils of different pH. *J. Sci. Food. Agric.* **37**: 961–968.

Schmidt, J.P. (1997) Understanding phytotoxicity thresholds for trace elements in land-applied sewage sludge. *J. Environ. Qual.* **26**: 4–10.

Smith, S.R. (1994) Effect of soil pH on availability to crops of metals in sewage sludge-treated soils. I. Nickel, copper and zinc uptake and toxicity to ryegrass. *Environ. Pollut.* **85**: 321–327.

Shuman, L.M. (1985) Fractionation method for soil microelements. *Soil Sci.* **140**: 11–22.

Sigmond, E. (1904) *Agricultural Chemistry*. Society for Natural Sciences, Budapest, (Hungarian).

Stigliani, W.M. (1988) Changes in valued "capacities" of soils and sediments as indicators of nonlinear and time-delayed environmental effects. *Environ. Monitor. Assessm.* **10**: 245–307.

Tessier, A., Campbell, P.G.C., and Bisson, M. (1979) Sequential extraction procedure for the speciation of particulate trace metals. *Anal. Chem.* **51**: 844–850.

Wagner, P. (1883) Beiträge zur Ausbildung der Düngerlehre. *Landwirtsch.Jahresblatt* **12**: 601–643.

SECTION C
CASE STUDIES

8

Effects of Mine Wastewaters on Freshwater Biota in Tropical Northern Australia

Scott J. Markich[1]*, Ross A. Jeffree*[2]*, David R. Jones*[3] *and Sven R. Sewell*[*4]

INTRODUCTION

Metals are ubiquitous in the environment. Many are essential for mediating biological functions in biota. Essential trace metals, such as Co, Cu, Fe, Mn, Ni, and Zn, are involved in many of the enzymatic and metabolic reactions that take place within an organism, either as a component or an activator of enzymes (Lehninger, 1982). However, some metals, such as Hg, Pb, and U, have no known biological function, and are therefore classified as non essential. All metals, whether essential or non-essential to an organism, become toxic beyond certain threshold concentrations (Depledge et al., 1994).

The sustained input of metals into aquatic environments has resulted in a range of ecological problems on local, regional, and global scales. Mining and processing of metal ores can be an important contributor to metal pollution in aquatic ecosystems. Due to the proximity of mining operations to freshwater environments, the latter are potentially at risk of metal pollution (Nriagu, 1990; United Nations, 1997). Indeed, the literature is replete with examples of ecological disturbance, or even total degradation, of freshwater environments due to the toxic effects of metals from mining operations (Jeffree and Williams, 1975; Kimmel et al., 1981; Fucik et al., 1991; He et al., 1998; Malmqvist and Hoffsten, 1999; Cherry

[1] Environment Division, Australian Nuclear Science and Technology Organization, Private Mail Bag 1, Menai, New South Wales 2234, Australia
[2] Environment Division, Australian Nuclear Science and Technology Organization, Private Mail Bag 1, Menai, New South Wales 2234, Australia
[3] Earth Water Life Sciences, P.O. Box 39443, Winnellie, Northern Territory 0821, Australia
[4] ERA, Ranger Mine, Locked Bag 1, Jabiru, Northern Territory 0886, Australia
[*] Present address. EcOz, PO Box 449, Christmas Island, Indian Ocean 6798.

et al., 2001). Although much attention has been focussed on trace metals by virtue of their intrinsic toxicity, it should also be recognized that an input of major ion salts can have an adverse impact on the chemical limnology of a pristine freshwater system (Hart et al., 1991). Fine particulate matter eroded from mined areas can also have a negative impact on receiving waters. Although not toxic per se, unless it contains elevated levels of metals, particulate matter can be detrimental to pelagic and benthic biota by virtue of exclusion of light and smothering of biota (Lloyd et al., 1987; Newcombe and McDonald, 1991; Stowar, 1997).

Two case studies from tropical northern Australia have been selected to illustrate contrasting mine site environmental management practices, particularly with respect to mining impacts on downstream aquatic biota. The first case study describes a decommissioned and rehabilitated copper-uranium mine (Rum Jungle), which caused severe downstream ecological impact due to acid mine drainage from the site. The second case study describes the extensive chemical and biological monitoring programs that have been developed and implemented at an operating uranium mine (Ranger) to ensure that no adverse downstream ecological impact occurs. The severe ecological impacts observed at Rum Jungle, for example, increased public awareness of environmental issues, and significantly contributed to the rigorous environmental management practices currently implemented at Ranger.

Both mine sites have been extensively studied over the last 25 years, and hence provide a long-term dataset to quantify both actual and predicted ecological effects in relation to measured contaminant (e.g. metal and/or pH) levels. This has led to an increasing number of studies investigating the mechanism(s) of metal-organism interactions (i.e. uptake, toxicity), particularly an understanding of biological responses in relation to aqueous metal speciation, including how water quality variables (e.g. pH, hardness, alkalinity, and dissolved organic matter) may influence trace-metal bioavailability. Such an approach enhances the capability of predicting the biological effects of metals across a range of water chemistry conditions. The measurement of metal speciation and bioavailability permits national numerical guidelines for protecting aquatic ecosystems to be potentially relaxed on a site-specific basis. This is discussed in detail with respect to the responses of aquatic biota to U.

Rum Jungle is a decommissioned copper-uranium mine (Fig. 8.1) operational from 1953 to 1971 (Watson, 1975). At the time of development, there was no legislative requirement for mining projects to undergo environmental impact assessment; therefore, the environmental damage that subsequently resulted from mining operations at Rum Jungle was not of significant concern at the time, but was of enormous concern later. The major environmental impact at Rum Jungle was the pollution of the

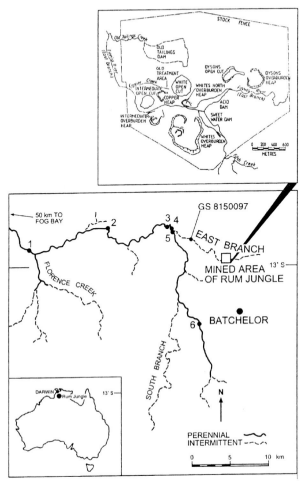

Fig. 8.1: Location map showing the Rum Jungle mine (inset) in relation to the Finniss River system, northern Australia. GS, Gauging station. Stream flow is northerly.

east branch and the main channel of the Finniss River (Fig. 8.1). This had the effect of extending the impact well beyond the boundaries of the actual mining and milling areas. The river has been contaminated by acid mine drainage (principally Cu, Zn, Mn, U, Al, sulfate, and acidity), a product of oxidation of sulfide minerals in the mine waste, originating from multiple sources (Davy and Jones, 1975). Between 1983 and 1985, the site was rehabilitated by the Australian Government at a cost of A$18 million (Allen and Verhoeven, 1986). This resulted in a substantial (ca. 70%) reduction in the measured annual loads of Cu, Zn, Mn, and sulfate entering the Finniss River (Richards et al., 1996; Lawton, 1998). These reductions in contaminant loads have been followed by an increase in the diversity and abundance of freshwater biota downstream from the mine (Jeffree and Twining, 2000; Jeffree et al., 2001).

Ranger is an operating uranium mine (Fig. 8.2), which commenced mining and milling operations in 1981. It is located a short distance upstream from the floodplain of Magela Creek (Fig. 8.2) and is surrounded by the World Heritage listed Kakadu National Park. In contrast to Rum Jungle, the Ranger mine is monometallic, with U being the only metal present in significant concentrations in the orebodies (Kendall, 1990). Uranium is of potential ecotoxicological significance in the mine wastewaters. The highest quality water stored on site may be released into Magela Creek during the wet season (December to April), subject to strict legislative requirements and regulations (NT DME, 1982). Indeed, over 57 acts of Parliament and associated regulations cover mining and

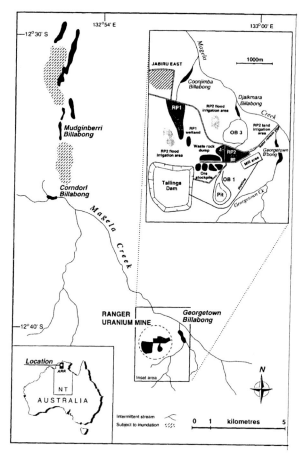

Fig. 8.2: Location map showing the Ranger uranium mine (inset), adjacent to Magela Creek, Northern Australia. OB, Orebody. RP, Retention pond. Stream flow is northerly.

milling operations at Ranger (Wedd, 1989). It is arguably the most regulated mine in Australia. Biological monitoring of Magela Creek and toxicity evaluation of the mine wastewaters has been undertaken on a continuous basis for some 20 years. Thus far, there have been no detectable impacts of mine wastewaters on aquatic ecosystems downstream of the Ranger mine.

The Rum Jungle and Ranger sites are located in a tropical monsoonal environment, which is characterized by a defined summer wet season, with average annual rainfall of about 1500 mm during this period. There is essentially no rainfall for the remainder of the year. During the wet season, stream flows are large, with the potential for substantial dilution of wastewater inputs from both mine sites. However, this type of rainfall pattern means that flow in many streams in the region is ephemeral, with base flow ceasing soon after the end of the wet season. For this reason, inputs of mine wastewaters can potentially have a greater impact at the end of the wet season as base flow recedes, and especially during the long dry season.

This region of northern Australia is also characterized by naturally acidic first flush waters at the start of each wet season. The rainfall at this time is acidic, with pH values as low as 4 having been measured in Kakadu National Park (Noller et al., 1990). The acidity comes from trace levels of organic acids (acetic and formic acid originating from biological activity) as well as nitric oxides produced by intense electrical thunderstorm activity that occurs at the start of the wet season. In addition, products of oxidation of naturally occurring acid sulfate soils, that underlie many of the low-lying coastal flood plain areas, may contribute to the acid load.

From an ecological perspective, the naturally occurring acidic first flush water after the dry season is particularly important as this has often caused extensive fish kills (Brown et al., 1983; Noller and Cusbert, 1985; Townsend et al., 1992; le Gras et al., 1999). The toxicity of this water is probably due to a combination of trace metals solubilized by the low pH, very low concentrations of dissolved oxygen and organic compounds originating from certain plant species. This toxicity is a natural part of the seasonal cycle that needs to be distinguished from mining impacts.

The acidic first flush can also solubilize metals that may be present in sediments downstream from mine sites. Thus, the initially ameliorated toxicity can be reexpressed. Such remobilization is especially likely to occur where precipitates of metal hydroxides have been deposited on the stream bed as a result of the interaction between acid mine drainage and inputs of alkaline groundwater into the stream. In the case of Rum Jungle, the low pH waters of the first flush mobilize metals from the river sediments into solution, resulting in very high trace metal concentrations in surface waters of the east branch and the main channel of the Finniss River.

RUM JUNGLE COPPER-URANIUM MINE

Background

The pollution at Rum Jungle and its effect on the proximate terrestrial environment and the Finniss River system were recognized towards the end of mining operations. The main causes of pollution are as follows (Allen and Verhoeven, 1986):

1. Three large overburden heaps (Whites, Intermediate and Dysons) and one smaller heap (Whites North) (Fig. 8.1) underwent bacteriologically catalyzed pyritic oxidation, which released dissolved trace metals (e.g. Cu, Zn, Mn and U), sulfate, and acid into the environment.
2. Two open cut pits (Whites and Intermediate) (Fig. 8.1), which were allowed to fill with water once the ore bodies were mined out, became polluted with trace metals and acid.
3. A Cu heap leach pile, constructed as an experiment to remove Cu from low-grade oxide and sulfide ore, continued to oxidize, releasing trace metals and acid into the environment.
4. Erosion of tailings from the Old Tailings Dam (Fig. 8.1), released unneutralized wastes and low levels of radioactivity to the river system.
5. Various areas of the site, including the old treatment plant site, the ore stockpile area, the Acid and Sweetwater dams (Fig. 8.1), as well as the outwash areas around the overburden heaps, became denuded of vegetation.

It was obvious by the mid-1960s that effluent from the treatment plant and leachate from the mine waste dumps were adversely affecting riparian vegetation at the mine site. More importantly, aquatic biota in both the east branch of the Finniss River (hereafter referred to as the East Branch), which flows through the mine site, and the Finniss River downstream of its confluence with the East Branch, were being severely impacted. The major pollutants in the East Branch and the Finniss River were Cu, Zn, Mn, sulfate, and acidity (Davy and Jones, 1975). Interestingly, U and Al were not considered to be major pollutants.

The Rum Jungle Rehabilitation Project was established in 1982 as a Federal Government funded project. The following remedial measures were undertaken to reduce the annual pollutant loads of Cu, Zn, and Mn being generated at the mine site (Allen and Verhoeven, 1986):

1. Waste dumps, which were the major contributors of trace metals, were covered to reduce infiltration of water and ingress of oxygen, thereby reducing the oxidation rate of the sulfides and the volume of water available to transport the soluble oxidation products.

2. Engineered runoff channels and erosion control banks were constructed on the dumps to reduce infiltration and consequent flushing of trace metals out of the dumps.
3. Tailings and residues from a Cu heap leach operation were placed in an open cut pit and covered to reduce acid production.
4. Acidic water in the open cuts was treated with lime to neutralise the pH and precipitate trace metals.
5. High flows in the East Branch (Fig. 8.1) were redirected through both open cuts, so that annual flushing with catchment runoff would minimize the build-up through time of acid and trace metals in the surface water layer of the pits (Ritchie, 1994; Zuk et al., 1994).

The annual contaminant loads and total water flow in the East Branch were measured at GS 8150097 (location shown in Fig. 8.1) during the pre- and post-remedial periods. Post-remedial monitoring of the waste rock dumps demonstrated reductions in both oxygen levels and temperatures inside the dumps, indicative of slowing in the oxidation rate of the sulfides (Ritchie, 1994).

Ecological and water quality studies in the Finniss River system were undertaken prior to, and after, remediation of the site. This river represents one of the few locations in the world where there is an extensive and high quality dataset, against which to assess ecological recovery of a river system following reduction of contaminant loads. The results of these investigations are described below.

Pre-remedial Baseline Studies

Contaminant loads and concentrations in water

Annual contaminant loads and total water flow, derived from water quality and flow gauge measurements at GS 8150097 (Fig. 8.1), entering the Finniss River between 1969/70 and 1973/74 are provided in Table 8.1. Over this period, the calculated annual loads of Cu, Zn, Mn, and sulfate delivered into the Finniss River system ranged from 44 to 106, 22 to 30, 46 to 110 and, 3300 to 13000 tonnes, respectively. Frequency distributions of the daily measured pH and total concentrations of Cu, Zn, and Mn in the East Branch from 1968 to 1974, are shown in Fig. 8.3 to provide an overview of water quality in the Finniss River system prior to remediation.

However, the annual sequence and chemical composition of the pollution were complex (Conway and Davy, 1975; Davy et al., 1975). Davy and Jones (1975) concluded that

- at the start of the wet season the pollution load is high in Cu relative to other pollutants;
- as the wet season progresses, the concentration of Mn increases relative to other pollutants; and

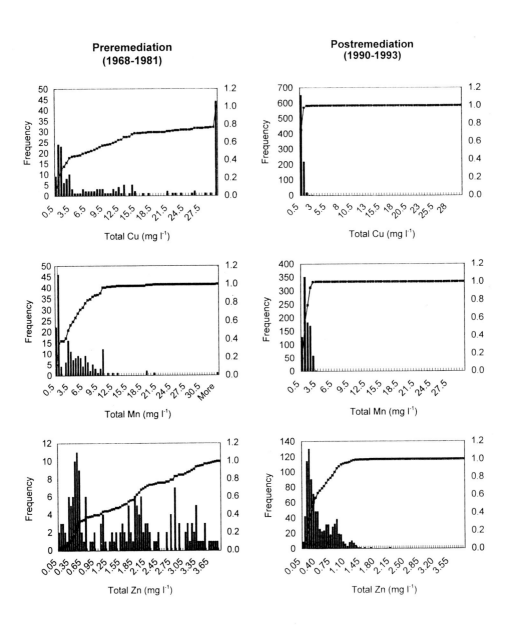

Fig. 8.3: Histograms of Cu, Mn and Zn water concentration frequencies in the East Branch of the Finniss River (GS 8150097) for preremediation (1968-1981) and post-remediation (1990-1993) monitoring periods.

Table 8.1: Annual loads (tonnes) of sulfate, Cu, Zn, and Mn entering the Finniss River system from the Rum Jungle Mine site[a]

Year	Annual flow (GL)	Sulfate (tonnes)	Cu (tonnes)	Zn (tonnes)	Mn (tonnes)
Pre-remediation					
1969/70	7	3300	44	—	46
1970/71	33	12000	77	24	110
1971/72	31	6600	77	24	84
1972/73	22	5500	67	22	77
1973/74	69	13000	106	30	87
Post-remediation					
1990/91	41	4000	15	7.4	31
1991/92	7	1260	3.8	2.7	9.1
1992/93	30	2696	12	3.9	25
1993/94	26	2281	13	5.3	18

[a] Based on acidified total metal concentrations in water measured in the East Branch at GS 8150097 (Fig. 8.1).

- at the beginning of the dry season, but when the East Branch has not yet ceased to flow into the Finniss River, Zn is preferentially absorbed by Finniss River sediments.

In the East Branch, at the end of the dry season when water flow ceased to the Finniss River, total concentrations of Cu in the surface waters of billabongs (permanent water bodies) were as high as 250 mg l^{-1} (Jeffree and Williams, 1975). In the Finniss River, the highest concentrations of contaminants (Cu, Zn, Mn, Co, Ni, U, sulfate, and acidity) occurred when convective thunderstorms caused early wet season flows in the East Branch and these coincided with low flows from the remainder of the Finniss catchment (Jeffree and Williams, 1980). During these "first flush" events, fish-kills in the Finniss River were observed to extend between 15 km and somewhat less than 30 km downstream of its confluence with the East Branch (Jeffree and Williams, 1975). These toxic events were most likely responsible for the general reduction over the subsequent dry season in fish diversity and abundance (Jeffree and Williams, 1975, 1980). When flow ceases from the East Branch, there is continuous flushing throughout the dry season in the perennial Finniss River.

Ecological impacts

In the East Branch downstream of the mine, no rooted or submerged plants were observed (Fig. 8.4) and live bank-side *Pandanus* were rare, although their dead stumps were present. Absence of the stabilizing effect of bank-side roots was associated with shallow gullying, gently sloping banks, and considerable deposits of sand along the banks, indicative of

Fig. 8.4: Photograph of a pool in the East Branch of the Finniss River prior to remediation (1974) showing the remains of bankside vegetation denuded by mine contaminants. Note the discoloured water.

erosion and consequent deposition. Fish were virtually absent from the pools of the East Branch, although physical and structural characteristics of these pools were similar to those at sites not exposed to pollution, which contained about seven species (Jeffree and Williams, 1975). A number of fish species were found in the unpolluted tributaries of the East Branch.

Many groups of Insecta, which are indicators of an unpolluted state, as well as the phyla Mollusca and Porifera, were absent in the East Branch downstream of the mine. However, some families of insects occurred abundantly; nearly all had an impervious cuticle and obtained atmospheric or dissolved oxygen through bubbles.

In the Finniss River, there was also an appreciable decline in fish diversity and abundance for at least 15 km downstream of the confluence with the East Branch. The total number of fish species collected (using >25 mm nets) in major billabongs over four sampling periods during the dry seasons of 1973 and 1974 is shown in Fig. 8.5. Sites in the Finniss River immediately downstream of the confluence with the East Branch showed lower numbers of species than those upstream or further downstream. There is a gradual recovery in both the number of species and individuals in the Finniss River with increasing distance downstream of the confluence with the East Branch. A comparable pattern of impact was also observed for the smaller fish species sampled in small embayments connected to the Finniss River (Jeffree and Williams, 1975).

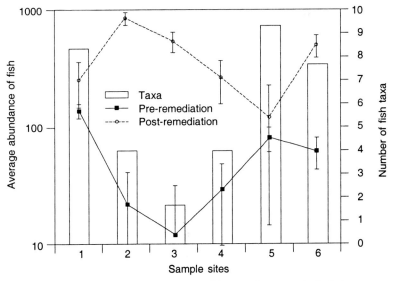

Fig. 8.5: Average abundances of seven fish species sampled during pre- and post-remediation, and the number of fish taxa caught at each sample site (see Fig. 7.1) prior to remediation (histograms).

Individual fish species that showed the greatest reduction in numbers in the zone of impact, relative to sites outside this zone, were the eel-tailed catfish (*Neosilurus* spp.), black-striped grunter (*Amniataba percoides*), black bream *(Hephaestus fuliginosus)*, and freshwater longtom (*Strongylura krefftii*). However, the abundance of the chequered rainbow fish (*Melanotaenia splendida inornata*) increased in the 15 km region downstream of the confluence with the East Branch during the dry season (Jeffree and Williams, 1975). This finding indicates that this species was relatively more tolerant of the pollutants present and/or that it could more rapidly recolonize less contaminated areas within in the polluted zone.

The distribution and diversity of macroinvertebrates were not specifically investigated during initial studies of the Finniss River system. However, the diet (gut contents) of the purple-spotted gudgeon (*Mogurnda mogurnda*), a species whose distribution, abundance, and size frequencies were not obviously affected by pollution, were studied in detail during the 1974 dry season (Jeffree and Williams, 1980). This study provided the following picture for the rather broad subset of organisms that the fish consumed:

1. Significant ($P \leq 0.05$) differences were observed between polluted and unpolluted zones of the river for eight out of 19 food types, considered individually. Five of these types were aquatic respirers: Pisces, Cladocera, Copepoda, and Chironomidae were more abundant in the polluted zone, but the decapod *Macrobrachium* was more abundant in the unpolluted zone. Two were the air respirers, Dytiscidae and Veliidae, and both were more abundant in the unpolluted zone. The respiratory mode of one food (Coleoptera larva), that was more abundant in the polluted zone, was indeterminate.

2. Hierarchical classification analysis was used to indicate the relationships between fish dietary samples from the polluted and unpolluted zones, when selected food sets were considered in combination. These analyses showed that the set of aquatic respirers showed the greatest segregation between polluted and unpolluted collections.

3. Discriminant function analysis was relatively ineffective in discriminating the zones using air respirers, but highly effective using aquatic respirers, and those sets of foods effective in discriminating the zones were largely different between the beginning and end of the dry season.

4. For those food sets which best differentiate between the polluted and unpolluted collections, the results of both the hierarchical and discriminant function analyses indicated greater heterogeneity among the collections from the polluted zone than from the unpolluted zone.

A difference in heterogeneity would follow from the importance of pollutants in determining the abundances of foods, when the effects of the pollutants are also variable, because of

1. changes in the concentrations of pollutants as the dry season progressed;
2. different distances downstream from the confluence of the East Branch with the Finniss River which would cause sediments, detritus, and algae to be differentially loaded;
3. occurrence of local unpolluted inflows, such as small tributaries joining the main stream, and throughflow seeping into the Finniss River from the banks and larger aquifers; and
4. entry of allochthonous material (e.g. leaf fall), which can be expected to influence the levels and chemical forms of pollutants.

Post-remedial Studies

Contaminant loads and concentrations in water and sediment

The annual contaminant loads and total water flow measured at GS 8150097 during the post-remedial period (1990/91-1993/94) are given in Table 8.1. Mean annual flow between the pre-remedial and post-remedial periods varied by 20–25% only, which simplifies subsequent comparisons, given the positive relationship between annual flow and load (Jeffree et al., 2001). During the post-remedial period, mean annual loads of Cu, Zn, Mn, and sulfate were reduced by a factor of 7, 5, 4 and 3, respectively, relative to pre-remediation (Table 8.1).

Prior to remediation, the distribution of pH values was positively skewed (Table 8.2), with a highly acidic median value of 4.2. Following remediation, the distribution becomes negatively skewed, i.e. there is a higher frequency of higher pH values. The median increased by more than two pH units to 6.3, indicative of a substantial reduction in the acidity of the mine effluents, although there is little change in the range of pH values (Table 8.2). Additionally, the distribution of Cu, Zn and Mn concentrations were negatively skewed due to the occurrence of some very high values (Fig. 8.3).

Prior to remediation, Cu concentrations in surface waters of the East Branch exhibited a maximum of 160 mg l^{-1}, with a median of 6.8 mg l^{-1} (Table 8.2). Following remediation, there was considerable reduction in the occurrence of the higher Cu concentrations, with a maximum of 2.7 mg l^{-1} and a median of 0.6 mg l^{-1} (Table 8.2; Fig. 8.3). There were small differences in the ranges of Zn water concentrations pre- and post-remediation, but higher concentrations declined markedly in occurrence following remediation, as indicated by a reduction in the median Zn concentration from 1.5 to 0.5 mg l^{-1}, and in the mode from 2.8

Table 8.2: Statistical characteristics of contaminant frequency distributions in the East Branch[a]

	Preremediation: 1968-1981 (1968-1974)					Postremediation: 1986-1995				
	pH	Cu	Zn	Mn	U	pH	Cu	Zn	Mn	U
Maximum	6.8	160	3.8	33.1	—	7.2	3.6	2.7	3.7	0.09
	(6.3)	(134)	(1.9)	(33.1)	(2.8)					
Minimum	3.1	0.13	0.03	0.01	—	3.4	0.01	0.02	0.01	0.01
	(3.1)	(1.0)	(0.3)	(1.0)	(0.02)					
Median	4.2	6.8	1.5	3.0	—	6.3	0.6	0.5	1.9	0.03
	(3.8)	(3.0)	(0.7)	(8.0)	(0.3)					
Mode	4.0	104	2.8	10	—	6.7	0.3	0.3	1.3	0.03
	(3.1)	(3)	(0.7)	(19)	(0.2)					
Skew	1.1	1.7	0.36	2.6	—	-0.6	1.4	1.1	0.5	-0.1
	(1.2)	(3.1)	(0.41)	(1.3)	(3.0)					
N	197	188	179	189	—	454	1615	1615	1615	21
	(23)	(13)	(5)	(14)	(12)					
WQG[b]							0.009	0.053	1.7	0.0035

[a] Metal concentrations are reported in mg l^{-1}.
[b] Water quality guideline (WQG) values for the protection of moderately disturbed freshwater ecosystems in Australia (ANZECC and ARMCANZ, 2000). Values for Cu and Zn were corrected for water hardness (i.e. a median water hardness of 280 mg l^{-1} as $CaCO_3$).

to 0.33 mg l^{-1} (Table 8.2; Fig. 8.3). For Mn, there is a more substantial reduction in the occurrence of higher concentrations following remediation, indicated by a reduction in the maximum concentration from 33.1 to 3.7 mg l^{-1}, the median from 3.0 to 1.9 mg l^{-1} and the mode from 10 to 1.3 mg l^{-1}. Based on a very limited dataset (Lawton, 1998; ANSTO, unpublished), median concentrations of U decreased 10-fold (from 310 to 30 µg l^{-1}) following remediation. Metal concentrations in the surface waters of the Finniss River typically decrease exponentially with increasing distance downstream of the confluence with the East Branch (Markich et al., 2002).

Although the maximum and median concentrations of Cu, Mn, Zn, and U have decreased substantially in the surface waters of the East Branch (and the Finniss River) following remediation of the Rum Jungle mine, the median concentrations still exceed the national water quality guidelines (ANZECC and ARMCANZ, 2000) for the protection of freshwater ecosystems (Table 8.2).

The majority of the metal load in mine drainage will ultimately be incorporated into stream sediment downstream of the source. As a result, the concentrations of contaminants in the sediment that are likely to damage benthic biota need to be considered. In July 1995 and August 1999, sediments at several control and impacted sites in the East Branch and the Finniss River were sampled and analyzed for metals. Median metal concentrations at each site are given in Table 8.3. For the control sites in the East Branch above the mine, and in the Finniss River above

Table 8.3: Metal concentrations in sediment from the Finniss River and East Branch (EB) compared with national guideline values[a]

Distance from EB (km)	Cu	Zn	Mn	U	Pb	Ni	Co	Fe
−15	18	14	200	19	17	9.1	8.6	7480
−0.5	21	13	198	22	16	10	10	6920
1	359*	126	262	149	118	152*	59	12900
4	284*	60	204	129	84	69*	29	10780
7	84	36	163	37	38	26	17	8100
10	40	21	166	25	14	17	11	7230
15	23	13	121	19	7.1	11	8.9	5740
30	7.9	5.8	55	6.8	4.6	4.0	4.1	3630
EB 2 km above mine	12	7.5	79	11	9.3	6.8	6.0	4890
EB below mine	483*	116	195	219	111	81*	28	16700
	(120–3500)	(24–240)	(154–2900)	(25–747)	(30–590)	(25–415)	(2.7–49)	(10500–269000)
SQG—low[b]	65	200	—	—	50	21	—	—
SQG—high[c]	270	410	—	—	220	52	—	—

[a] Median (and range) values are reported in (mg kg^{-1} dry weight).

[b] Sediment quality guideline value below which there is a low probability of biological effect (ANZECC and ARMCANZ, 2000). Sediment metal concentrations that exceed this value, but are lower than the SQG-high value (see below) are indicated in bold.

[c] Sediment quality guideline value above which there is a high probability of biological effect (ANZECC and ARMCANZ, 2000). Sediment metal concentrations that exceed this value are indicated in bold and an asterisk(*).

the confluence with the East Branch, the concentrations of Cu, Zn, Pb, and Ni are not considered to be of ecotoxicological concern (Table 8.3): that is, the measured concentrations were below their guideline values, indicating a low probability of biological effect (ANZECC and ARMCANZ, 2000). Sediment guideline values are not currently available for Mn, U, Co, and Fe in Australia, or elsewhere, due to a lack of toxicity data for benthic aquatic biota.

Sediment concentrations of Cu, Zn, U, Pb, Ni, Co, and Fe in the East Branch downstream of the mine and in the Finniss River immediately downstream of the confluence with the East Branch, were substantially higher than in the control sites (Table 8.3). As expected, these metal concentrations declined with increasing distance downstream of the mine, approaching background (or control) concentrations some 30 km downstream of the East Branch confluence (Table 8.3). The sediment concentrations of Cu and Ni in the East Branch and in the Finniss River up to 4 km downstream of the East Branch confluence exceeded the upper national sediment guideline values (Table 8.3), indicating a high probability of biological effect. For example, the median measured concentrations of Cu and Ni in the sediments of the East Branch were about two-fold higher than those at the upper sediment guideline value (Table 8.3). The results for Ni are somewhat unexpected, since this metal was never identified as being a contaminant of potential ecotoxicological concern from the Rum Jungle mine. Sediment concentrations of Pb exceeded the lower sediment guideline value in the East Branch and in the Finniss River up to 4 km downstream of the East Branch confluence by twofold (Table 8.3).

Ecological Impacts

Field studies that commenced in the early 1990s indicated that there has been a small recovery in the riparian vegetation of the East Branch following the remediation of the mine site, and associated reductions in annual contaminant loads and surface water concentrations. Using image analysis of vegetation patterns in the Finniss River system, Stratford (1994) still found considerable detriment proximal to the East Branch.

In May 1993 (early dry season), Jackson and Ferris (1998) conducted a study of macroinvertebrate community composition and associated water quality in the East Branch that was very similar in design to pre-remedial investigations in its choice of sites and sampling methodologies, but more quantitative in nature. This study showed:

1. Sites exposed to mine pollution had only a few dominant taxa with comparatively low average numbers of individuals. These were adult coleopterans hemipterans and dipterans.

2. Ephemeroptera, odonata, corbiculids, hyriids, planorbids, and thiarids were not found in polluted habitats. However, trichopterans and palaemonid prawns were found in the lowest reaches of the East Branch, immediately upstream of its confluence with the Finniss River.

3. A clear distinction between polluted and unpolluted reference sites of the East Branch was noted with respect to the average number of individual macroinvertebrates, irrespective of taxon; that is, a significantly ($P \leq 0.01$) greater number were sampled at unpolluted sites.

4. During this sampling period, the highest concentrations of metals measured in the surface waters of the East Branch, at sites receiving effluent from the mine, were 20.3 mg l^{-1} for Cu and 82.8 mg l^{-1} for Zn. These concentrations decreased with increasing distance downstream of the mine.

Jackson and Ferris (1998) concluded that there have been no apparent increases in the number of macroinvertebrate taxa in the section of the East Branch receiving effluent from the mine, following remediation of the Rum Jungle mine site. However, there does appear to have been some ecological improvement in the lowest reaches, indicated by the presence of freshwater atyid shrimps.

In a more comprehensive study, Edwards (2001) evaluated spatial and temporal patterns in macroinvertebrate community composition in relation to seasonal changes in metal pollution in the East Branch. Benthic macroinvertebrates, together with physicochemical variables, were collected from August 1994 to August 1995, to include both a wet and a dry season. A marked seasonal trend in the physicochemistry of the East Branch surface waters identified three distinct levels of ecological impact. The most significant impact occurred late in the dry season at sites closest to the mine. All other sites and sampling times were regarded as being of medium level impact, with the reference sites being minimally impacted. In agreement with the study of Jackson and Ferris (1998), Edwards (2001) found that sites in the East Branch furthest from the mine showed some degree of ecological improvement or recovery late in the dry season, with community composition very similar to the reference sites.

Ferris et al. (1995) sampled benthic algae in the East Branch (1993-1995) during the period of recessional flow that occupies the early dry season. An obvious, and roughly, exponential gradient of pollution (trace metals, sulfate, and acidity) develops during this time (Lawton, 1998). At the site of lowest pH (3.0) and highest metal concentrations (e.g. 30 mg Cu l^{-1}), the species richness was <10% for diatoms, and 50% for non-diatoms, relative to control sites. Collectively, the algal species richness in the East Branch downstream of the Rum Jungle mine site was, on average, <50% of that typically found at control sites. The changing community of benthic diatoms reflects the gradient of pollution in a statistically significant way (Ferris and Vyverman, 1996) such that it is possible to predict approximately where a sample was collected on the basis of the species composition. Furthermore, Ferris et al. (1995) found that the relative contribution of three major algal groups (blue-green algae, green algae, and diatoms) to the overall number of taxa present at a site changes with position along the pollution gradient. Blue-green algae were absent from the most polluted end of the gradient, leaving only green algae and a few species of diatoms. Overall, the quantitative data on benthic diatoms provide a potentially useful basis for measuring any further improvement in the ecosystem health of the East Branch, based on primary producers at the base of the food chain.

Markich et al. (2002) sampled freshwater bivalves (*Velesunio angasi*) in 1996 from the Finniss River system at 10 sites *a priori* exposed and non-exposed to mine pollution. Secondary ion mass spectrometry (SIMS) was used to measure Cu, Mn, Zn, U, Ni, Co, Pb, and Fe/Ca ratios across the annual shell laminations of the longest-lived bivalves found at each site, with the aim of evaluating the ability of the shells to archive measured annual metal inputs and their temporal patterns. At sites not exposed to mine pollution, relatively constant and similar (baseline) signals were found for all metals in the shell laminations of *V. angasi*, dating as far back as 1965. At sites contaminated by mine pollution, relatively constant, but variably elevated, signals were evident for Cu, Mn, Zn, Ni, and Co in the shell, which extended back only to the end of remediation (1986). Since remediation, the temporal patterns of Cu, Zn, and Mn observed in the shells at the most polluted sites reflected those of the measured annual dissolved loads in the surface waters. The average concentrations of Cu, Mn, Zn, Ni, and Co in the shells decreased (3–13 fold) with increasing distance downstream of the mine site, until concentrations characteristic of the nonexposed sites were reached. This geographic pattern of decline in pollution signal in the shell with increasing distance downstream of the pollution input is consistent with the pattern established for water and sediment chemistry. Overall, the SIMS results supported the proposition that the shells of *V. angasi* can be used as archival indicators of metal pollution in surface waters of the Finniss River over their lifetime.

Copper is generally considered to be of greatest ecotoxicological significance, particularly in the East Branch. Indeed, the median Cu concentration (600 μg l^{-1}; Table 8.2) in the East Branch substantially exceeds that which causes 50% mortality in a range of Australian tropical freshwater crustaceans, molluscs and fish under laboratory conditions (see review by Markich and Camilleri (1997)). For example, at 4 μg Cu l^{-1}, 50% of freshwater shrimps (*Caradina* sp.) died after four days, whereas 50% of chequered rainbowfish (*Melanotaenia splendida inornata*) and fly-specked hardyhead (*Craterocephalus stercusmuscarum*) died when exposed to 60 and 17 μg Cu l^{-1}, respectively, for four days.

In addition to Cu, Al is present at very high concentrations (40-90 mg l^{-1}) in surface waters of the East Branch, up to about 1 km downstream of the mine (J. Ferris, unpublished). As the pH of the surface water reaches about 5.0, Al is precipitated as aluminium hydroxide (an obvious white floc at the sediment surface). The maximum measured concentration of Al (90 mg l^{-1}) exceeds the national water quality guideline (ca. 1 μg l^{-1} at pH <6.5) for the protection of freshwater ecosystems guideline (ANZECC and ARMCANZ, 2000) by a factor of 90,000. In the only study on the toxicity of Al to Australasian freshwater biota, Camilleri et al. (2002) reported an average 4-day LC$_{50}$ value of 460 μg Al l^{-1} for a tropical freshwater fish (purple spotted gudgeon (*M. mogurnda*)) at pH 5.0. Aluminium concentrations in the East Branch just below the mine are at least two orders of magnitude above this value, indicating that this metal is of ecotoxicological concern.

Within the Finniss River, ecological studies conducted following remediation have shown the following environmental improvements over the first 15 km downstream of the confluence with the East Branch:

1. The indicator fish species, black eel-tailed catfish (*Neosilurus ater*), banded grunter (*Amniataba percoides*), and archer fish (*Toxotes chatareus*), which were virtually absent, have returned.
2. Fish community structure has returned to natural levels of diversity and abundance (Zuk et al., 1994; Jeffree et al., 2001).
3. Ecotoxicologically sensitive species of Crustacea (atyid shrimps, e.g. *Caradina nilotica*) occur within impacted regions of the river where no living specimens were found prior to remediation (J. Twining, unpublished).

It is important to note that some ecological recovery is occurring in the Finniss River even though the concentrations of Cu, Zn, and Ni in the sediment, and Cu, Zn, U, Co, and Al in the surface waters, often substantially exceed their respective national guideline values for the protection of aquatic ecosystems. Guidelines for metals in aquatic ecosystems are typically based on total concentrations (see below).

However, it is well recognized that the biological effects of metals in water and sediment are critically dependent on their physicochemical form, or speciation, and not their total metal concentration (Mudroch et al., 1999; Brown and Markich, 2000). As a specific example, the total concentration of Cu (26 μg l^{-1}) in surface waters of the Finniss River, approximately 15 km downstream of the confluence with the East branch, was shown to exceed the hardness-corrected national guideline value (5.5 μg l^{-1} at a hardness of 150 mg l^{-1} as CaCO$_3$) by a factor of 5. After filtering the water sample through a 0.4 μm membrane filter, the Cu concentration was reduced to 14 μg l^{-1}. Geochemical speciation modeling, using a model such as HARPHRQ (Brown et al., 1991), was then used to determine the speciation of Cu in the sample (i.e. pH, 7.4; temperature, 24°C; dissolved oxygen, 90% saturation; dissolved organic carbon, 6.2 mg l^{-1}; conductivity, 270 μS cm^{-1}; hardness, 150 mg l^{-1} as CaCO$_3$). Dissolved inorganic Cu, which is a reasonable initial estimate of bioavailable Cu (Markich et al., 2001), was calculated to be 2.5 μg l^{-1}, a 10-fold reduction from the total concentration, but most importantly is below the Cu guideline value. Chemical measurement techniques may also be used (Markich et al., 2001). Overall, a consideration of metal speciation is essential when determining what fraction of the total concentration metal is actually bioavailable.

Physiological and/or genetic tolerance of freshwater biota to metals such as Cu and Zn in the waters of the Finniss River is an alternative mechanism to explain their partial recovery since remediation of the Rum Jungle mine, despite metal concentrations in the sediments and water that exceed national guideline values. Gale et al. (2003) found that black striped rainbowfish (*Melanotaenia nigrans*) from the East Branch have acquired tolerance to the elevated Cu (and possibly other metal) concentrations in the surface waters. The 4-day LC_{50} value of Cu was eight-fold higher for *M. nigrans* living in the East Branch (1120 µg l^{-1}) than at reference sites (135 µg l^{-1}). Fish from the East Branch accumulated significantly ($P \leq 0.05$) less (up to 50%) Cu in all tissues compared to that of reference fish, when experimentally exposed to low and elevated (×10) Cu concentrations. Based on dissimilar allozyme frequencies and reduced heterozygozity in fish from the East Branch, the mechanism of Cu tolerance may be genetically based. The selection of allozyme genotypes less sensitive to inhibition by Cu may allow fish from the East Branch to survive Cu concentrations that exceed the capacity of the exclusion mechanism.

RANGER URANIUM MINE

Background

Over the past two decades there has been a major research effort to gather environmental data in the Alligator Rivers Region (Fig. 8.2). This is largely as a consequence of the Australian Government's decision (Fox Inquiry; Fox et al., 1977) to permit the mining and milling of two (Narbalek and Ranger) of the four extensive deposits of U known to exist in this area. The Narbalek deposit was mined out and has recently been rehabilitated (Supervising Scientist, 2001).

The Ranger deposit, located adjacent to Magela Creek (Fig. 8.2), is currently being mined by a conventional, simultaneous mining/milling, open-cut operation. There are two economic orebodies (Orebody 1 and 3) jointly containing about 125,000 tonnes of U_3O_8 at an average grade of 0.32% (Supervising Scientist, 2001). Mining of Orebody 1 was completed in 1994. Mining of Orebody 3 commenced in 1997. Mining and milling operations are expected to continue until about 2006. In 1996, the Ranger mine was the third largest producer of uranium oxide in the western world, accounting for 10% (4149 tonnes) of the world's production (ERA, 2001). Mining of U at the Jabiluka deposit (90,400 tonnes of U_3O_8 at an average grade of 0.43%), located approximately 20 km to the north of the Ranger mine, is yet to commence. It is proposed that ore from the Jabiluka deposit will be transported to the existing Ranger site for milling (Kinhill and ERA, 1998).

As a consequence of heavy wet season rainfall, one of the most important environmental issues associated with the operational phase of U mining at the Ranger mine has been the management of excess water that accumulates within the mine retention ponds (Fig. 8.2). The options for water management are severely constrained as a result of the location of the project area upstream of the Magela Creek floodplain, which forms part of the World Heritage listed Kakadu National Park. The Ranger mine is the most tightly regulated mine in Australia with respect to discharges of water from the site, and is required to adhere to detailed environmental requirements determined by government authorities (NT DME, 1982).

Magela Creek is an ephemeral stream consisting of a series of braided sandbed channels with sandy levees and a series of billabongs and channels that expand onto a seasonally flooded black-soil floodplain (Nanson et al., 1993). High peak flows occur in response to intense storm events in the upper catchment. High flow conditions (20–2000 m^3 s^{-1}) during this period impart a uniform and common water chemistry to Magela Creek (Humphrey et al., 1990). Water quality is very good, reflecting the essentially pristine nature of the catchment (Table 8.4). For this reason, the potential for impact of wastewater on aquatic ecosystems downstream of the Ranger mine is of particular concern. Hence, any mine wastewater releases need to be carefully controlled to minimize ecological detriment.

The process water circuit (tailings dam, pit 1, plant and retention pond (RP) 3) at the Ranger mine operates in a closed loop, with all water that enters being retained within it. A second type of water is defined by the term "restricted release zone" (RRZ). This includes runoff from the process plant area, and runoff and seepage from the economic and low-grade ore stockpiles. Seepage and surface runoff water from the RRZ drains into RP2. There is legislative provision for the release of water from RP2. However, careful operation of the water management system at the mine has meant that this water has never been released.

Water entering Magela Creek from the mine lease comes from catchments that are located outside the RRZ. This discharge occurs via the Georgetown and Djalkmara billabongs (Fig. 8.2) and only during, and immediately after, the wet season. The discharge of excess wastewaters from RP4 (which is located outside the RRZ because its catchment is waste rock), should it be necessary, is permitted subject to strict, long-established, release criteria. Additionally, an application needs to be made for state government (ministerial) approval prior to any such release (NT DME, 1982). These criteria include stream flow rate (>20 m^3 s^{-1} during the wet season), maximum allowable chemical addition, additional annual load limits, and prerelease biological testing (NT DME, 1982; ERA, 2000).

The geochemistry of the waste rock is such that only U (and its radioactive progeny), Mg, Mn, and S (as SO$_4$) in runoff and seepage are likely to limit the discharge of mine wastewaters into Magela Creek (Noller, 1991). From the point of view of solids management, waste is defined as material containing less than 0.02% U. Waste rock can be stockpiled outside the RRZ.

Table 8.4: Fresh surface water chemistry of Magela Creek during the wet season

Parameter	Units	Available sources[a] (1972-1992)			
		Mean[b]	95% CI	Range	N
pH		6.0	5.9-6.1	4.2-7.0	2540
Temperature[c]	°C	28.0	27.6-28.5	25.1-32.5	2230
Conductivity	µS cm^{-1}	17	16-19	6.1-75	1065
Turbidity	NTU	3.2	2.9-3.7	0.6-61	1210
Suspended solids	mg l^{-1}	7.1	6.9-7.6	1.2-146	855
Na	mg l^{-1}	0.99	0.89-1.1	0.36-9.1	1455
K	mg l^{-1}	0.37	0.31-0.45	0.08-2.7	1455
Ca	mg l^{-1}	0.45	0.43-0.48	0.09-1.6	1455
Mg	mg l^{-1}	0.60	0.58-0.63	0.26-2.8	1455
Cl	mg l^{-1}	1.3	1.1-1.6	0.20-17	1455
SO$_4$	mg l^{-1}	0.24	0.14-0.38	0.10-9.0	1530
HCO$_3$	mg l^{-1}	2.6	2.4-2.9	0.5-11	1455
NO$_3$	µg l^{-1}	15	12-19	3.1-110	1365
PO$_4$	µg l^{-1}	7.0	6.0-8.4	3.0-45	1365
Fe	µg l^{-1}	90	87-96	25-870	740
Al	µg l^{-1}	70	67-75	12-220	235
Mn	µg l^{-1}	9.7	8.6-11	1.0-57	875
Cu	µg l^{-1}	0.70	0.68-0.73	0.11-1.6	255
Zn	µg l^{-1}	0.30	0.28-0.34	0.18-1.4	300
U (as UO$_2$)	µg l^{-1}	0.11	0.091-0.13	0.02-2.3	1755
Pb	µg l^{-1}	0.076	0.073-0.080	0.033-0.13	255
Cd	µg l^{-1}	0.013	0.010-0.018	0.010-0.054	55
Total organic		4.1[d]	4.0-4.3[d]	0.3-8.9[d]	1365
carbon	mg l^{-1}	10.7[e]	10.5-11.1[e]	2.2-94[e]	
Dissolved		3.7[d]	3.6-3.9[d]	0.2-8.0[d]	915
organic carbon	mg l^{-1}	8.9[e]	8.7-9.3[e]	0.7-54[e]	

[a] Data were obtained from refereed journal articles (Hart and McGregor, 1980; Hart et al., 1982, 1987a, 1987b, 1992a, 1992b; Morley et al., 1985), government and corporate reports (Morley, 1981; Walker and Tyler, 1982; NT DTW, 1983; Humphrey and Simpson, 1985; Cusbert et al., 1992), as well as unpublished material from routine monitoring programs conducted by Ranger Uranium Mine and Federal (Supervising Scientist) and State (Northern Territory Department of Mines and Energy) government authorities.

[b] Geometric mean values.

[c] Bottom water temperature; considered to be more representative of water temperatures experienced by benthic organisms.

[d] Low–median flow conditions (i.e. < 20 m^3 s^{-1}).

[e] High flow conditions (i.e. > 20 m^3 s^{-1}).

Uranium is the primary element of potential ecotoxicological concern. Uranium concentrations (as UO$_2$) in wastewaters (particularly RP4) during the wet season have ranged from 15 to 190 µg l^{-1}, with a median of *ca.* 50 µg l^{-1} (a factor of ca. 450 greater than for surface water in Magela Creek) since 1992 (Markich, 1998; ERA, 1999, 2000).

Whilst chemical analysis of the surface water of Magela Creek indicates a slight chemical signature from the mine, no deleterious effects on aquatic biota in Magela Creek downstream of the mine have been detected since operations commenced. This record stands in marked contrast to the legacy of substantial downstream impacts from the Rum Jungle mine, and is a testament to the very high standards of environmental management on the Ranger leases. This case study describes the use of multi-tiered biological testing and monitoring (using local invertebrate and vertebrate species) to evaluate impacts downstream of a mine site, and the research that is being done to develop locally applicable numerical guidelines for solutes (especially U) present in mine water. The framework described should be applicable to all types of mining operations throughout the world.

Biological Assessment of Mine Wastewaters

General

During the wet season, the quality of the surface water is regularly monitored in Magela Creek, upstream and downstream of the mine site, as well as in the retention ponds (RP1, RP2, and RP4) as part of the statutory water quality monitoring program. This is complemented by nonstatutory, laboratory-based, prerelease toxicity testing of RP4 water. Creekside toxicity testing has also been conducted during periods of water release from RP4 since the 1991/92 wet season. Similar approaches have been effectively used in the United States (e.g. Naugatack River (Mount et al., 1986) and Skeleton Creek (Norberg-King and Mount, 1986)).

The laboratory and creekside toxicity tests are complemented by measurements of metals and radionuclides in the flesh of the freshwater mussel, *Velesunio angasi*, collected from a billabong downstream of the mine, to assess the bioaccumulation of metals and radionuclides over time. Baseline metal concentrations in the flesh of *V. angasi* were determined throughout the Magela Creek catchment between 1980 and 1982 (Allison and Simpson, 1989) so that future concentrations may be compared. Apart from being a suitable indicator species for biomonitoring over the longer term, *V. angasi* is also a food source for the traditional (Aboriginal) owners of Kakadu National Park. The annual sampling and analysis of mussels for metals and radionuclides forms the only part of the biological work at the mine that is specified as a statutory (regulatory) requirement in the Authorisation to Operate (NT DME, 1982). To date, there is no evidence to indicate that metal or radionuclide concentrations in mussel flesh have increased over time (e.g. ERA, 1996, 1997, 1998, 1999, 2000). Furthermore, metal concentrations in the flesh of *V. angasi* from Magela Creek, downstream of the Ranger mine, are consistent with other minimally polluted waters (see review by Markich (1996)).

As part of a "whole ecosystem" approach, a program has been developed and trialed since 1994 to assess the health of aquatic ecosystems comprising the Magela Creek catchment (Corbett, 1997; Sewell et al., 1997). This approach involves comparing the numbers of key trophic levels occupied by organisms in billabongs and riparian areas at test and control sites. It provides a tool for monitoring the health of aquatic ecosystems over long periods to determine if there are any significant ($P \leq 0.05$) chronic impacts from mine wastewater releases. The whole ecosystem approach represents the highest tier in the biological systems approach for assessing mining impacts at Ranger. To date, this approach has not provided any evidence of deleterious impacts on the aquatic ecosystems downstream of the Ranger mine.

Pre-release laboratory toxicity testing of mine wastewaters

The management of excess wastewater on site during the wet season requires the controlled release of water from RP4 in most years. Water from RP4 receives runoff and seepage from a waste rock dump and typically contains U, Mg, and sulfate at levels 500, 250, and 2000 times higher, respectively, than the receiving waters of Magela Creek (see reviews by Markich (1998) and Klessa (2000)). The minimum dilution requirements of RP4 water is initially determined using pre-release toxicity testing. The actual rate of wastewater discharge into Magela Creek is determined after consideration of water chemistry (Magela Creek and RP4 water), the flow rate of Magela Creek, and the results of biological testing. The rationale for this approach has been discussed by Brown (1986). In short, because the impact of complex and varying wastewaters on aquatic environments cannot be predicted from their individual constituents (cf. water quality standards regulated by chemical analysis), testing the toxicity of mine wastewaters on aquatic biota is desirable.

An important pre-requisite for the biological assessment of mine wastewaters is the identification of biota in Magela Creek potentially at risk. In a comprehensive discussion on the use of freshwater biota to assess mine wastewaters in the Alligator Rivers Region, Humphrey and Dostine (1994) concluded that soft-bodied and gill-breathing organisms (or their life stages) are most at risk from solutes present in the water.

A detailed description of the developments that led to the present use of toxicity testing is provided by Johnston (1991). The principal features of the approach are as follows:

1. Direct measurement of the change in toxicological responses (lethal and sublethal) of local biota, selected from widely different taxa and trophic levels, to the actual wastewater (RP4) as it is diluted with receiving (Magela Creek) water.

2. Dilution of wastewater required to render it harmless can be used as a control parameter to regulate its release. However, this must follow the application of a "safety" factor to account for the possible occurrence of large undetected risks, additivity of undetected effects, and protection of species other than those tested.

The suitability of 19 local freshwater organisms was assessed for their potential use in the pre-release tests. Several organisms were found to be useful, based on sensitivity, representation from different trophic levels (e.g. vertebrate predator, invertebrate predator and invertebrate herbivore), and suitability for laboratory culture (see review by Holdway (1992a)). Three of these, [*Hydra viridissima* (green hydra), *Moinodaphnia macleayi* (water flea) and *M. splendida inornata* (chequered rainbowfish)] (Table 8.5), are currently used to assess the toxicity of RP4 waters following established protocols (Bywater et al., 1991; Hyne et al., 1996; Markich and Camilleri, 1997).

Table 8.5: Local freshwater organisms currently used in pre-release toxicity tests with Ranger RP4 water

Organism	Test endpoint	Test duration (days)
Green hydra *Hydra viridissima* (Cnidaria: Hydrozoa)	Population growth	4
Water flea *Moinodaphnia macleayi* (Crustacea: Branchiopoda)	Adult survival Brood size	7[a]
Chequered rainbowfish *Melanotaenia splendida inornata* (Chordata: Osteichthyes)	Embryo survival	6[a]

[a] As the endpoints of these tests are defined by biological events (i.e. production of third brood and hatching of eggs) rather than a pre-determined exposure time, the typical test duration may be extended by up to two days.

Pre-release toxicity tests compare organism response in a control water, obtained from a pristine billabong located upstream of the mine, with various dilutions (typically, 0% (control), 0.3%, 1.0%, 3.2%, 10%, and 32%) of test (RP4) water. From these tests, the maximum concentration at which no observed effect (NOEC) and the lowest concentration at which an effect is observed (LOEC), relative to the controls, are determined from statistical analysis (one-way analysis of variance) of the test results obtained prior to the release of RP4 water.

It is evident from the data obtained in January 1997 (Table 8.6) that for all three test organisms none of the RP4 dilutions tested produced an observable effect (i.e. LOEC > 32%). However, RP4 dilutions tested in December 1998 showed an observable effect at 10% for all organisms, indicating that the toxicity of RP4 water may vary between wet seasons (Table 8.6). Furthermore, Holdway (1992a) has found that the toxicity of RP4 water to a range of freshwater organisms may vary considerably within a single wet season.

Table 8.6: Selected results from pre-release toxicity testing of RP4 water

Organism	Test start date	NOEC (%)	LOEC (%)
Green hydra	22/1/96	32	>32
(H. viridissima)	14/1/97	32	>32
	11/12/98	3.2	10
Water flea	21/1/96	32	> 32
(M. macleayi)	14/1/97	32	> 32
	18/12/98	3.2	10
Chequered	18/1/96	1	3.2
Rainbowfish	14/1/97	32	> 32
(M. splendida inornata)	11/12/98	3.2	10

[a] NOEC: no observed effect concentration.
[b] LOEC: lowest observed effect concentration.

The January 1996 results from toxicity tests with RP4 water demonstrate that not all organisms are equally sensitive (Table 8.6). The LOEC for the chequered rainbowfish was found to be 3.2%, whilst it was >32% for the green hydra and the water flea. Holdway (1992a) found that no one species was always the most sensitive to RP4 dilutions, highlighting the need to use a battery of test organisms for assessing the toxicity of whole effluent.

Creekside toxicity testing of Magela Creek water

The Ranger mine is currently trialing a field-based toxicity testing program to accompany the laboratory-based tests. The aim of the field-based program of testing is to check the efficacy of pre-release tests and provide an early warning of impacts caused by the mine wastewater releases. This is achieved by quantifying and assessing the relative toxicity of Magela Creek water, both upstream and downstream, before, during, and after the release of wastewater from RP4. Such a program is required to achieve the overall environmental protection objective, i.e. no observable effects upon local aquatic organisms, in a comprehensive and sensitive biological monitoring program. Development of the procedures used for the creekside monitoring program have been described by Humphrey et al. (1990, 1995a) and Boyden et al. (1995). In addition to providing early warning of unexpected toxic effects, the creekside program provides verification of the adequacy of release standards derived from the results of the laboratory-based prerelease toxicity tests, which employ equally sensitive endpoints.

Work stations have been established on the banks of Magela Creek, one using water upstream of the mine (control site) and another about 5 km downstream of the mine. At each site, two local freshwater organisms (*Amerianna cumingii* (freshwater snail) and *Melanotaenia nigrans* (black-striped rainbowfish)) (Table 8.7) are exposed to a constant flow of creek water in bench-mounted test containers using a pump and header (settling) tank arrangement. The test using *A. cumingii* assesses reproduction (egg production) of the adults and as well as the survival of the embryos and juveniles produced from the eggs (Lewis, 1992; Humphrey et al., 1995b). The test using *M. nigrans* determines larval survival (Boyden et al., 1995) (Table 8.7).

Table 8.7: Local freshwater organisms currently used in creekside monitoring of RP4 releases

Organism	Test endpoint	Test duration (days)
Snail	Adult egg production	4
Amerianna cumingii (Mollusca: Gastropoda)	Embryo/juvenile survival	15[a]
Black-striped rainbowfish *Melanotaenia nigrans* (Chordata: Osteichthyes)	Larval survival	4

[a] As the endpoint of this test is defined by a biological event (escape of juveniles from egg masses) rather than a pre-determined exposure time, the typical test duration may vary by up to two days.

The results of a number of tests conducted from 1991/92 to 1995/96 are given in Fig. 8.6. For the fish tests, larval survival at both sites was found to be between 80% and 95%, with relatively low variability. The differences between the upstream and downstream sites is also plotted; the mean value of the difference is approximately zero, and in all cases the measured differences from this mean value are not statistically significant ($P > 0.05$). Periods when RP4 water were discharged, are indicated in Fig. 8.6. These data show that such releases had no effect on the ability of larval *M. nigrans* to survive in waters downstream of the mine. Snail data show that there is considerable natural variability in the egg production rate of *A. cumingii*. Nevertheless, variability at the downstream site is similar to that at the upstream site, and the difference in response is not statistically significant ($P > 0.05$). Again, the discharge of water from RP4 (at the indicated times) had no detectable effect on the reproductive rate of *A. cumingii*. The above results are consistent with those from the last several wet seasons (1996/97-2000/01) (Supervising Scientist, 2001).

The results of the creekside tests verified the pre-release laboratory toxicity tests. These results are no surprise given that the rate of release

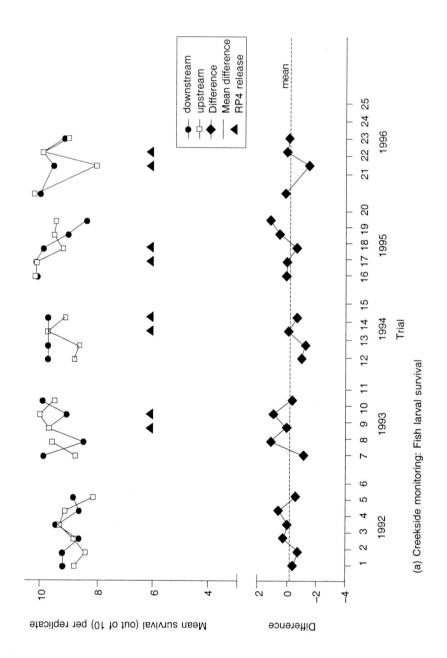

(a) Creekside monitoring: Fish larval survival

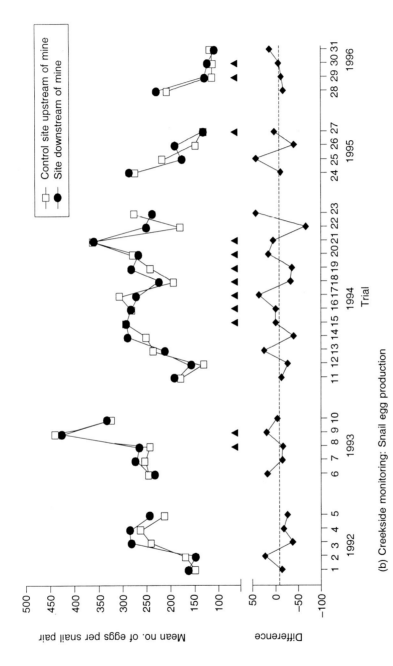

(b) Creekside monitoring: Snail egg production

Fig. 8.6: Results of creekside toxicity tests from 1991/92 to 1995/96 using (a) fish larval survival and (b) snail reproduction. Data are shown for animals exposed to Magela Creek water from sites upstream and downstream of the Ranger mine. Differences between upstream and downstream responses are also shown. Adapted from Johnston and Needham (1999) with permission.

of water from RP4 was limited by the requirements imposed by the chemical standards (NT DME, 1982) and by the dilution required on the basis of the prerelease toxicity tests. The monitoring data, therefore, confirm the adequacy of the controls imposed on the discharge of water from the mine site.

A practical in situ technique used to assess the effects of mine wastewater releases has been the measurement of the reproductive responses of freshwater mussels (*V. angasi*) held in mesh-covered containers buried along the creek edges (Humphrey et al., 1990). Adverse effects detected were of short duration only, and localized, being confined to a "mixing zone" immediately downstream of the discharge point.

Monitoring the community structure of macroinvertebrates and fish in Magela Creek

Macroinvertebrate communities have been sampled from a number of sites in Magela Creek at the end of substantial wet season flows, on an annual basis from 1988 to the present, with the aim of developing a monitoring technique to detect any impact from mining. Design and methodology of the research project have been gradually refined over this period to meet the needs of cost efficiency and improved ability to confidently attribute any observed changes to mining impact (Humphrey et al., 1990, 1995a).

Macroinvertebrate data for the 10-year period 1987/88–1996/97 have been analysed for a site in Magela Creek upstream of the Ranger mine (control) and a site about 5 km downstream (identical to those used for the creekside tests) to illustrate comparative changes in community structure (Johnston and Needham, 1999). Results are expressed in terms of dissimilarity (Bray-Curtis) measures: metrics that quantify the degree to which community structure differs between two samples, sites or sampling occasions. Dissimilarity measures range from 0 (the taxa and relative abundances of two samples are identical) to 1 (the taxa and relative abundances of two samples have nothing in common). The results concluded that year to year changes in the dissimilarity index between the upstream and downstream sites are smaller than the natural variability that occurs at the control site. Hence, any change in community structure that has occurred downstream from the mine has not been ecologically significant and may have been due to natural variability.

Additionally, studies of fish community structure in several billabongs have shown no evidence of mining related impacts since 1994 (Johnston and Needham, 1999). In one study, fish in two billabongs, one on Magela Creek downstream of the Ranger mine and the other on a separate "control" catchment (Nourlangie Creek, a tributary of the South Alligator River in Kakadu National Park), have been monitored by a visual counting technique, using a canoe with a transparent bow. Multivariate measures of the dissimilarity between these two fish communities indicated that although there were differences between streams, the differences remained relatively constant over the study period.

In summary, the biological monitoring program at Ranger has shown that operation of the mine has had no detectable impact on a range of sensitive indicators of ecological health, including survival of larval fish, reproduction of freshwater snails, and community structure of macroinvertebrates and fish.

Biological Response of Freshwater Biota to U

Surface water

Uranium is of greatest potential ecotoxicological concern in mine wastewaters that may be released from RP4 into Magela Creek during the wet season. Worldwide literature on the toxicity of U to freshwater biota has been largely derived from studies over the last two decades on the potential impacts of U mining in the Alligator Rivers Region of Australia's Northern Territory (see reviews by Markich and Camilleri (2002) and Markich (1998)). Data are available on the toxicity of U to 19 freshwater organisms from the Magela Creek catchment, including unicellular algae, hydra, bivalves, water fleas, prawns, and fishes. Organisms tested cover a range of trophic levels. Studies have exposed test organisms to U in natural and synthetic Magela Creek water. Synthetic water simulates the chemical composition of Magela Creek water during the wet season and provides a simple, standard, and well-defined water for toxicity testing with a variety of organisms. The use of synthetic water overcomes the complexity and variability of natural water and permits a more detailed understanding to be developed of the mechanisms of metal-organism interactions. This approach greatly improves the capability for predicting the biological effects of metals in fresh surface waters of varying chemistry.

The toxicity of U to the selected local test organisms varies considerably (Markich and Camilleri, 1997; Franklin et al., 2000). For example, the LOEC for the unicellular green alga, *Chlorella* sp. (13 μg l^{-1}; 3-day population growth) is 90-fold lower than that for the purple-spotted gudgeon (fish), *M. mogurnda* (1140 μg l^{-1}; 4-day larval survival) in synthetic Magela Creek water under similar physicochemical conditions. With respect to the three test organisms currently used to assess the toxicity of RP4 water prior to release (Table 8.5), the water flea (LOEC = 19 μg l^{-1}; 7-day reproduction) is typically more sensitive to U in natural Magela Creek water than green hydra (LOEC = 170 μg l^{-1}; 4-day population growth) (Markich and Camilleri, 1997). The toxicity (6-day embryo survival) of U to the chequered rainbowfish has not been determined.

However, Bywater et al. (1991) reported a 4-day LC_1 (analogous to a LOEC) and LC_{50} of 880 and 2660 µg U l^{-1}, respectively, for juvenile (7-day old) chequered rainbowfish. For the creekside toxicity tests, indications are that the snail is more sensitive to U (LOEC = ~170 µg l^{-1}; 4-day egg production) (Lewis, 1992) than the black-striped rainbowfish (LOEC = 320-370 µg l^{-1}; 4-day larval survival) (Bywater et al., 1991; Holdway, 1992b).

The LOEC values for all organisms exceed the guideline value of 5 µg U l^{-1} recommended to protect freshwater biota in Magela Creek (NT DME, 1982). This value was based on a maximum allowable addition of 3.8 µg U l^{-1} to the receiving waters of Magela Creek (allowing up to 1.2 µg U l^{-1}), which forms part of the Authorisation to Operate. More importantly, this value is consistent with the guideline value of 5 µg U l^{-1} derived by Markich and Camilleri (1997) and Hogan et al. (2002) using available local freshwater toxicity data. Based on the available data, it would appear that the existing guideline values are protective for the range of species tested.

Sediment

In contrast to the extensive development of water-only toxicity tests with local aquatic organisms, scant attention has been given to the development of sediment toxicity tests. Sediments in Magela Creek (particularly the backflow billabongs) may serve as a source and a sink of dissolved contaminants such as U. A sediment toxicity test employing a local chironomid, *Chironomus crassiforceps*, has been developed for assessing the bioavailability of U and Cu in the Magela Creek flood plain. (Peck et al., 2002).

Uranium concentrations in the sediments of Mudginberri Billabong, located in the main channel of Magela Creek 12 km downstream of the Ranger mine, have remained relatively constant [3.9 ± 0.5 mg kg^{-1} (mean ± 95% confidence limit); range, 2-6 mg kg^{-1}] since 1984 (ERA, 2000). This result indicates no cumulative impact from mining on the Ranger lease over this period. Furthermore, Noller and Hart (1993) found that U concentrations in Magela Creek sediments (corrected for particle size and % organic carbon) were similar between upstream and downstream sites. They concluded that transport of U by Magela Creek water was most likely associated with the filterable or fine colloidal material (< 0.45 µm) and not with the suspended matter (>0.45 µm).

A sediment guideline value for U does not currently exist in Australia or elsewhere due to a lack of toxicity data. However, measured concentrations of Cr, Cu, Pb, and Zn in Magela Creek sediments (Thomas and Hart, 1984; ERA, 2000) are below their respective national sediment guideline values (ANZECC and ARMCANZ, 2000) (Table 8.8).

Table 8.8: Concentrations of metals in sediments (mg kg^{-1} dry weight) from Magela Creek compared with national guideline values

	U	Cr	Cu	Pb	Zn
Mean	4.1	41	23	14	25
95% CI	3.1–5.1	36–46	19–27	12–16	20–30
Range	1.4–7.5	25–65	9–37	9–20	10–54
SQG–low[a]	--	80	65	50	200

[a] Sediment quality guideline value below which there is a low probability of biological effect (ANZECC and ARMCANZ, 2000).

Overall, the concentrations of metals, including U, in Magela Creek sediments downstream of the mine are likely to have minimal impact on aquatic biota, except for perhaps in extreme physicochemical conditions, such as naturally-occurring acidic first flush events that may cause the release of metals from sediments into the surrounding water. Given that all evidence indicates that the levels of metals present have not been significantly ($P > 0.05$) increased by mining at Ranger, any such release would be expected to be part of the natural cycle of exposure for the local aquatic organisms.

Speciation and Bioavailability of U to Freshwater Biota

Guidelines for metals in aquatic ecosystems

Guidelines for metals in aquatic ecosystems are typically based on total (i.e. dissolved and particulate) concentrations (Gardiner and Zabel 1989; CCREM 1991; US EPA 1995a, 1995b, Roux et al., 1996). However, it is well recognized that biological effects of metals are critically dependent on their physicochemical form, or speciation, and not total metal concentration (Brown and Markich, 2000; Markich et al., 2001). Metals in aquatic systems exist in a variety of physicochemical forms, including the free metal ion (M^{z+}) and metal complexes with organic and inorganic ligands in dissolved, colloidal and/or particulate forms (Pickering, 1995). Although the relationship between metal speciation and biological effect is complex, metals present as the free metal ion (e.g. UO_2^{2+} or Cu^{2+}) or as weak complexes, that are able to dissociate at a cell or gill membrane, are generally considered more bioavailable than metals in strong complexes or adsorbed to colloidal and/or particulate matter (Hamelink et al., 1994). Guidelines based on total metal concentrations will generally be overprotective, since only a fraction of the total metal concentration in natural waters will be generally bioavailable, especially in samples containing appreciable concentrations of suspended particulate matter and/or dissolved organic matter (DOM) (Markich et al., 2001).

In a world first, metal speciation and bioavailability are incorporated into the Australian and New Zealand water quality guidelines for the protection of aquatic ecosystems, using a risk-based, decision tree approach (Markich et al., 2001). The decision tree includes three increasingly complex levels of analysis: (1) acid-extractable metal concentration in an unfiltered sample; (2) acid-extractable metal concentration in an filtered sample, and (3) concentrations of specific metal species (e.g. geochemical modeling) or groups of metal species (e.g. chemical measurement) and/or biological testing. The complexity of analysis for a water sample depends on guideline conformity at each level of the decision tree. The measurement of metal speciation and bioavailability permits guidelines for protecting aquatic ecosystems to be potentially relaxed on a site-specific basis. This signifies a move to further develop chemical indicators so that they better reflect biological effects.

Factors affecting metal speciation and bioavailability

Metal speciation and bioavailability in fresh surface waters may be strongly influenced by a variety of physicochemical variables, including hardness (i.e. primarily Ca and/or Mg concentration), alkalinity, pH and natural dissolved organic matter (including fulvic and humic acids) (Campbell and Stokes, 1985; Meador, 1991; Markich and Jeffree, 1994; Erickson et al., 1996). The effects of these water quality variables on the toxicity of U to freshwater organisms from Magela Creek have been investigated.

Dissolved organic matter and pH. Markich et al. (2000) found that valve movement responses (measured in terms of the duration of valve opening, DVO) of the freshwater mussel, *V. angasi*, exposed to U in synthetic Magela Creek water were highly dependent ($P \leq 0.001$) on the pH and/or the concentration of dissolved organic matter (expressed in the form of a model fulvic acid (FA), comprising aspartate, citrate, malonate, salicylate, and tricarballyate in defined ratios). For a given model FA concentration, the toxicity (2-day EC_{50}) of U to *V. angasi* decreased exponentially as the pH increased from 5.0 to 6.0 (Fig. 8.7). Similarly, for a given pH, the toxicity of U to *V. angasi* decreased exponentially as the model FA concentration increased from 0 to 7.9 mg l^{-1}.

In the absence of model fulvic acid, the toxicity of U to *V. angasi* decreased by a factor of about 5, with an increase in pH from 5.0 to 6.0 (e.g. the EC_{50} decreased from 556 µg l^{-1} at pH 6.0 to 103 µg l^{-1} at pH 5.0). Additionally, in the presence of the maximum concentration of model FA (7.9 mg l^{-1}), the toxicity of U to *V. angasi* at pH 5.0 and 6.0 was reduced by about a factor of 2. For example, the EC_{50} increased from 556 µg U l^{-1} at pH 6.0 to 1080 µg U l^{-1} at pH 6.0 + 7.9 mg l^{-1} FA. The toxicity of U to *V. angasi* at a given total U concentration was greatest at pH 5.0 in the absence of model FA, and least at pH 6.0 with the maximum concentration (7.9 mg l^{-1}) of model FA (Fig. 8.7).

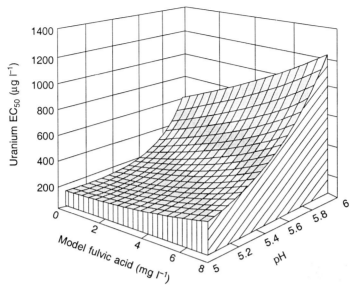

Fig. 8.7: Three-dimensional surface plot showing how the toxicity (EC_{50}) of U to the freshwater bivalve *V. angasi* varies as a function of pH and model fulvic acid concentration.

Speciation of U at pH 5.0 and pH 6.0 + 7.9 mg l^{-1} FA was calculated using the geochemical speciation model HARPHRQ (Fig. 8.8). In summary, at pH 5.0, in the absence of model FA, UO_2^{2+} (43–58%) and UO_2OH^+ (26–36%) were the dominant uranyl species predicted to form (Fig. 8.8a). Relative proportions of these two species decreased with increasing total U concentration; such decreases were offset by increases in the relative proportions (0–12%) of polymeric uranyl species (i.e. $(UO_2)_2(OH)_2^{2+}$ and $(UO_2)_3(OH)_5^+$). In contrast, proportions of UO_2^{2+} (2–4%) and UO_2OH^+ (8–12%) formed a minor contribution to total U concentration at pH 6.0 + 7.9 mg^{-1} FA (Fig. 8b). As expected, predicted speciation of U was altered by addition of model FA, with the formation of three organic uranyl species ($UO_2(OH)Cit^{2-}$, UO_2Mal and $UO_2(OH)Mal^-$, where Cit is citrate and Mal is malonate). The predicted concentrations of UO_2^{2+} and UO_2OH^+ at pH 5.0 were in close agreement (2-5% difference) with those measured using time-resolved laser-induced fluorescence spectroscopy. A more detailed discussion of U speciation under these conditions is given by Markich (1998).

No previous studies have determined the effects of fulvic and/or humic acids on U uptake by, or toxicity to, aquatic organisms. However, with respect to the effect of citrate, one of the organic acids used in the model FA, results of this study appear to be consistent with those of Yong and

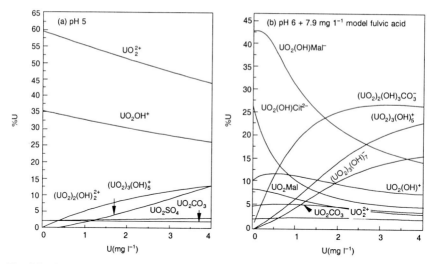

Fig. 8.8. Predicted speciation (% distribution) of U in synthetic Magela Creek water at (a) pH 5.0 in the absence of model fulvic acid, and (b) pH 6.0 in the presence of the maximum concentration of model fulvic acid (7.9 mg l⁻¹). Uranyl species comprising <2% are not shown.

Macaskie (1995), who found that the formation of uranyl-citrate reduced the toxicity of U to the bacterium *Citrobactor* at pH 7.0.

Markich et al. (2000) also compared the valve movement response (DVO) of *V. angasi* to U in natural and synthetic Magela Creek water (matched in terms of major and trace element concentrations, pH (5.5) and dissolved organic carbon concentration (6.4 mg l⁻¹), to test the practical use of the latter in predicting the potential toxicity of U to freshwater organisms in Magela Creek. The results showed that there was no significant ($P > 0.05$) difference in the mean DVO of *V. angasi* exposed to U in both waters (Table 8.9). The above results strongly indicate that the valve movement response of *V. angasi* to U in Magela Creek water can be reliably estimated using customized synthetic water. This approach may ultimately improve risk assessment models for the protection of aquatic ecosystems downstream of the Ranger mine. Therefore, the risk to *V. angasi* in surface waters of Magela Creek during releases of mine wastewaters containing U is predicted to be greatest at pH 5.0 in the presence of very low FA concentrations (FA is seldom absent) and least at pH 6.0 in the presence of high FA concentrations. Overall, the results emphasize the potential biological effects of U in a specific freshwater system with variable concentrations of FA, where relatively small changes in seasonal pH (5–6) occur.

Using multiple linear regression analysis, combined with geochemical speciation modeling, Markich et al. (2000) provided evidence to show that under the prescribed experimental conditions the biological response

Table 8.9: Comparative toxicity of *V. angasi* to U in synthetic and natural Magela Creek water at pH 5.5 and 6.4 mg l^{-1} DOC (5.7 mg l^{-1} FA)[a]

Biological endpoint	Synthetic water	Natural water
NOEC	275	285
LOEC	292	301
EC_{50} (95% CI)	358 (335–381)	367 (343–391)

[a] Values are reported in μg l^{-1}. NOEC: no observed effect concentration. LOEC: lowest observed effect concentration.

(BR) of *V. angasi* to U, was related to the activity of particular U species (i.e. BR $\propto 1.86 \times UO_2^{2+} + UO_2OH^+$) and not total U concentration. These results indicate that UO_2^{2+} has nearly a 2-fold higher binding affinity than UO_2OH^+ does at the cell membrane surface.

Results on the effects of pH on the valve movement response of *V. angasi* to U apparently differ with the results of Franklin et al. (2000), who showed that toxicity (3-day EC_{50}; population growth) of the unicellular green alga *Chlorella* sp. to U in synthetic Magela Creek water increased by a factor of 1.8 (78 to 38 μg U l^{-1}) when pH increased from 5.7 to 6.5 in the absence of dissolved organic matter. Speciation of U in this concentration range is predicted to be dominated (70-93%) by the polymeric uranyl hydroxy carbonate species, $UO_2(OH)_3CO_3^-$, with only a very small (<4%) proportion occurring as the free uranyl ion (UO_2^{2+}). Given the small proportion of UO_2^{2+} at both pH values, the effect of H^+ itself may have been responsible for reducing the toxicity of U to *Chlorella* sp. at pH 5.7. Indeed, Franklin et al. (2000) found that intracellular U was 2-fold lower at pH 5.7 than at 6.5. No other studies have reported on the effect of pH on the toxicity of U to freshwater organisms.

Hardness. The influence of water hardness on the toxicity of U to freshwater organisms in Magela Creek is potentially important since Ca and Mg concentrations in RP4 are typically a factor of 25 (11 mg l^{-1}) and 250 (150 mg l^{-1}) higher, respectively, than Magela Creek water (Table 8.4; *ca.* 520 mg l^{-1} as $CaCO_3$ in RP4 water compared to 3-4 mg l^{-1} as $CaCO_3$ in Magela Creek). Riethmuller et al. (2000) found that the toxicity (4-day EC_{50}) of U to the green hydra *H. viridissima* in synthetic Magela Creek water (pH 6.0 ± 0.2 and constant alkalinity, 4 mg l^{-1} as $CaCO_3$) decreased by a factor of 2 (114 to 219 μg U l^{-1}) with a 50-fold increase in water hardness (6.6 to 330 mg l^{-1} as $CaCO_3$). In contrast, the authors found no significant ($P > 0.05$) effect of water hardness on the toxicity (4-day LC_{50}) of U to the purple spotted gudgeon, *M. mogurnda* (i.e. the 95% confidence limits overlapped; 1750 ± 115 μg U l^{-1} at 6.6 mg l^{-1} as $CaCO_3$ and 1645 ± 145 μg U l^{-1} at 330 mg l^{-1} as $CaCO_3$).

Charles et al. (2002) found that a 50-fold increase in water hardness (8-400 mg l^{-1} as $CaCO_3$), added as Ca and Mg sulfate, resulted in a five-fold decrease in the toxicity of U to the tropical freshwater alga *Chlorella* sp. (i.e. an increase in the 3-day EC_{50} from 56 µg U l^{-1} at 8 mg l^{-1} as $CaCO_3$ to 270 µg U l^{-1} at 400 mg l^{-1} as $CaCO_3$). Few other studies have determined the effects of increasing water hardness on the toxicity of U to freshwater organisms (Parkhurst et al., 1984 (fish); Poston et al., 1984 (water flea); Barata et al., 1998 (water flea)). However, because these studies confounded the effects of increasing water hardness with increasing alkalinity and pH, the true effects of water hardness (i.e. Ca and/or Mg concentration) on U toxicity could not be discerned.

General Discussion

The two case studies presented describe two contrasting, but complementary, biological assessment approaches required to
1. minimize downstream impact by a new or currently operating mine; and
2. measure the degree of ecological recovery of a river system seriously impacted by historical, or existing, mining activity.

In the case of a new mine, one of the more essential requirements will be to obtain a reliable baseline data set for water quality that accounts for seasonal variation, such as naturally acidic first flush events or other significant annual occurrences. If this is not done, subsequently measured impacts can incorrectly be attributed to the mining operation. The first level of comparison of the measured baseline values would be with the current water quality guidelines, usually for ecosystem protection. However, these guidelines may be based on data that are not applicable to local (site-specific) conditions. For example, a river may already be impacted by the weathering products from the surface expression of an orebody. For cases such as this, the baseline monitoring program should include biological testing, since for routinely measured physicochemical parameters (i.e. major ions, metals and nutrients) do not provide an overall measure of toxic response. In particular, they will not account for seasonal changes in the presence of, for example, naturally occurring toxic organic compounds.

Ideally, local organisms should be used for biological testing. However, few mining companies would have the resources available in the first instance to develop such procedures for local organisms. A practical alternative would be a battery of toxicity tests using "standard" organisms (e.g. water fleas, fish, algae) to compare the toxic response in a synthetic culture water with test water from the site (i.e. the "water-effect ratio"; US EPA (1994)). Results from this assessment will provide an indicative assessment of toxicity, which will provide a firmer base than physicochemical measurements alone. Tests using the same group of "standard" test organisms could then be done on a seasonal basis following commencement of the mine, and the response compared with that obtained for the baseline studies. This approach would provide additional confidence in the event that the before and after distributions of test results were the same.

It is possible that a mine may wish to discharge water from the site. In this case, it may be able to be simply demonstrated (using chemical analysis) that the dilution factors are so large that there will be a negligible increase in the concentration of solutes downstream of the mixing zone. However, if the increase is significant, then in the first instance, the regionally applicable ecosystem protection criteria should be used as a benchmark. If the increment of concentration is greater than this, or regulatory concerns are such to question the application of "generic" protection criteria, then toxicity testing using local organisms may be required to establish the no observable effects dilution level for the wastewater in question.

In the event that a river system has already been seriously impacted by discharge from an existing mine, or sites on which mining is no longer taking place, then a combination of physicochemical parameters and biological indicators should be used to map the downstream extent of adverse impact. In this case, generic ecosystem protection guidelines could be used as a starting benchmark. However, it is very important to recognize that using measured concentrations of solutes alone, especially from one-off or "snapshot" river profiles, may give a misleading impression owing to the potential for considerable variability in the ratio of flows in discharges from the mine site and the stream. Measurements of biological indicators, incorporating species abundance, diversity and community composition, are likely to provide a better integrated picture of the effect of impacts over a period of time. In the event that remedial measures are undertaken, the upstream and downstream results from the original surveys can be used to track, and quantitatively document, recovery of the system.

One of the potential problems associated with the traditional biological indicators approach, is the difficulty in obtaining appropriately matched control habitat environments. This may not be an issue if the river channel has very similar limnological characteristics upstream, and extending downstream below the (serious) impact zone. However, in many situations this will not be the case, and so control sites in a suitable unpolluted river, located as close as possible, will be required. However, there is no guarantee that there will be a good match between the species types present in similar physical environments in the two rivers. In this event, it would be more appropriate to use a trophic-level approach, where the occupancy of the various available niches in the system are compared. This approach is being developed in Northern Australia by Corbett (1997).

ACKNOWLEDGMENTS

We are grateful to Jeff Evans and Jeremy Thomson for preparing some of the figures. Paul Brown, John Bennett, and Andrew Jackson provided useful comments on an earlier manuscript.

REFERENCES

Allen, C.G. and Verhoeven, T.J. (1986) *The Rum Jungle Rehabilitation Project.* Final Project Report, Northern Territory Department of Mines and Energy, Darwin, Australia.

Allison, H.E. and Simpson, R.D. (1989) *Element Concentrations in the Freshwater Mussel, Velesunio angasi, in the Alligator Rivers Region.* Technical Memorandum 25, Supervising Scientist for the Alligator Rivers Region, Canberra, Australia.

ANZECC and ARMCANZ (2000) *Australian and New Zealand Guidelines for Fresh and Marine Water Quality.* Australia and New Zealand Environment and Conservation Council and the Agriculture and Resource Council of Australia and New Zealand, Canberra, Australia.

Barata, C., Baird, D.J., and Markich, S.J. (1998) Influence of genetic and environmental factors on the tolerance of *Daphnia magna* to essential and non-essential metals. *Aquat. Toxicol.* **42**: 115–137.

Boyden, J., Pidgeon, B., and Humphrey, C.L. (1995) *ERISS Protocol for the Creekside Monitoring of Magela Creek Waters II. Larval Black-striped Rainbowfish, Melanotaenia nigrans, Survival test.* Internal Report 181. Supervising Scientist, Darwin, Australia.

Brown, P.L., Haworth, A., Sharland, S.M., and Tweed, C.J. (1991) *HARPHRQ: An Extended Version of the Geochemical Code PHREEQE.* Nirex Safety Series Report 188. United Kingdom Atomic Energy Authority, Harwell Laboratory, Oxon.

Brown, P.L. and Markich, S.J. (2000) Evaluation of the free ion activity model of metal-organism interaction: Extension of the conceptual model. *Aquat. Toxicol.* **51**: 177–194.

Brown, T.E., Morley, A.W., Sanderson, N.T., and Tait, R.D. (1983) Report of a large fish kill resulting from natural acid water conditions in Australia. *J. Fish Biol.* **22**: 335–350.

Brown, V.M. (1986) Development of water quality criteria from ecotoxicological data. In *Water Quality Management: Freshwater Ecotoxicology in Australia*, Hart, B.T. (Ed.). Chisholm Institute of Technology, Melbourne, Australia, pp. 107–117.

Bywater, J.F., Banaczkowski, R., and Bailey, M. (1991) Sensitivity to uranium of six species of tropical freshwater fishes and four species of cladocerans from northern Australia. *Environ. Toxicol. Chem.* **10**: 1449–1458.

Camilleri, C., Markich, S.J., Noller, B.N., Turley, C.J., Parker, G., and van Dam, R. Silica reduces the toxicity of aluminium to a tropical freshwater fish. (2002) *Chemosphere*, in press.

Campbell, P.G.C. and Stokes, P.M. (1985) Acidification and toxicity of metals to aquatic biota. *Can. J. Fish. Aquat. Sci.* **42**: 2034–2049.

CCREM (1991) *Canadian Water Quality Guidelines.* Canadian Council of Resource and Environment Ministers, Inland Waters Directorate, Environment Canada, Ottawa, Canada.

Charles, A.L., Markich, S.J., Stauber, J.L., and De Filippis, L.F. (2002) The effect of water hardness on the toxicity of uranium to a tropical freshwater alga (*Chorella* sp.). *Aquat. Toxicol.*, in press.

Cherry, D.S., Currie, R.J., Soucek, D.J., Latimer, H.A., and Trent, D.J. (2001) An integrative assessment of a watershed impacted by abandoned mined land discharges. *Environ. Pollut.* 111: 377–388.

Conway, N.F. and Davy, D.R. (1975) The pollution cycle at Rum Jungle: Controlling factors, basic mechanisms and degree. In *Rum Jungle Environmental Studies*, Davy, D.R. (Ed.). Australian Atomic Energy Commission, Sydney, Australia, AAEC/E365.

Corbett, L. (1997) Environmental monitoring at the Ranger mine. In *Proceedings of the 22nd Annual Environmental Workshop*. Minerals Council of Australia, Canberra, Australia, pp. 240–255.

Cusbert, P.J., leGras, C.A., Hunt, C.W., and Akber, R.A. (1992) *Magela Creek Water Quality During 1991/92 Wet Season: An Interim Report*. Internal Report 67. Supervising Scientist for the Alligator Rivers Region, Canberra, Australia.

Davy, D.R. and Jones, J. (1975) Annual inputs of pollutants to the Finniss River system. In *Rum Jungle Environmental Studies*, Davy, D.R. (Ed.). Australian Atomic Energy Commission, Sydney, Australia, AAEC/E365.

Davy, D.R., Jones, J., and Lowson, R.T. (1975) Water quality in the Rum Jungle area. In *Rum Jungle Environmental Studies*, Davy, D.R. (Ed.). Australian Atomic Energy Commission, Sydney, Australia, AAEC/E365.

Depledge, M.H., Weeks, J.M., and Bjerregaard, P. (1994) Heavy metals. In *Handbook of Ecotoxicology*, Vol. 2, Calow, P. (Ed.). Blackwell Scientific, London. pp. 79–105.

Edwards C.A. (2001) *Effects of the Remediated Rum Jungle Mine on the Macroinvertebrate Community Composition in the East Branch of the Finniss River, Northern Territory*. M.Sc. Dissertation, University of Technology, Sydney, Australia.

ERA. (1996) *Environmental Annual Report: Ranger Mine*. Energy Resources of Australia, Jabiru, Australia.

ERA (1997) *Environmental Annual Report: Ranger Mine*. Energy Resources of Australia, Jabiru, Australia.

ERA (1998) *Environmental Annual Report: Ranger Mine*. Energy Resources of Australia, Jabiru, Australia.

ERA (1999) *Environmental Annual Report: Ranger Mine*. Energy Resources of Australia, Jabiru, Australia.

ERA (2000) *Environmental Annual Report: Ranger Mine*. Energy Resources of Australia, Jabiru, Australia.

ERA (2001) *Annual Report*. Energy Resources of Australia Ltd, Jabiru, Australia.

Erickson, R.J., Benoit, D.A., Mattson, V.R., Nelson, H.P., and Leonard, E.N. (1996) The effects of water chemistry on the toxicity of copper to fathead minnows. *Environ. Toxicol. Chem.* 15: 181–193.

Ferris, J.M. and Vyverman, W. (1996) Tropical benthic diatoms in a pollution gradient associated with acid rock drainage. In *Proceedings of the INTECOL V International Wetland Conference*, The International Association of Ecology, Perth, Australia, p. 120.

Ferris, J.M., Vyverman, W., and Ling, H.U. (1995) Benthic algae in a pollution gradient associated with acid mine drainage. In *Proceedings of the Second Annual Conference of the Australasian Society for Ecotoxicology*, Australasian Society for Ecotoxicology, Sydney, Australia, p. 16.

Fox, R.W., Kelleher, G.G., and Kerr, C.B. (1977) *Ranger Uranium Environmental Inquiry. Second Report*. Australian Government Publishing Service, Canberra, Australia.

Franklin, N., Stauber, J.L., Markich, S.J., and Lim, R.P. (2000) pH-dependent toxicity of Cu and U to a tropical freshwater green alga (*Chlorella* sp.). *Aquat. Toxicol.* 48: 275–289.

Fucik, K.W., Herron, J., and Fink, D. (1991) The role of biomonitoring in measuring the success of reclamation at a hazardous waste site. In *Aquatic Toxicology and Risk Assess-*

ment, American Society for Testing and Materials, Philadelphia. *ASTM 1124*. pp. 212–220.

Gale, S.A., Smith, S.V., Lim, R.P., Jeffree, R.A., and Petocz, P. (2003) Mechanism of copper tolerance in a population of black-banded rainbowfish (*Melanotaenia nigrans*). *Aquat. Toxicol.*, in press.

Gardiner, T. and Zabel, T. (1989) *United Kingdom Water Quality Standards Arising from European Community Directives—An Update*. Final Report 0041, Water Research Centre, Swindon, United Kingdom.

Hamelink, J.L., Landrum, P.F., Bergman, H.L., and Benson, W.H. (Eds). (1994) *Bioavailability: Physical, Chemical and Biological Interactions*. Lewis Publishers, Boca Raton.

Hart, B.T., Bailey, P., Edwards, R., James, K., Swadling, K., Meredith, C., McMahon, A., and Hortle, K. (1991) Biological effects of saline discharges to streams and wetlands: A review. *Hydrobiologia* 210: 105–144.

Hart, B.T., Currey, N.A., and Jones, M.J. (1992b) Biogeochemistry and effects of copper, manganese and zinc added to enclosures in Island Billabong, Magela Creek, Northern Australia. *Hydrobiologia* 230: 93–134.

Hart, B.T., Davies, S.H., and Thomas, P.A. (1982) Transport of iron, manganese, cadmium, copper and zinc by Magela Creek, Northern Territory, Australia. *Water Res.* 16: 605–612.

Hart, B.T. and McGregor, R.J. (1980) Limnological survey of eight billabongs in the Magela Creek catchment, Northern Territory. *Aust. J. Mar. Freshwater Res.* 31: 611–626.

Hart, B.T., Noller, B.N., le Gras, C., and Currey, N.A. (1992a) Manganese speciation in Magela Creek, Northern Australia. *Aust. J. Mar. Freshwater Res.* 43: 421–441.

Hart, B.T., Ottaway, E.M., and Noller, B.N. (1987a) Magela Creek system, northern Australia. I. 1982-1983 Wet-season water quality. *Aust. J. Mar. Freshwater Res.* 38: 261–288.

Hart, B.T., Ottaway, E.M., and Noller, B.N. (1987b) Magela Creek system, Northern Australia. II. Material budget for the floodplain. *Aust. J. Mar. Freshwater Res.* 38: 861–876.

He, M., Wang, Z., and Tang, H. (1998) The chemical, toxicological and ecological studies in assessing the heavy metal pollution in Le An River, China. *Water Res.* 32: 510–518.

Hogan, A., van Dam, R., Camilleri, C., and Markich, J.J. (2002). Toxicity of uranium to the tropical alga *Chlorella* sp. for the derivation of a site specific trigger value for Magela Creek, northern Australia. In *Interact 2002: Programme and Abstract Book*, Warne, M. and Hibbert, B. (Eds). The Royal Australian Chemical Institute, The Australasian Society of Ecotoxicology and the International Chemometrics Society, Sydney, Australia, pp 157.

Holdway, D.A. (1992a) Control of metal pollution in tropical rivers in Australia. In *Pollution in Tropical Aquatic Systems*, Connell, D.W. and Hawker, D.W. (Eds). CRC Press, Boca Raton, pp. 231–246.

Holdway, D.A. (1992b) Uranium toxicity to two species of Australian tropical fish. *Sci. Total Environ.* 125: 137–158.

Humphrey, C.L., Bishop, K.A., and Brown, V.M. (1990) Use of biological monitoring in the assessment of effects of mining wastes on aquatic ecosystems of the Alligator Rivers Region, tropical northern Australia. *Environ. Monitor. Assess.* 14: 139–181.

Humphrey, C.L. and Dostine, P.L. (1994) Development of biological monitoring programs to detect mining waste impacts upon aquatic ecosystems of the Alligator Rivers Region, Northern Territory, Australia. *Mitt. Int. Ver. Theor. Angew. Limnol.* 24: 293–314.

Humphrey, C.L., Faith, D.P., and Dostine, P.L. (1995a) Baseline requirements for assessment of mining impact using biological monitoring. *Aust. J. Ecol.* 20: 150–166.

Humphrey, C.L., Lewis, B.F., Brown, I., and Suggit, J.L. (1995b) *ERISS Protocol for the Creekside Monitoring of Magela Creek Waters I. Freshwater Snail, Amerianna cumingii, Reproduction and survival test*. Internal Report 180. Supervising Scientist, Darwin, Australia.

Humphrey, C.L. and Simpson, R.D. (1985) *The Biology and Ecology of Velesunio angasi (Bivalvia: Hyriidae) in the Magela Creek, Northern Territory*. Open File Record 38. Supervising Scientist for the Alligator Rivers Region, Canberra, Australia.

Hyne, R.V., Rippon, G.D., Hunt, S.M., and Brown, G.H. (1996) *Procedures for the Biological Toxicity Testing of Mine Waste Waters using Freshwater Organisms*. Supervising Scientist Report 110. Supervising Scientist, Canberra, Australia.

Jackson, S. and Ferris, J.M. (1998) Macroinvertebrate ecology of the Finniss River East Branch at the beginning of the 1993 dry season. In *Rum Jungle Rehabilitation Project, Monitoring Report 1988-1993*, Kraatz, M. (Ed.). Northern Territory Department of Lands, Planning and Environment, Darwin, Australia, pp. 77–83.

Jeffree, R.A. and Twining, J.R. (2000) Contaminant water chemistry and distribution of fishes in the East Branch, Finniss River, following remediation of the Rum Jungle uranium/copper mine site. In *Contaminated Site Remediation: From Source Zones to Ecosystems*, Johnston, C.D. (Ed.). Centre for Groundwater Studies, Wembley, Australia, pp. 51–56.

Jeffree, R.A., Twining, J.R., and Thomson, J. (2001) Recovery of fish communities in the Finniss River, northern Australia, following remediation of the Rum Jungle uranium/copper mine site. *Environ. Sci. Technol.* **35**: 2932–2941.

Jeffree, R.A. and Williams, N.J. (1975) Biological indicators of pollution of the Finniss River system, especially fish diversity and abundance. In *Rum Jungle Environmental Studies*, Davy, D.R. (Ed.). Australian Atomic Energy Commission, Sydney, Australia, AAEC/E365.

Jeffree, R.A. and Williams, N.J. (1980) Mining pollution and the diet of the purple-striped gudgeon *Mogurnda mogurnda* Richardson (Eleotridae) in the Finniss River, Northern Territory, Australia. *Ecol. Monogr.* **50**: 457–485.

Johnston, A. (1991) Water management in the Alligator Rivers Region: A research view. In *Proceedings of the 29th Congress of the Australian Society of Limnology*, Hyne, R.V. (Ed.). Supervising Scientist for the Alligator Rivers Region, Canberra, Australia, pp. 10–34.

Johnston, A. and Needham, R.S. (1999) *Protection of the Environment Near the Ranger Uranium Mine*. Supervising Scientist Report 139. Environment Australia, Canberra, Australia.

Kendall, C.J. (1990) Ranger uranium deposits. In *Geology of the Mineral Deposits of Australia and Papua New Guinea*, Hughes, F.E. (Ed.). The Australasian Institute of Mining and Metallurgy, Melbourne, Australia, pp. 799–805.

Kimmel, W.G., Miller, C.A., and Moon, T.C. (1981) The impact of a deep-mine drainage on the water quality and biota of a small hard-water stream. *Proc. Pennsyl. Acad Sci.* **55**: 137–141.

Kinhill and ERA (1998) *The Jabiluka Project. The Jabiluka Mill Alternative*. Kinhill Engineers and Energy Resources of Australia, Sydney, Australia.

Klessa, D.A. (2000) *The Chemistry of Magela Creek. A Baseline for Assessing Change Downstream of Ranger*. Supervising Scientist Report 151. Supervising Scientist, Canberra, Australia.

Lawton, M.D. (1998) Surface water quality and hydrology: East Finniss River 1988/89-1992/93 and mine site surface water quality 1989-91. In *Rum Jungle Rehabilitation Project, Monitoring Report 1988-1993*, Kraatz, M. (Ed.). Northern Territory Department of Lands, Planning and Environment, Darwin, Australia, pp. 17–28.

le Gras, C., Klessa, D.A., Cusbert, P., Pidgeon, B., and Boyden, J. (1999) Natural acidification processes leading to fish mortality in a tropical billabong of Northern Australia. In *Proceedings of the EnviroTox '99 International Conference*. The Royal Australian Chemical Institute and the Australasian Society for Ecotoxicology, Geelong, Australia, p. 35.

Lehninger, A.L. (1982) *Principles of Biochemistry*. Worth Publishers, New York.

Lewis, B.F. (1992) *The Assessment of Seven Northern Territory Gastropod Species for Use as Biological Monitors of Ranger Uranium Mine Retention Pond Waters*. Open File Record 100. Supervising Scientist for the Alligator Rivers Region, Canberra, Australia.

Lloyd, D.S., Koenings, J.P., and LaPerriere, J.D. (1987) Effects of turbidity in fresh waters of Alaska. *North Am. J. Fish. Manage.* **7**: 34–45.

Malmqvist, B. and Hoffsten, P.O. (1999) Influence of drainage from old mine deposits on benthic macroinvertebrate communities in central Swedish streams. *Water Res.* **33**: 2415–2423.

Markich, S.J. (1996) *Element Concentrations in the Soft Tissues of a Freshwater Bivalve Species from the Mammy-Johnson River, NSW*. Australian Nuclear Science and Technology Organisation, Sydney, Australia, ANSTO/C485.

Markich, S.J. (1998) *Effects of Biological and Physicochemical Variables on the Valve Movement Responses of Freshwater Bivalves to Mn, U, Cd or Cu*. Ph.D. Dissertation, University of Technology, Sydney, Australia.

Markich, S.J. (2002) Uranium speciation and bioavailability in aquatic systems: An overview. *The Scientific World Journal* **2**: 707-729.

Markich, S.J., Brown, P.L., Batley, G.E., Apte, S.C., and Stauber, J.L. (2001) Incorporating metal speciation and bioavailability into water quality guidelines for protecting aquatic ecosystems. *Australas. J. Ecotoxicol.* **7**: 109-122.

Markich, S.J., Brown, P.L., Jeffree, R.A., and Lim, R.P. (2000) Valve movement responses of *Velesunio angasi* (Bivalvia: Hyriidae) to Mn and U: An exception to the free ion activity model. *Aquat. Toxicol.* **51**: 155–175.

Markich, S.J. and Camilleri, C. (1997) *Investigation of Metal Toxicity to Tropical Biota: Recommendations for Revision of the Australian Water Quality Guidelines*. Supervising Scientist Report 127. Supervising Scientist, Canberra, Australia.

Markich, S.J. and Jeffree, R.A. (1994) Absorption of divalent trace metals as analogues of calcium by Australian freshwater bivalves: An explanation of how water hardness reduces metal toxicity. *Aquat. Toxicol.* **29**: 257–290.

Markich, S.J., Jeffree, R.A., and Burke, P.T. (2002) Freshwater bivalve shells as archival indicators of metal pollution from a copper-uranium mine in tropical northern Australia. *Environ. Sci. Technol.* **36**: 821-832.

Meador, J.P. (1991) The interaction of pH, dissolved organic carbon, and total copper in the determination of ionic copper and toxicity. *Aquat. Toxicol.* **19**: 13–32.

Morley, A.W. (1981) *A Review of Jabiluka Environmental Studies*. Vol. 1. Pancontinental Mining, Sydney, Australia.

Morley, A.W., Brown, T.E., and Koontz, D.V. (1985) The limnology of a naturally acidic tropical water system in Australia. I. General description and wet season characteristics. *Verh. Int. Ver. Theor. Angew. Limnol.* **22**: 2125–2130.

Mount, D.I., Norberg-King, T.J., and Steen, A.E. (1986) *Validity of Effluent and Ambient Toxicity Tests for Predicting Biological Impact, Naugatuck River, Waterbury, Connecticut*. United States Environmental Protection Agency, Washington, DC, EPA/600/8–86/005.

Mudroch, A., Azure, J.M., and Mudroch, P. (Eds). (1999) *Manual of Bioassessment of Aquatic Quality*. Lewis Publishers, New York.

Nanson, G.C., East, T.J., and Roberts, R.G. (1993) Quaternary stratigraphy, geochronology and evolution of the Magela Creek catchment in the monsoon tropics of northern Australia. *Sediment. Geol.* **83**: 277–302.

Newcombe, C.P. and MacDonald, D.D. (1991) Effects of suspended sediment on aquatic ecosystems. *North Am. J. Fish. Manage.* **11**: 72–82.

Noller, B.N. (1991) Non-radiological contaminants from uranium mining and milling at Ranger, Jabiru, Northern Territory, Australia. *Environ. Monitor. Assess.* **19**: 383-400.

Noller, B.N., Currey, N.A., Ayers, G.P., and Gillett, R.W. (1990) Chemical composition and acidity of rainfall in the Alligator Rivers Region, Northern Territory, Australia. *Sci. Total Environ.* **91**: 23–48.

Noller, B.N. and Cusbert, P.J. (1985) Mobilisation of aluminium from a tropical floodplain and its role in natural fish kills: A conceptual model. In *International Conference of Heavy Metals in the Environment*, Lekkas, T.D. (Ed.). CEP Consultants, Athens, pp. 700-702.

Noller, B.N. and Hart, B.T. (1993) Uranium in sediments from the Magela Creek catchment, Northern Territory, Australia. *Environ. Technol.* **14**: 649–656.

Norberg-King, T.J. and Mount, D.I. (1986) *Validity of Effluent and Ambient Toxicity Tests for Predicting Biological Impact, Skeleton Creek, Enid, Oklahoma*. United States Environmental Protection Agency, Washington, DC, EPA/600/8–86/002.

Nriagu, J.O. (1990) Global metal pollution: Poisoning the biosphere. *Environment* **32**: 6–11, 28–33.

NT DME (1982) *Ranger Uranium Mine General Authorisation, A82/3* (Annex C, as amended 1994). Northern Territory Department of Mines and Energy, Darwin, Australia.

NT DTW (1983) *Alligator Rivers Region Regional Surface Water Quality Monitoring, November 1978-April 1981*. Northern Territory Department of Transport and Works, Darwin, Australia.

Parkhurst, B.R., Elder, R.G., Meyer, J.S., Sanchez, D.A., Pennak, R.W., and Waller, W.T. (1984) An environmental hazard evaluation of uranium in a Rocky Mountains stream. *Environ. Toxicol. Chem.* **3**: 113–124.

Peck, M.R., Klessa, D.A., and Baird, D.J. (2002) A tropical sediment toxicity test using the dipteran, *Chironomus crassiforceps* to test metal bioavailability with sediment pH change in tropical acid-sulfate sediments. Environ. Toxicol. Chem. **21**: 720-728.

Pickering, W.F. (1995) General strategies for speciation. In *Chemical Speciation in the Environment*, Ure, A.M. and Davidson, C.M. (Eds). Blackie Academic, London, pp. 9–32.

Poston, T.M., Hanf, R.W., and Simmons, M.A. (1984) Toxicity of uranium to *Daphnia magna*. *Water Air Soil Pollut.* **22**: 289–298.

Richards, R.J., Applegate, R.J., and Ritchie, I.A.M. (1996) The Rum Jungle rehabilitation project. In *Environmental Management in the Australian Mineral and Energy Industries: Principals and Practices*, Mulligan, D.R. (Ed.). University of New South Wales Press, Sydney, Australia, pp. 530–555.

Riethmuller, N., Markich, S.J., Parry, D., and van Dam, R.A. (2000) *The Effect of True Water Hardness and Alkalinity on the Toxicity of Cu and U to Two Tropical Australian Freshwater Organisms*. Supervising Scientist Report 155. Supervising Scientist, Canberra, Australia.

Ritchie, A.I.M. (1994) The waste-rock environment. In *Short Course Handbook on Environmental Geochemistry of Sulfide Mine-Wastes*, Blowes, D.W. and Jambour, J.L. (Eds). Vol. 22. Mineralogical Association of Canada, Waterloo, Canada, pp. 133–161.

Roux, D.J., Jooste, S.H., and MacKay, H.M. (1996) Substance-specific water quality criteria for the protection of South African freshwater ecosystems: Methods for derivation and initial results for some inorganic test substances. *S. Afr. J. Sci.* **92**: 198–206.

Sewell, S.R., Shinners, S.H., and Corbett, L.K. (1997) Biological monitoring to assess the impacts of mining operations on aquatic ecosystems. In *Short Course on Water Analysis and Monitoring*. Australian Centre for Minesite Rehabilitation Research and Northern Territory Department of Mines and Energy, Darwin, Australia.

Stowar, M. (1997) *Effects of Suspended Solids on Benthic Macroinvertebrate Fauna Downstream of a Road Crossing, Jim Jim Creek,]Kakadu National Park*. Internal Report 256. Environmental Research Institute of the Supervising Scientist, Darwin, Australia.

Stratford, M.L. (1994) *Image Analysis of Riparian Vegetation Patterns: Rum Jungle Mining Operations*. B.Env.Sc. (Hons.). University of Wollongong, Wollongong, Australia.

Supervising Scientist (2001) *Supervising Scientist. Annual Report 2000-01.* Supervising Scientist, Darwin, Australia.

Thomas, P.A. and Hart, B.T. (1984) *Textural Characteristics and Heavy Metal Concentrations in Billabong Sediments from the Magela Creek System.* Technical Memorandum 9. Supervising Scientist for the Alligator Rivers Region, Canberra, Australia.

Townsend, S.A., Boland, K.T., and Wrigley, T.J. (1992) Factors contributing to a fish kill in the Australian wet/dry tropics. *Water Res.* **26**: 1039–1044.

United Nations (1997) *Comprehensive Assessment of the Freshwater Resources of the World.* United Nations Economic and Social Council, New York.

US EPA (1994) *Interim Guidance on the Determination and Use of Water-Effect Ratios for Metals.* United States Environmental Protection Agency, Washington, DC, EPA-823-B-94-001.

US EPA (1995a) *Great Lakes Water Quality Initiative Criteria Documents for the Protection of Aquatic Life in Ambient Water.* United States Environmental Protection Agency, Washington, DC, EPA-820/B-95-004.

US EPA (1995b) Stay of federal water quality criteria for metals. *Fed. Reg.* **60**: 22228-22237.

Walker, T.D. and Tyler, P.A. (1982) *Chemical Characteristics and Nutrient Status of Billabongs of the Alligator Rivers Region, Northern Territory.* Open File Record 27. Supervising Scientist for the Alligator Rivers Region, Canberra, Australia.

Watson, G.M. (1975) Managerial history of the Rum Jungle project. In *Rum Jungle Environmental Studies*, Davy, D.R. (Ed.). Australian Atomic Energy Commission, Sydney, Australia, AAEC/E365.

Wedd, M. (1989) The regulation of uranium mining in the Northern Territory. In *Proceedings of the 11th North Australian Mine Rehabilitation Workshop*, Northern Territory Department of Mines and Energy, Darwin, Australia. pp. 13–18.

Yong, P. and Macaskie, L.E. (1995) Role of citrate as a complexing ligand which permits enzymatically-mediated uranyl ion bioaccumulation. *Bull. Environ. Contam. Toxicol.* **54**: 892–899.

Zuk, W.M., Jeffree, R.A., Levins, D.M., Lowson, R.T., and Ritchie, A.I.M. (1994) From Rum Jungle to Wismut: Reducing the Environmental Impact of Uranium Mining and Milling. *Trans. Am. Nucl. Soc.* **70** (Suppl. 1): 935–940.

9

Inplace Inactivation and Natural Ecological Restoration Technologies (IINERT)

W.R. Berti[1] and J.A. Ryan[2]

INTRODUCTION

Because of the many shortcomings of current remediation alternatives, there is a growing need for the development of alternative technologies that are low cost, low input, and environmentally benign, yet provide equivalent protection to human health and the environment. Many innovative remediation techniques being currently developed focus on exploiting or altering the soil chemistry to either remove contaminants from the soil or reduce their solubility and bioavailability while leaving them in the soil. Some of the more promising innovative remediation alternatives include plant-based or "phytoremediation" techniques such as phytoextraction and phytostabilization (i.e. inplace inactivation). In phytoextraction, the contaminant is gradually removed from the soil by plant uptake and harvesting (or, in the case of mercury, by volatilization). Conversely, inplace inactivation is a site-stabilization technique in which materials are applied to the soil to alter the soil contaminant chemistry, making the contaminant less water-soluble, less mobile, and less bioavailable. Inplace inactivation does not affect the total contaminant concentration, but reduces the risk of harm to a target organism (e.g. humans) making the contaminant unable to biologically interact with the organism. Inplace inactivation is based on fundamental soil chemistry, plant biology, agricultural practices, and experience with the restoration of drastically disturbed mine and roadside lands and construction sites. Inplace inactivation appears to be a relatively simple, low-cost, low-input method, and should prove adaptable to a wide range of contaminated sites.

---

[1] DuPont Central Research & Development, Newark, DE USA
[2] U.S. Environmental Protection Agency National Risk Management Laboratory, Cincinnati, OH USA

Background

Heavy-metal contamination [e.g. lead (Pb), cadmium (Cd), chromium (Cr), copper (Cu) and zinc (Zn)] in soils is widespread worldwide. Unlike organic compounds that can be degraded, heavy metals can only be covered or buried to prevent contact, be picked up and recycled or moved to a safer location, or be transformed into a less toxic form. They never go away. The most common remedy for Pb-contaminated soils, for example, has been to mix them with chemical binders such as portland cement and to relocate them to landfills, safely away from receptors. Portland cement works by increasing the particle size and imparting the resulting material with a high buffering capacity in the alkaline pH range. The large particles and alkaline pH buffering capacity that result from the stabilization process reduce the amount of contaminant that is extracted by laboratory leaching methods, including regulatory tests such as the Toxicity Characteristic Leaching Procedure (TCLP) (US EPA, 1990) and the Simulated Precipitation Leaching Procedure (SPLP) (US EPA, 1995). Soil washing is another remedy in which soils are subjected to intense dry and wet processing to remove size and density fractions containing the contaminants.

This process concentrates the contaminants in the water or fine particles and leaves large particles relatively clean. The larger particles produced by the soil washing process can be reused as fill material onsite. The water and fine particles in which the contaminants reside must be further treated and the solids recycled or properly disposed, usually by landfilling.

The chemical form of heavy metals in soils is an important consideration in determining the hazard to human health and the environment. Certain chemical forms of some heavy metals are very toxic; for other forms of heavy metals, particularly certain naturally occurring forms, the toxicity can be quite low. Nature itself provides hints for other solutions to remediate heavy metals in soils in addition to landfilling, covering, or washing. Populations living in or near natural Pb outcroppings often have lower blood and tissue Pb levels than do populations living in areas where Pb paint is used (USEPA, 1990). This is very likely because many natural mineral forms of Pb have very low bioavailability. They do not dissolve in the human digestive system when ingested, but rather pass though unabsorbed without causing harm. From these types of observations, scientists have postulated that if Pb contamination in soils can be converted to the less toxic forms, it might be safe to leave in place.

One potential method to reduce the hazard of heavy metals in soils is by chemically and physically manipulating the soil to convert the forms of the contaminants from those of greater hazard (i.e. high water solubility, high mobility, high bioavailability) to those of lesser hazard. The term "inplace inactivation" (US EPA, 1996a) has been coined to describe this process of chemically and physically inactivating contaminants, both in soil and other materials found at the earth's surface. Other names for this strategy include "phytostabilization", "agronomic stabilization", "in situ stabilization", and "phytorestoration". In this strategy, no actual reduction in pollutant concentration occurs. The risk reduction is provided by chemical and physical processes in the soil so that the soil can remain in place. Chemicals and materials that appear to be most promising for inplace inactivation include phosphates, mineral fertilizers, iron oxyhydroxides, other minerals, biosolids, and limestone. Conversion of Pb to less toxic forms has been demonstrated in soils amended with safe additives using common agricultural techniques (Berti and Cunningham, 1994; Rabinowitz, 1993; Ruby et al., 1994; Mench et al., 1994). To complement the use of soil amendments, a rich plant growth in treated areas will help hold the soils in place by preventing erosion, help reduce rain impact and water infiltration, and provide an effective barrier against actual contact with soil. In some cases, plant roots may absorb contaminants to further prevent off-site migration or leaching. Incorporating soil amendments and growing plants using existing agronomic techniques are more natural ways of restoring the ecology of a soil when compared with other remediation technologies such as soil excavation and landfilling, soil excavation and washing, and soil capping. Importantly, this agriculture-based technique should be less likely to impair the soil's potential for sustaining plant growth after treatment and should be relatively environmentally benign when compared with many conventional remediation practices.

Current Developments

Much of the work on inplace inactivation has involved Pb-contaminated soils. Research has shown that raising the soil pH, increasing organic matter, adding metal oxyhydroxides, and increasing certain anions (especially phosphate) can decrease Pb mobility and lower the characteristic hazard (as measured using the TCLP) by an order of magnitude (Berti and Cunningham, 1997). Figure 9.1 shows the effects of amending soil with phosphate fertilizer such as KH_2PO_4 and iron oxyhydroxides (FeOOH applied as Iron Rich, a coproduct in the manufacturing of white pigment at the DuPont Co. Edgemoor, DE, facility). The TCLP of untreated soils was well above the critical level of 5 mg Pb L^{-1}. The addition of phosphate and FeOOH, however, significantly reduced the TCLP-extractable Pb from soils without drastically altering the final pH of the TCLP solution.

A six-step sequential chemical extraction of these soils resulted in a "fingerprint" of the soil-Pb (Fig. 9.2) which appears to give valuable

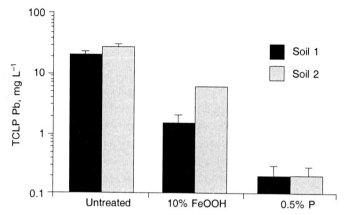

Fig. 9.1: TCLP Pb from untreated soils 1 and 2 and after treating with 0.5% P and 10% FeOOH. Total Pb in soils 1 and 2 are 1200 and 2400 mg kg^{-1}, respectively. Please note that the Y-axis is log-scaled in the figure Berti and Cunningham (1997).

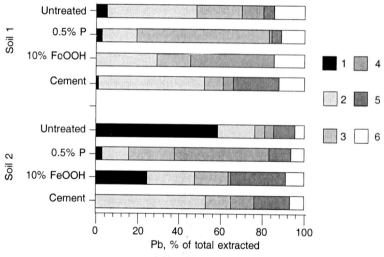

Fig. 9.2: Sequential chemical extraction of Pb from untreated soils 1 and 2 and after treating with 0.5% P, 10% FeOOH, and portland cement. Fraction number: 1: = exchangeable; 2: = carbonates; 3: = Mn oxides; 4: = organic; 5: = Fe oxides; 6: = residual Berti and Cunningham (1997).

information on the potential of Pb for leaching, plant uptake, and mammalian bioavailability through soil ingestion (Berti and Cunningham, 1997). In the untreated soils the majority of the Pb was extracted in the first two fractions, which represent the more soluble or available forms of Pb. Application of amendments, particularly phosphate and FeOOH, caused the Pb to shift from forms with high relative mobility and greatest potential hazard (Fractions 1 and 2) to those with low or no relative mobility (Fractions 3-6).

Reductions in contaminant solubility or mobility, as measured by the TCLP or sequential chemical extraction, provide useful information regarding risk. However, the primary concern of soil Pb is the direct ingestion of soil by children and the amount of Pb that subsequently is absorbed into the bloodstream, which is referred to as *bioavailability*. However, the definition of bioavailability varies substantially depending upon the receptor. This is discussed in detail in Chapters 2 and 3. *Absolute bioavailability* (ABA) is the fraction of the total dose (i.e. amount of soil Pb ingested) that is absorbed into the bloodstream. Absolute bioavailability of Pb in ingested food and water is currently assumed to be 50%, whereas the value for Pb in ingested soil is 30%. *Relative bioavailability* (RBA) is the ratio of ABA of a compound in a dose to its absolute bioavailability in a reference dose (e.g., the bioavailability of lead in soil relative to lead in food/water would be 60%).

Like contaminant solubility and mobility, bioavailability also can be reduced through the application of certain materials to soils. Results of an immature swine soil-dosing study at the University of Missouri and a weaning Sprague-Dawley rat dosing study at the U.S. Department of Agriculture-Agricultural Research Service (USDA-ARS) indicate a significant reduction in soil Pb bioavailability as a result of adding phosphate to a soil from Joplin, MO (total soil-Pb about 4500 $\mu g\ g^{-1}$). Treatments were applied to soils in laboratory and field experiments.

Measuring Bioavailability for Determining Risk

Animal dosing studies, such as the swine and rat studies reported here, have been the method of choice to measure bioavailability. Readers can refer to Chapter 3 for an excellent overview of other techniques of measuring bioavailability. As a test method to evaluate bioavailability for use in risk assessment and developing remediation treatment alternatives, however, animal studies may have major limitations. Animal dosing studies are expensive, complex, time consuming, and may have ethical concerns (Hrudey et al., 1996). The physiologically-based extraction test (PBET) (Ruby et al., 1996) is a quick chemical extraction to serve as an alternative to animal studies. It has been used to determine the bioaccessibility of Pb, As, and other soil contaminants. Bioaccessibility, as used here, is defined as the solubility of soil Pb in the simulated stomach solution of the PBET relative to the total Pb in the soil. Soil-Pb bioaccessibility determined in this way has been shown to correlate well with that using a Sprague-Dawley rat model (Ruby et al., 1996) and a swine model (Medlin, 1997) (R^2 = 0.88, Fig. 9.3).

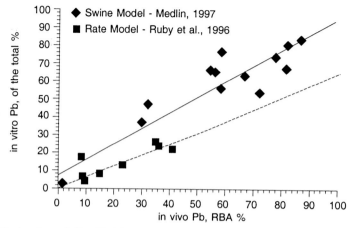

Fig. 9.3: In vitro relative Pb bioaccessibility measured using the PBET procedure at pH 1.5 vs. EPA swine model blood Pb relative bioavailability (diamonds: from Medlin, 1997) and in vitro relative Pb bioavailability measured using the PBET procedure at pH 2.5 vs. Sprague-Dawley rat model blood-Pb RBA (squares; from Ruby et al., 1996).

An interesting feature of Fig. 9.3 is the wide range of soil-Pb bioavailabilities measured in the two animal models on two different sets of soils. In the swine model, for example, (RBA) of Pb ranged from 1 to 87%, or less than 1 to 44% Pb (ABA). The overall average for Pb RBA in the swine model for the 15 soils was 53% (Ruby et al., 1996), or 27% Pb ABA, which compares well with the EPA default Pb ABA of 30%.

Soil-As bioaccessibility has also been shown to correlate well with soil-As bioavailability using a *Cynomolgus* monkey model (Ruby et al., 1996), a rabbit model (Ruby et al., 1996), and a swine model (Medlin, 1997) ($R^2 = 0.54$ overall, $n = 10$; $R^2 = 0.75$ for swine model data only, $n = 7$; Fig. 9.4). The correlations between in vitro and in vivo studies for As are not as good as those demonstrated for Pb. However, both of these comparisons indicate that a simple soil test for As, Pb, and perhaps other soil contaminants may be valuable for determining exposure from direct soil ingestion.

The Solubility/Bioavailability Research Consortium (SBRC) is working to develop the PBET method to determine the bioavailability of hazardous substances in contaminated soils (Ruby et al., 1997). The consortium is a collaborative effort among academics, consultants, regulators, and the regulated community. The initial research of the consortium will focus on Pb and As for the purpose of refining human exposure estimates. Other inorganic constituents such as beryllium (Be), Cd, Cr, and Hg will be considered for method development and validation once tests for Pb and As have been established.

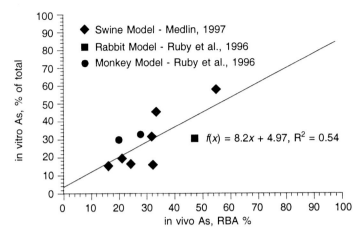

Fig. 9.4: In vitro relative As bioaccessibility measured using the PBET procedure at pH 1.5 vs. EPA swine model blood-As RBA (diamonds: from reference 12) and in vitro relative Pb bioavailability measured using the PBET procedure at pH 2.5 vs. rabbit (squares) and *Cynomolgus* monkey model (circles) blood-Pb RBA (squares: Ruby et al. (1996)).

The hazard presented by Pb at a residential site is often determined for young children (age 7 and younger), using the Integrated Exposure and Uptake Biokinetic model (IEUBK) (US EPA, 1994 a, b). This model considers several parameters, such as a child's nutritional status, exposure frequency and rate, and Pb bioavailability to predict the percentage of children in an exposed population who may have a blood Pb level greater than 10 μg dL^{-1}. Children with blood Pb levels above this value are considered to be at risk of suffering from Pb-related health problems. The IEUBK model uses a default value of 30% for the absolute bioavailability of soil-Pb. This may, however, overestimate or underestimate Pb bioavailability for a given site, as indicated in Fig. 9.3. A change in the default value of Pb bioavailability in the IEUBK model may change the percentage of children with blood Pb levels above 10 μg dL^{-1} (Fig. 9.5). A goal is that no more than 5% of exposed children will have blood Pb levels above the critical level, as predicted by the IEUBK model.

For a soil with 2000 mg Pb kg^{-1} and an absolute Pb bioavailability of 30% (60% relative bioavailability), the IEUBK predicts that over 50% of the exposed children will be above the critical blood Pb level. If Pb is closer to 50% absolute bioavailability, the number of children at risk may approach 80% of the population. When the Pb bioavailability is 10%, or is reduced to 10% by inplace inactivation, the percent of children who are at risk from soil Pb may be only about 10%. Although this hypothetical scenario still exceeds the goal for the percent of children above the critical blood Pb level, it indicates that default assumptions on Pb bioavailability may yield incorrect predictions concerning children at risk at a given site.

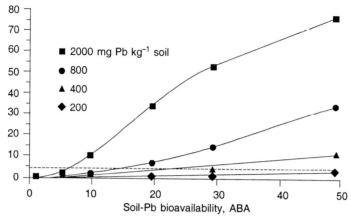

Percentage of children with blood-Pb greater than 10 ug dL⁻¹

Fig. 9.5: Influence of soil-Pb bioavailability on the percentage of children with blood-Pb levels greater than 10 μg dL⁻¹, as determined using the IEUBK model (US EPA, 1996a,b).

The Technical Review Workgroup for Lead (TRW - U.S. Environmental Protection Agency) has developed an approach to determine risks associated with nonresidential exposures to Pb in soil. The focus of the approach is to estimate fetal blood Pb levels in pregnant women who may be exposed to Pb-contaminated soils at existing or potential work sites (US EPA, 1996). Figure 9.6 shows the possible effects of varying soil-

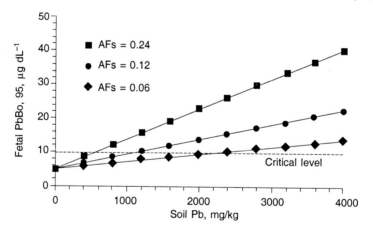

Fig. 9.6: Potential influence of soil-Pb bioavailability (AFs) on fetal blood Pb (PbB) concentrations in response to pregnant females exposed to total soil-Pb concentrations, as determined using the risk estimation algorithm from Technical Review Workgroup for Lead (US EPA, 1996). The critical level is the PbB concentration that results in a risk of Pb poisoning to the fetus.

Pb absorption (i.e. bioavailability) factors (Afs) on fetal blood Pb concentrations as a function of total soil Pb. This interim approach is recommended for use pending further development and evaluation of an integrated exposure biokinetic model for adults.

Advantages and limitations of Inplace Inactivation

Costs

Any site to which conventional remediation techniques can be applied is a candidate for inplace inactivation. In general, inplace inactivation appears to be less expensive and creates fewer operational hazards than the more conventional remedial methods. Specific capitalization costs are difficult to determine, but, in general, farming operation costs are low. Equipment for farming operations (e.g. fertilizing, plowing, etc.) is generally readily available. The overall remedial costs should be substantially lower than those associated with traditional methods. Because inactivation processes are generally inexpensive and easy to implement, they may also serve as interim measures for reducing risk prior to subsequent remediation by other techniques.

Plants as a tool for inactivation

Plant uptake and hazards to humans or animals who may consume the plants may be greatly reduced with inplace inactivation. Plants may sequester pollutants in their roots without translocating them to aboveground material. In these cases, additional degrees of inactivation are established.

Inplace inactivation, however, is not a panacea for contaminated soil. Plants are living organisms with constraints that are often in conflict with the nature of the pollutant and the site. Plants have certain pH and nutrient requirements and require sufficient water, air, and sunlight. Soil amendments such as limestone, phosphate, other plant nutrients, ash, biosolids, and metal (Fe/Mn) oxyhydroxides can be useful in stabilizing metals and establishing and growing plants when used at the proper rates. They are most effectively applied and incorporated prior to planting. Vegetative growth requires that the soil be friable enough to allow plant roots to take hold. Local seasonal and rainfall conditions can have an impact on the difficulty of establishing and maintaining plant growth. However, sites initially hostile to plants can be converted to encourage reasonable plant growth through the use of proper soil amendments, soil preparation, and plant establishment techniques.

Plant vegetative covers used to decrease erosion may require time to become fully established (e.g. one to three growing seasons). The establishment of a vegetative cover depends on the plant, climate, and soil. However, techniques such as the use of fast-growing grasses, mulch, and landscape fabric are available to prevent erosion until full plant cover is established. Permanent establishment of selected plant species may be delayed because of seasonal changes in growth rates or suboptimum climatic conditions, particularly in the late fall and winter. Plant establishment during summer months may require irrigation.

How practical is inplace inactivation?

The more the site resembles a farmer's field, the less expensive and more easily applied this technology will be. Compared with the equipment needed for many conventional remediation practices, the supplementary equipment needed for inplace inactivation (e.g. basic farming equipment) is easier to transport to remote sites because construction access roads are not necessary. Debris surrounding a site may impose limitations on the use of farming implements. However, the utilization of downsized equipment, small tractors and plows or rototilling equipment, may alleviate these concerns. Plants can be planted in a wide variety of unfavorable conditions by using technologies such as hydroseeding. Abandoned buildings and lots (even in more polluted areas) are often quickly covered with weeds and scrub trees, demonstrating the adaptability of certain plant species. The site requirements needed to grow plants, except for the time of year and short-term weather conditions, may be less rigorous than those for other remediation technology requirements.

In general, knowing the type and extent of contamination at a site plays a crucial role in

- assessing the feasibility of inplace inactivation,
- determining the proper type and amount of soil amendments to be added, and
- selecting the proper plant species to be used.

Plows have been designed to have a maximum plowing depth of up to 1m. Soil auguring equipment is also available that can deliver and mix materials to a depth of 30 m or higher.

For inplace inactivation, soil amendments are selected based on their ability to stabilize the pollutants and compatibility with plant growth. Plant species are selected based on their ability to tolerate site conditions and maximize plant growth and ground cover. Information on the productivity of soil and the amendments necessary to optimize its productivity can be obtained with the help of government agencies and agricultural colleges and universities.

Field Studies and Results

Inplace inactivation is a nascent technology, but has a sound technical basis stemming from the work done in the mining and smelting industries to revegetate drastically disturbed land (Mumshower, 1994; Sopper, 1993). It has the potential to develop into a viable remediation option in cases where pollutants
- are relatively nonleachable and
- pose little eminent risk to human health or the environment.

This strategy is currently being used to remediate zinc-, lead-, and nickel-contaminated soil at a Superfund site in Palmerton, Pennsylvania, USA. Metal stabilization has been accepted in a Record of Decision (ROD) at the Superfund site. Inplace inactivation is also undergoing field trials in Joplin, Missouri, USA. This area was the site of Pb and Zn mining and smelting activities for about 100 years, beginning in the mid 1800s. These operations resulted in elevated soil Pb levels across much of the county (about 6500 ha). Soil Pb levels in the area vary considerably, ranging approximately from 1100 to 5350 mg kg^{-1} at a depth of 1-8 cm, and from 2000 to 4800 mg kg^{-1} at a depth of 8-15 cm. At residential homes with the most severe Pb contamination or where the blood Pb level in children indicated a health risk, the contaminated soil around the homes is being excavated and replaced. This plan of remediation will continue house by house until the property surrounding approximately 2500 homes at risk is remediated (M. Doolan, 1997, personal communication). A workgroup of the IINERT team is demonstrating inplace inactivation practices that can be implemented easily at a site in Joplin, including some practices that can be accomplished by the homeowners themselves. In this way, they can take steps to help protect the health of their families. It is hoped that this demonstration will also raise public and regulatory support for inplace inactivation of other contaminated sites.

In early 1997, the workgroup installed a field validation study at Joplin to evaluate the effectiveness of amendments including phosphate fertilizers (triple superphosphate), phosphoric acid, and high-Fe byproducts such as Compro and Iron Rich to reduce Pb bioavailability. Extensive preliminary research to characterize the site and select treatments was performed prior to field implementation. Several rounds of sampling and testing of soil from the Joplin site and treated field plots has been conducted to date. Soil Pb bioavailability in swine was reduced by 25-30% in treated soils as compared with that in untreated ones. (Casteel et al., 2001). In weaning rats, Pb bioavailability in treated soils was reduced about 50% compared with that in untreated ones (Q. Xue, J.G. Hallfrisch, R.L. Chaney and S.L. Brown, personal communication). Analysis of plants

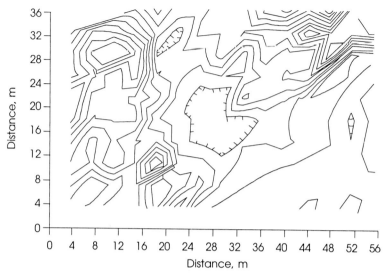

Fig. 9.7: Joplin field research site soil Pb-concentration (in $\mu g \ g^{-1}$) isolines, sampled Fall 1996. Soil samples collected on a 3.6 x 3.6 m^2 grid.

from the site has shown as high as an 80% reduction of Pb in the plant tops as a result of treatments when compared with untreated plots (Brown et al, 1998). Additional plant and soil sample collections are ongoing. Information on the progress of this research can be found on the IINERT RTDF (Remediation Technology Development Forum) web site at *http:// www.rtdf.org/iinert.htm*.

Cost Benefit Analysis

An economic analysis comparing inplace inactivation techniques to some currently practiced remediation techniques shows that both phytoremediation techniques are considerably less expensive than their current counterparts (Table 9.1). These comparisons are based on a hypothetical site that is covered with some rubble and is contaminated with >0.2% Pb (w/w) to a depth of 30 cm. The soil exceeds US regulatory TCLP limits and must be disposed of in a hazardous waste landfill. The costs are adjusted to reflect an annual discount rate of 12% and inflation rate of 3%.

The table includes the theoretical cost of inplace inactivation. It also includes the calculated costs of several currently practiced site-remediation techniques. Decontamination of the site requiring excavation, stabilization, and off-site disposal of the contaminated material assumes removal of the soil to 30 cm, stabilization of the contaminated soil with cement, and final placement in a hazardous waste landfill. In soil washing, however,

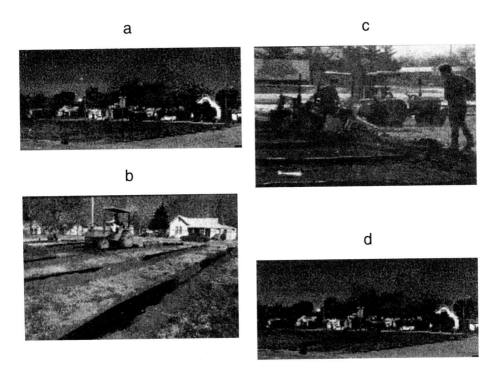

Fig. 9.8: Joplin, MO, IINERT field research site: (a) May 1996 prior to start of work; (b) and (c) March 1997 during the establishment of treatment plots by USDA, and d) May 1998, showing the plots and vegetative growth 14 months after installation. All plots were physically separated with plastic barriers to minimize soil and treatment movement among neighboring plots, as shown in (b) and (c).

the fine material is separated and landfilled instead of the whole soil. Common site-stabilization practices include asphalt and soil capping. In these techniques, a layer of either asphalt or uncontaminated soil is placed over the contaminated area to prevent environmental exposure of the contaminants and to restrict water infiltration into the contaminated profile.

Of the site decontamination techniques, solidification and stabilization off-site are the most expensive, with costs exceeding US$1.5 million per hectare excavated to a depth of 30 cm. This technique remediates a site in one year, but requires expensive landfill space and does not include extensive restoration of the site. Soil washing to remove the fine soil fractions is the second most expensive decontamination technique (US$790,000 per hectare to 30 cm).

Table 9.1: Descriptions and costs of remediation alternatives for a Pb contaminated site.[a]

Alternative	Description	Net Present Cost —US$ per ha—
Site Decontamination		
Excavation, stabilization, and off-site disposal[b]	The soil is excavated to 30 cm, stabilized with cement, and placed in a hazardous waste landfill. No backfilling with uncontaminated soil, or revegetation, is included	1,600,00
Soil washing[c]	The fine material (assumed to comprise 20% of the soil) is separated from the soil at a soil washing facility, stabilized with cement, and placed in a hazardous waste landfill.	790,000
Phytoextraction	Soil containing 0.14% Pb (w/w) is decontaminated to a final Pb concentration of 0.04% (w/w) for residential landuse. An assumed plant biomass of 40 ton/ha-year containing 1% Pb (w/w, dry weight basis) is produced for 10 years. Na₄EDTA is used at an equal molar ratio to the Pb removed. Plant material is landfilled as hazardous waste following on-site moisture reduction.	279,000
Site Stabilization		
Asphalt capping	This cap includes a base layer of 20 cm, a subgrade layer of 25 cm, and a top layer of 4 cm asphalt. This site can be used as a parking lot.	160,000
Soil capping	A cap of 60 cm of uncontaminated soil is placed over the contaminated area. No other geotextiles or barriers are included. A layer of vegetation is established on the surface to stabilize the cap and reduce water infiltration.	130,000
Inplace inactivation	The soil is amended with lime and nitrogen and potassium fertilizers for plant growth. Triple super phosphate (90 ton ha⁻¹) and Iron Rich (400 ton ha⁻¹) are applied to inactivate Pb. The site is seeded with grass and mowed four times per year for 30 years.	60,000

a. Costs summarized from Cunningham and Berti, 2000.
b. Estimate based on a cost of US$207 to landfill a metric ton of soil, not including stabilization.
c. Estimate based on a cost of US$226 to landfill a metric ton of soil, including stabilization.

Site-stabilization techniques are generally less expensive than decontamination techniques. Of the site stabilization techniques considered in this comparison, asphalt capping is the most expensive, costing US$160,000. Asphalt capping effectively prohibits contact of the soil with the environment. However, it limits land use to parking lots or similar functions. A less expensive yet similar technique is soil capping with a 60 cm thick layer of uncontaminated soil covering the contaminated area. Vegetation is established to stabilize the soil cap. The cost per hectare for soil capping is approximately US$140,000. Compared to these two stabilization techniques, inplace inactivation is an economically attractive alternative, costing approximately US$53,000 per hectare. In this scenario, inplace inactivation includes site preparation, plowing, application of amendments to inactivate Pb, lime and fertilizer application, planting, and mowing (4 times per year). Soil amendments to inactivate soil Pb include 90 tonnes ha^{-1} triple superphosphate fertilizer, and about 400 tonnes ha^{-1} Iron Rich. All site-stabilization techniques require some annual maintenance, such as asphalt patching or repairs, or mowing or reseeding (soil capping and inplace inactivation).

When compared with commonly practiced remediation techniques, inplace inactivation is the least expensive alternative. However, this technique may also be less invasive to the site and more quickly promote the restoration of a healthy ecosystem.

Research and Development—Future Needs

Several areas of research and development have been identified by the IINERT Action Team:

- Develop a more thorough understanding of the factors that control contaminant bioavailability to humans, which should include the biological, chemical, and physical factors that affect bioavailability.
- Develop and validate simple techniques that can be used to assess soil-metal bioavailability to humans. These simple techniques should be well correlated to appropriate human or animal (e.g. pigs and rats) model surrogates.
- Develop correlations between soil components (i.e. metal species, nonmetal-containing components) and the soil-metal bioavailability which determine the short and long-term stabilities of soil-metal components.
- Develop treatment technologies and processes for the additions of materials to metal-contaminated soils that induce the formation of less bioavailable metal forms, providing a practical approach to inplace inactivation.
- Develop and validate simple techniques that can be used to evaluate environmental hazards for both soil contaminants and for various remediation options.

CONCLUSIONS

Reducing metal availability and maximizing plant growth through inplace inactivation may prove to be an effective method of in situ soil-metal remediation on industrial, urban, smelting, and mining sites. In addition, these stabilization techniques can occur as part of a treatment train with other remediation methods, including those now under development, the most intriguing of which is "biomining" the available fraction of metal pollutants with plants.

ACKNOWLEDGEMENTS

We thank all the people who have contributed to the IINERT RTDF, especially R. Blanchar, S. Brown, J. Carter, S. Casteel, R. Chaney, M. Doolan, J. Drexler, A. Green, D. Hesterberg, D. Mosby, G. Pierzynski, M. Ruby, and J. Yang.

REFERENCES

Berti, W.R., and Cunningham, S.D. (1994) Remediating soil with green plants. In *Trace Substances, Environment and Health*, Cothern, C.R., (Ed.). : Northwood Science Reviews, pg. 43–51.

Berti, W.R., and Cunningham, S.D. (1997) Inplace inactivation of soil-Pb in Pb contaminated soils. *Environ. Sci. Technol.* **31**: 1359–1364.

Brown, S.L., Xue, Q., Hallfrisch, J. G., and Chaney, R.L. (1998) *RTDF IINERT: Progress Report. US Department of Agriculture - Agriculture Research Service*, Beltsville, MD, USA.

Casteel, S.W., Evans, T.J., Cowart, R.P., Yang, J., and Mosby, D. (2001) Effects of phosphorus treatment on soil lead bioavailability. *Agronomy Abstracts*.

Cunningham, S. D., Berti, W. R. 2000. Phytoextraction and phytostabilization: Technical, economic, and regulatory considerations of the soil-lead issue. In N. Terry and G. Banuelos (ed.) *Phytoremediation of Contaminated Soil and Water: Symposium Proceedings of 4th International Conference on the Biogeochemistry of Trace Elements*, Berkeley, California, June 1997, 359-376. CODEN: 68INA7 CAN **132**: 22–455 AN 1999: 705825 CAPLUS

Hallifrisch. J., Veillon, C., Hill, A.D., Patterson, K., Chaney, R., Xue, Q., Brown, S., and Ryan, J.A. (2001) Effect of soil Pb inactivation treatments on bioavailability of smelter contaminated soil to rat. *Agronomy Abstracts*.

Hrudey, S.E., Chen, W., and Rousseaux, C.G. (1996) *Bioavailability in environmental risk assessment*. CRC Lewis Publishers, New York, pg. 51.

Ma, Q.Y., Logan, T.J., Trama, S.J. (1995) Lead immobilization from aqueous solutions and contaminated soils using phosphate rocks. *Environ. Sci. Tech* **29**: 1118–1126.

Medlin, E.A. 1997. An in vitro method for estimating the relative bioavailability of lead in humans. MS thesis. University of Colorado, Boulder, Colorado, USA.

Mench, M.J., Didier, V.L., Loffler, M., Gomez, A., and Masson, P. (1994) A mimicked in-situ remediation study of metal-contaminated soils, with emphasis on cadmium and lead. *J. Environ. Qual.* **23**: 58–63.

Munshower, F.F. (1994) *Practical Handbook of Disturbed Land Revegetation*. Lewis Publishers. Ann Arbor, MI, USA p. 265.

Rabinowitz, M.B. (1993) Modifying soil lead bioavailability by phosphate addition. *Bull. Environ. Contam.* **51**: 438–444.

Ruby, M.V., Davis, A., and Nicholson, A. (1994) In situ formation of lead phosphates in soils as a method to immobilize lead. *Environ. Sci. Technol.* **28**: 646–654.

Ruby, M. V., Davis, A., Schoof, R., Eberle, S., Sellstone, C. M. (1996) Estimation of lead and arsenic bioavailability using a physiologically based extraction test. *Environ. Sci. Technol.* **30**: 422–430.

Ruby, M.V., Weis, C., Christensen, S., Post, G., Brattm, B., Drexler, J., Chappell, B., Ramaswami, A., Berti, W., Jensen, R., Carpenter, M., and Edwards, D. (1997) *Solubility/ bioavailability research consortium: Statement of purpose.* Exponent Environmental Group, Inc., Boulder, CO.

Sopper, W.E. (1993) *Municipal Sludge Use in Land Reclamation.* Lewis Publishers Ann Arbor, MI, USA pp. 163.

US Environmental Protection Agency. (1990) *Hazardous waste management system; identification and list of hazardous waste; toxicity characteristic revision; final rule,* Part II. Fed. Reg. 55(61): 11 798–11 877 & Part V. Fed. Reg. 55(126): 26 986–26 998.

US Environmental Protection Agency. (1994a) *Guidance manual for the Integrated exposure uptake biokinetic model for lead in children.* EPA/540/R-93/081, PB93–963510. U.S. Environmental Protection Agency, Office of Emergency and Remedial Response: Research Triangle Park, NC.

US Environmental Protection Agency. (1994b) Technical support document: parameters and equations used in the integrated exposure uptake biokinetic model for lead in children (v0.99d).1 U.S. Environmental Protection Agency, Office of Emergency and Remedial Response: Research Triangle Park, NC.

US Environmental Protection Agency. (1996a) *IINERT* Soil-Metals Action Team. Research and Development. Solid Waste and Emergency Response. EPA 542-F-96-101D.

US Environmental Protection Agency (1996b) *Recommendations of the Technical Review Workgroup for Lead for an Interim Approach to Assessing Risks Associated with Adult Exposures to Lead in Soil. Adult Lead Risk Assessment Committee,* Technical Review Workgroup for Lead. New York.

10

An Assessment of the Revegetation Potential of Acidic Basemetal Tailings using Metal-Tolerant Grass Species and Lime

W. J. Morrell[1], N. Bolan[2], P.E.H. Gregg[2] and R.B. Stewart[2]

INTRODUCTION

The establishment of vegetation on pyrite-bearing mine wastes is often problematic due to the tendency of these materials to rapidly oxidize and acidify on exposure to air and water (Johnson et al., 1994). Although the acidification of growing media and the ensuing onset of acid mine drainage (AMD) is a problem associated most commonly with pyrite-rich colliery spoil (Caruccio, 1975), these processes also impact mine wastes derived from gold and base-metal sulfide deposits (Harries, 1997; Table 10.1). In addition to low pH and high bioavailability of phytotoxic metals, metalliferous wastes typically exhibit multiple nutrient deficiencies and poor physical growing conditions, which may further hinder plant establishment (Bradshaw and Chadwick, 1980). An approach commonly used to establish vegetation on metalliferous mine-wastes is to cap them with composite covers of clay, rock, and soil which serve to physically isolate contaminants and to provide growing media for plants (Table 10.1).

The base-metal tailings derived from the Tui Mine in New Zealand (Fig. 10.1) have remained largely devoid of vegetation since their abandonment more than 25 years ago. Although the tailings deposit is relatively small in terms of both volume (100,000 m^3) and surface area (< 2 ha), in situ capping of the tailings will be a costly undertaking due to the site's poor accessibility. Direct vegetation of the tailings, although unlikely to dramatically improve the quality of AMD emanating from the tailings (see Pang (1995) for discussion of AMD impact associated with

---

[1] Centre for Mined Land Rehabilitation, University of Queensland, Brisbane, 4072, Australia
[2] Institute of Natural Resources, Massey University, Palmerston North, New Zealand

Table 10.1: Rehabilitation strategies for metalliferous mining wastes from selected Australian and New Zealand mine sites.

Minesite	Ore	Toxicity	Revegetation strategies/comments	Reference
Rum Jungle (NT Australia)	Cu, U[238]	Al, Cu, Co, Fe, Al, Cu,	Tailings were back-filled into an open cut pit and capped with low permeability cover. Waste rock dumps were recontoured and capped with clay/rock and soil cover to control AMD.	Harries and Ritchia, 1988; Bell et al., 1989
Captains Flat (NSW Australia)	Au, Cu, Pb, Zn	Cu, Pb, Zn	Trials indicated poor establishment of grass cover on tailings treated with lime (50t/ha) and fertilizer. Tailings were subsequently recontoured and capped with a composite clay/rock and soil cover.	Craze, 1977, 1979; Bell et al., 1989; Harries, 1997
Mt Isa (QLD Australia)	Ag, Cu, Pb, Zn	As, Cu, Pb, Zn	A variety of tailings ameliorates including fly ash, sewage sludge, and straw mulch have been trialed on site with limited success. Rehabilitation strategies are still under review but are likely to involve capping.	Bell et al., 1989
Mt Lyell (TAS Australia)	Cu	Fe, Cu, Zn, Pb	Tailings were historically dumped into local river system. Tailings rehabilitation strategies are still under review. Proposed strategies to rehabilitate waste rock dumps include recontouring, and the use of HDPE membrane and capping.	Koehnken, 1998; Harries, 1997
Tui Mine (New Zealand)	Au, Cu, Pb, Zn	As, Cu, Pb, Zn	Rehabilitation strategies for the tailings are still under review. Research has looked at use of metal tolerant and hyperaccumulating plants, tailings amelioration using lime and organic matter. Capping or codisposal of tailings at adjacent minesite remain viable options.	Morrell 1997; Bennett et al., 1998

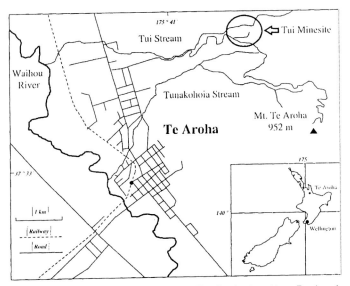

Fig. 10.1: Location of the Tui Minesite, Te Aroha, New Zealand.

the Tui minesite), may provide a cost-effective means of physically stabilizing the surface tailings in an environment prone to high intensity rainstorms and geological instability (Arand, 1986).

At minesites where the isolation of acid-generating wastes through burial or capping is impractical, a common means of facilitating plant establishment is to treat such wastes with high rates of acid-neutralizing materials and fertilizer (Pulford, 1991). Lime ($CaCO_3$) is one of the most commonly used liming agents. In addition to neutralizing acid and minimizing the bioavailability of toxic metals, liming improves nutrient availability, provides a source of calcium, reduces metal toxicity, and inhibits pyrite oxidation (Smith and Sobek, 1978). The use of lime to ameliorate acid-generating materials has been widely adopted in the coal mining industry and forms the focus of a considerable body of literature (Costigan et al., 1984; Ziemkiewicz and Skousen, 1995). Comparatively little research, however, has been published on the use of lime for amending strongly acidic, base-metal tailings for direct vegetation purposes (Maclean and Dekker, 1976; Burt and Caruccio, 1986).

In the United Kingdom, the use of metal-tolerant grass populations is also widely accepted as a potentially rapid and economical way of revegetating metal-contaminated materials (Smith and Bradshaw, 1979; Johnson et al., 1994). Although observations of metal-tolerant flora date back to the 16th century, Bradshaw (1952) was one of the first researchers to moot the idea of using these plants to revegetate metal-contaminated minesites (Gemmell, 1977). To date, over 50 species of plants capable of evolving metal-tolerant ecotypes have been identified (Antonovics et al., 1971; Eslick, 1995). Much research has focused on the use of metal-tolerant cultivars of *Festuca rubra* and *Agrostis capillaris* (Darmer, 1973; Johnson et al., 1977; Smith and Bradshaw, 1979; Bradshaw and McNeilly, 1981; Wong, 1982; Whitely and Williams, 1993, Johnson et al., 1994).

This chapter reports the outcomes of a glasshouse study that investigates the combined use of lime and two metal-tolerant grass cultivars to revegetate the Tui mine tailings. Change in bioavailability of metals as reflected by 0.02 M $CaCl_2$ extraction is also presented.

MATERIALS AND METHODS

Trial Design

To investigate the interaction of various liming rates and metal tolerance, a randomized factorial experiment was conducted using tailings collected from the Tui minesite. Four liming rates and five plant taxa were assessed in the trial, including a low fertility legume (*Lotus corniculatus*), two metal-tolerant grass species (*Agrostis capillaris* var. Parys Mountain and *Festuca rubra* var. Merlin) and two nonmetal-tolerant grass species (*A. capillaris* var. Egmont and *F. rubra* var. Lobi). An additional treatment was included for *L. corniculatus*, which consisted of inoculating the lime-amended tailings with rhizobium. The trial comprised 100 pots, including four replicates of each treatment and an additional four amended but unsown treatment pots that were destructively sampled during the course of the trial to monitor the effect of liming on soil characteristics.

Tailings

A composite sample (500 kg) of tailings was obtained from surface material (0 – 20 cm) collected at several locations on the Tui tailings dam. Prior to use the samples were air-dried, sieved to less than 4 mm, uniformly mixed and stored at 4°C. The highly variable geochemical characteristics of the tailings precluded the use of intact cores for plant growth trials.

Liming Rates and Treatment Preparation

The tailings were amended with four rates of finely ground, analytical-grade lime (nil, 8.25, 16.5 and 112 Mg ha^{-1}). The highest liming rate (112 Mg ha^{-1}) was based on the theoretical acid generating potential (TAGP) of the tailings. This rate represents the amount of lime required to neutralize the acid potentially produced by the oxidation of remnant sulfide minerals within the Tui tailings. The TAGP of the Tui tailings was calculated stoichiometrically using Fe and Cu data to estimate the quantities of pyrite (FeS_2) and chalcopyrite ($CuFeS_2$) in the tailings (Morrell et al., 1996).

The low and intermediate liming rates (8.25 and 16.5 Mg ha^{-1} respectively) were based on the pH buffering capacity of the tailings, with the intermediate rate (1 pH-6) representing the amount of lime required to raise the pH of the tailings to 6.0. The lower of these rates (½ pH-6) was arbitrarily set at 50% of the 1 pH-6 rate.

After lime addition, the tailings samples were moistened to 70% field capacity (–5 kPa) and left to equilibrate in open plastic bags at 25°C for 3 weeks. The tailings were then transferred into open pots, which were sown with the equivalent of 100 viable seeds. The viability was tested prior to the commencement of the trial. Moisture contents were maintained throughout the incubation period and trial via the addition of distilled water and a plant nutrient solution (Middleton, 1973). At 4 and 6 weeks post sowing, the trial pots were flushed with distilled water (4 pore volumes) to mitigate surface salt accumulation.

Tailings Characterization and Monitoring

Several chemical analyses were conducted on the tailings immediately prior to amelioration. Parameters measured included pH and electrical conductivity (EC) (using tailings:distilled water ratios of 1:2.5 and 1:5 respectively (Blakemore et al., 1987), total elemental concentrations as assessed by hydrofluoric (HF) acid digestion followed by inductively coupled plasma-atomic emission spectrometry measurements (ICP-AES), and 25% v/v nitric acid (Smith and Bradshaw, 1979) and 0.1 M hydrochloric acid (Viets and Bowen, 1965) extractable Fe and Cu concentrations estimated using flame atomic absorption spectrometry (FAAS). Several nutrient parameters were also assessed including exchangeable Ca, K, Mg and Na, Olsen P (Blakemore et al., 1987) and calcium phosphate extractable sulfate (Searle, 1979). Particle size analysis of the tailings was carried out using the wet pipette method described by Thomas (1973).

During the trial, the chemical characteristics of the tailings were monitored periodically by assessing pH, EC, 0.1 M HCl and 0.02 M calcium chloride ($CaCl_2$) extractable (Hoyt and Nyborg, 1972) As, Cd, Cu, Fe, Pb, Mn, and Zn and 1 M KCl (Blakemore et al., 1987) and 0.02 M $CaCl_2$ extractable Al levels in the unsown treatment pots. At the end of the growing period, pH conditions were assessed in all pots to examine variability between the sown and unsown treatment pots.

Seed Germination

To assess germination trends and initial seedling survival rates, emergent seedlings exhibiting fully expanded green cotyledons were counted 1, 2, and 4 weeks after sowing.

RESULTS AND DISCUSSION

Characterization

Physical characterization of the tailings indicates that it provides a satisfactory physical medium for plant growth. The predominantly fine sand to silt-sized (0.002 – 0.25 mm) tailings exhibit adequate water holding capacity (19.3% at - 5 kPa), moderate bulk density (1.3 Mg/m^3), and low impedance to root penetration (Morrell, 1997).

In terms of their chemical characteristics, however, the untreated tailings are an extremely inhospitable growing medium. The tailings exhibit low pH (2.3), high EC (1.26 dS/m), low pH buffering capacity (1.7 Mg $CaCO_3$/ 1000 Mg tailings/pH unit), high concentrations of phytotoxic metals, and multiple nutrient deficiencies (Table 10.2). They contain negligible amounts of organic matter (0.06%, Baskaran et al., 1996) and comprise predominantly quartz with minor kaolinite (Morrell et al., 1996).

Table 10.2: Selected total elemental concentrations of Tui Tailings as determined by ICP-AES following HF digestion.

Element	Concentration $(mg\ kg^{-1})$
Al	4090
As	185
Ca	71
Cd	5.2
Co	2.7
Cu	250
Fe	13400
K	875
Mg	93
Mn	5
Na	118
P	91
Pb	13400
S	9427
Zn	662

The unamended tailings contained about 70 mg kg^{-1} and 1800 mg kg^{-1} of 0.1 M HC1 extractable Cu and Fe, respectively. Appreciably higher total concentrations of Cu and Fe were observed within the tailings (Table 10.2). From these values, the amounts of pyrite and chalcopyrite, which are the primary acid-generating minerals, were calculated. The theoretical acid generating potential (TAGP) of the tailings, as measured from the contents of these minerals, is estimated to be 43.2 Mg $CaCO_3$/1000 Mg tailings. Pyrite accounts for more than 98% of this potential acidity in the tailings which exhibit negligible acid neutralizing potential (Morrell, 1997).

GERMINATION AND YIELD RESPONSES

Nil Lime Treatment

None of the species germinated in the nil lime treatment. The failure of all taxa to germinate in this treatment was not surprising given the extremely acidic nature of the medium. The pH of the nil lime treatment was found to remain low throughout the duration of the trial, dropping only slightly over the 11-week trial period from a preincubation level of 2.3 to about 2.1 in the unsown treatment pot (Fig. 10.2). Similar pH values of 2.2 were found in the sown trial pots at the end of the 8-week growing period.

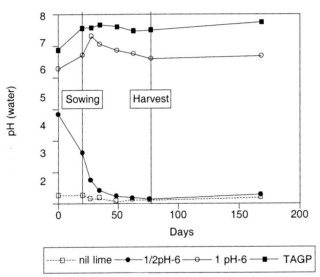

Fig. 10.2: Changing pH of unsown pots treated with various rates of lime (nil = 0, ½ pH-6 = 8.25, 1 pH-6 = 16.5, TAGP = 112 Mg $CaCO_3$ ha^{-1}).

Germination failure and plant mortality at low pH (<4.0) is generally attributed to enzyme deactivation and phytotoxicity associated with high hydrogen ion activity and/or the high bioavailability and release of toxic metal cations (e.g. Al^{3+} or Cu^{2+}) into the soil solution (Gemmell, 1977; Nriagu, 1984). Aluminium toxicity is likely to have played a significant inhibitory role in the nil lime treatment in terms of germination. High levels of 1 M KCl and 0.02 M $CaCl_2$ extractable Al were observed in the unamended tailings (Figs. 10.3 and 10.4). On Day 28 of the trial, for example, the concentration of 0.02 M $CaCl_2$ extractable Al in the unsown treatment pot was over 100 times the level known to adversely effect white clover growth (Edmeades et al., 1983; Hume et al., 1988). It is likely that Cu, Fe, and Zn, which were present in high readily available concentrations in the tailings, also inhibited seed germination (Fig. 10.3).

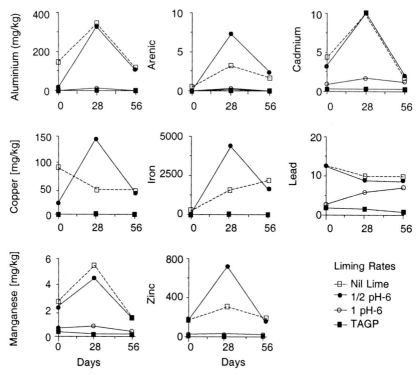

Fig. 10.3:   Concentrations of various metals extracted by 0.02 M CaCl₂ at different liming rates.

½ pH-6 treatment

In this treatment, four out of the five plant taxa failed to germinate despite the addition of the equivalent to 8.25 Mg of lime per hectare. Of the 5% of *Lotus corniculatus* seed that germinated, none of the emergent seedlings survived for more than a few days. Near total germination failure on this treatment is again attributed to the extreme acidity of the amended tailings at the time of sowing. Although the pH of the tailings initially rose to 4.9 following lime addition, the pH of the tailings decreased to 3.6 by the end of the 3-week incubation period (Fig. 10.2). The pH continued to decrease during the course of the trial to the point where by at Day 50, no marked pH difference was evident between the nil and ½ pH-6 lime treatments (Fig. 10.2).

Although lime addition at the ½ pH-6 rate resulted in the initial reduction of most metal concentrations in the tailings, by Day 28 of the trial, the quantities of 0.02 M CaCl₂ extractable Cu, Fe, and Zn was at least twice that found in the nil lime treatment (Fig. 10.3). Calcium chloride extractable As concentrations were also higher in the ½ pH-6 treatment (3 cf. 7 mg kg⁻¹), as were the levels of 1 M KCl extractable Al and 0.1 M HCl extractable Cu, Fe, and Zn (Fig. 10.4). The metal concentration trends

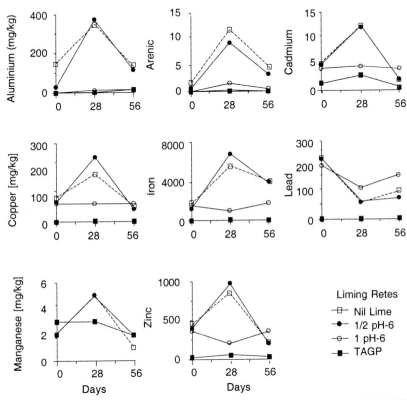

Fig. 10.4: Concentration of various metals extracted by 0.1 M HCl and 1 M KCl (aluminium only) at different liming rates.

observed in this study are consistent with the results of an unpublished incubation study in which the addition of low rates of lime to the Tui tailings was found to exacerbate growing conditions. This was attributed to the decrease in pH and significant increase in the bioavailability of Fe and sulfate in the tailings (Morrell, 1997). Although this contradicts the widely accepted belief that the addition of neutralizing agents generally inhibits the process of sulfide oxidation and therefore reduces metal toxicity (Geidel, 1979; Walder and Chavez, 1995), other studies have shown that low application rates of lime may in fact accelerate abiotic sulfide oxidation (Backes *et al.*, 1986; Burt and Caruccio, 1986) and potentially exacerbate growing conditions.

1 pH-6 treatment

In contrast to the nil and ½ pH-6 liming rates, "satisfactory" germination and subsequent seedling growth was obtained in the 1 pH-6 treatment. Although marked germination differences were observed between the three plant species, little variation was noted between the metal-tolerant and nonmetal-tolerant grass cultivars (Fig. 10.5(a)). Germination rates of close to 100% were attained for both varieties of *A. capillaris* within a week of sowing. The survival rate of *A. capillaris* seedlings was low, however, with only about 50% of the Parys Mountain seedlings and 35% of the Egmont seedlings surviving through to Week 4 of the trial (Fig. 10.5(a)). Although lower germination rates were attained for both varieties of *F. rubra* (≈60%) and *L. corniculatus* (≈40%), these species demonstrated comparatively high survival rates (Fig. 10.5a).

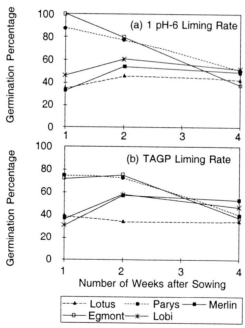

Fig. 10.5: Percentage germination at various periods after sowing at (a) 1pH-6 liming rate and (b) TAGP liming rate. Plants failed to germinate and/or establish on the nil and ½ pH-6 lime treatments.

As anticipated, the application of lime at a rate of 16.5 Mg ha^{-1} was found to increase pH of the tailings to approximately 6. During the 3-week incubation period, however, the pH of the amended tailings continued to rise, reaching a maximum of 7.3 (Fig. 10.2). A gradual decrease in pH, however, was noted during the course of the 8-week growing period. The pH of the planted pots (6.6 ± 0.1) measured at the end of this period was found to compare well with the pH observed in the unsown treatment pot (6.6).

At the 1 pH-6 liming rate, the metal-tolerant varieties of *A. capillaris* (Parys Mountain) and *F. rubra* (Merlin) were found to significantly outyield their nonmetal-tolerant counterparts by factors of 10 and 4 respectively (Fig. 10.6).

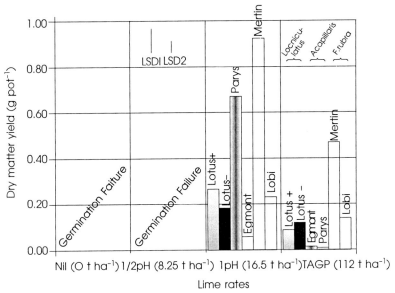

Fig. 10.6: Dry matter yields of different plant taxa at various liming rates LSDI = least significant difference for grass species (5%); LSD2 = least significant difference for *Lotus corniculatus* only (5%).

Although the dry matter yields of *L.corniculatus* were comparable to those of the nonmetal-tolerant grass species on the 1 pH-6 treatments, the overall performance of this species was poor (Fig. 10.6). The plant shoots were stunted and exhibited symptoms of chlorosis consistent with that of N deficiency. Although significantly higher yields of *L.corniculatus* were obtained from the pots inoculated with rhizobium at the 1 pH-6 liming rate, the reversal of this trend at the higher liming rate (TAGP) and the lack of obvious root nodulation suggest that this response is not associated with the fixation of nitrogen by rhizobia (Fig. 10.6). The root systems of the *L. corniculatus* were characterized by short, over-thickened lateral and primary roots. These symptoms, although indicative of Al toxicity, are also potentially attributable to a number of heavy-metal toxicities and nutrient deficiencies.

The concentrations of most metals extracted by 0.02 M $CaCl_2$ were relatively low in the 1 pH-6 treatment (Fig. 10.3), indicating their low bioavailability. The level of 0.02 M $CaCl_2$ extractable Al, at the time of sowing, for example, was about 10 mg kg^{-1} in the 1 pH-6 treatment compared with 170 mg kg^{-1} in the nil lime treatment. Calcium chloride extractable Al levels remained comparatively low throughout the trial as did Cu and Zn levels. The concentrations of 1 M KCl Al and 0.1 M HCl extractable As, Cu, Fe, and Zn were also appreciably lowered by the application of lime at the 1 pH-6 rate (Fig. 10.4).

TAGP treatment

Germination trends for this treatment were comparable to those observed at the 1 pH-6 liming rate. Poorer germination results (80% cf. 100%), however, were attained for both *A. capillaris cultivars*, which indicated that germination conditions were poorer in this treatment (Fig. 10.5).

The trend in pH changes for the TAGP treatment (112 Mg ha^{-1}) was almost parallel to that of the 1 pH-6 treatment (Fig. 10.2). The pH of the unsown treatment pot, however, increased to 7.5 during the 3-week incubation period where it remained for the entire monitoring period. Comparison of the pH of the planted pots (7.6 ± 0.1) with that of the equivalent unsown treatment pot (7.5) again indicated that there was good agreement between the unsown and sown trial pots.

The addition of 112 Mg ha^{-1} lime (i.e., TAGP rate) had a highly significant (P < 0.001) adverse impact on plant yield compared to the 1 pH-6 rate. Average herbage yields of all five plant taxa decreased notably at the TAGP liming rate (Fig. 10.6). The taxa demonstrating the greatest negative response to this higher rate of lime addition was the metal-tolerant variety of *A.capillaris* (Parys Mountain). Dry matter yields of this taxa decreased from about 0.7g pot^{-1} at the 1 pH-6 rate to about 0.01g pot^{-1} at the higher liming rate (Fig. 10.6).

Although the two highest lime rates had no significant effect (P = 0.328) on the relative yields of *A. capillaris* and *F. rubra*, a strongly significant (P > 0.001) interaction was identified between lime treatment and plant metal tolerance. This interaction is clearly evidenced in Fig. 10.6, where at the highest liming rate (112 Mg ha^{-1}), a marked decrease in the relative performance of the metal-tolerant plants in relation to the nonmetal-tolerant taxa is evident. This interaction is most marked for *A. capillaris* and suggests that the growth of this species is adversely affected by high pH conditions. In contrast to *A. capillaris*, no dramatic decrease in dry matter yields was noted for *F. rubra* at the TAGP liming rate, and the metal-tolerant variety of *F. rubra* again significantly outyielded the nonmetal-tolerant variety by a ratio of more than 3:1. The reduction in yield of *F. rubra* var. Merlin at the higher liming rate was, however, significant (P < 0.05).

The highest rate of lime application had highly significant (P < 0.001) and detrimental effect on the mean yields of *L. corniculatus*. At the TAGP rate, yields of Lotus were only about 50% of that found at the 1 pH-6 rate (Fig. 10.6). Although *L. corniculatus* treatments performed better than the two varieties of *A. capillaris*, the vigor and health of the legume grown on this treatment was again extremely poor.

The decrease in relative yield differences between metal-tolerant and nonmetal-tolerant cultivars of the two respective grass species and the reduced availability of A1, Cu, Fe, Pb, and Zn in the TAGP treatment indicates that the overall reduction in dry matter yield at the higher lime rate (112 t ha^{-1}) was not caused by metal toxicity (Figs. 10.3 and 10.4).

Plant Nutrition

Potential problems affecting plant growth associated with high pH are well recognized and include enzyme inactivation; induced B, Cu, Fe, Mg, Mn, Zn, and P deficiencies; and accelerated N loss through leaching and volatilization (Doubleday, 1972; MacLean and Dekker, 1976; Gemmell, 1977; Pulford, 1991). In an attempt to explain the variation between species performance and the reduction of yield on the TAGP treatment, foliar analysis was conducted on selected herbage derived from the two "successful" treatments that yielded herbage (Table 10.3). Nutrient data for Ca, K, Mg, N, and P, and estimated total uptake values and nutrient status classifications (based on data for *Trifolium repens* and *Lolium perenne* grown in soil) are presented in Table 10.3. The low quantities of dry matter produced by both *A. capillaris* cultivars at the TAGP liming rate precluded analysis of this material.

Phosphorus

Foliar analysis of the herbage indicated that P was deficient in seven of the eight treatments (Table 10.3). Analyses of the tailings sampled from the unsown treatment pots confirmed that the total P (<77 mg kg^{-1}) and Olsen P (<4.1 mg kg^{-1}) content of the tailings remained low throughout the trial despite the addition of high rates of soluble P fertilizer. Most plants grown on the tailings also showed some visual symptoms of P deficiency.

Table 10.3: Nutrient content and estimated total nutrient uptake by herbage from selected treatments. Uptake (mg kg^{-1}) = mean pot yield (g) × nutrient (%) × 10,000. Critical values based on figures for *Lolium perrene* and *Trifoliam repens* grown in soils (adapted from McLaren and Cameron, 1996).

	L. corniculatus		F. rubra Merlin		F. rubra Lobi		A. capillaris Parys	Egmont
	1 pH-6	TAGP	1 pH-6	TAGP	1 pH-6	TAGP	1 pH-6	pH-6
Phosphorus (%)	0.10	0.08	0.34	0.27	0.25	0.12	0.27	0.27
Critical P (%)	0.3	0.3	0.3	0.3	0.3	0.3	0.8	0.3
Uptake (mg kg^{-1})	181	97	3104	1259	787	168	1491	164
Nitrogen (%)	3.71	6.67	4.39	4.54	4.61	4.91	4.53	4.29
Critical N (%)	4.4	4.4	4.0	4.0	4.0	4.0	4.0	4.0
Uptake (mg kg^{-1})	6680	8001	39734	21551	14738	6877	24940	2577
Calcium(%)	0.78	1.65	0.39	0.64	0.73	1.07	0.41	0.61
Critical Ca (%)	0.3	0.3	0.2	0.2	0.2	0.2	0.2	0.2
Uptake (mg kg^{-1})	1401	1984	3548	3054	2323	1502	2250	368
Magnesium (%)	0.20	0.45	0.11	0.21	0.16	0.25	0.12	0.15
Critical Mg (%)	0.13	0.13	0.15	0.15	0.15	0.15	0.15	0.15
Uptake (mg kg^{-1})	355	544	1032	1017	525	354	660	89
Potassium (%)	1.40	1.06	3.06	2.28	2.22	1.18	4.07	3.28
Critical K (%)	1.7	1.7	1.7	1.7	1.7	1.7	1.7	1.7
Uptake (mg kg^{-1})	2512	1271	27693	10849	7101	1648	22391	1967

Herbage derived from the metal-tolerant cultivar of *F. rubra* (Merlin) grown on the 1 pH-6 liming rate was the only material found to contain satisfactory levels of P (Table 10.3). This observation is important as it was this treatment that yielded the greatest levels of dry matter, which strongly suggests that P was the nutrient governing plant growth on the tailings. Phosphorus is considered to be the primary limiting nutrient on many mining wastes (Smith and Bradshaw, 1979; Bradshaw and Chadwick, 1980).

Deficiencies of P were most marked in herbage originating from tailings amended with the TAGP rate of lime (Table 10.3). This result probably reflects the tendency of phosphate to form low solubility precipitates such as tricalcium phosphate ($Ca_3(PO_4)_2$) at high pH. Research by Doubleday (1972), Gemmell (1981), and Costigan et al. (1982) indicates that the addition of heavy rates of lime to pyritic mine wastes has variable impact on the P nutrition of pastoral plant species. One common finding of these studies, however, was that the performance of legumes was more adversely affected by the application of high rates of lime than that of any of the grass species investigated. Doubleday (1972) found that even moderate rates of lime addition (rates sufficient to attain a pH of 6.5) adversely affected dry matter yields on some spoils and reduced the growth response of *Lolium perenne* to phosphate.

Nitrogen

In contrast to P levels, herbage N levels were considered to be adequate at both the 1 pH-6 and TAGP liming rates (Table 10.3). *L. corniculatus* herbage grown on the lower of the two rates, however, was found to contain less than 3.8% N, a level considered to indicate deficiency. This analytical finding is consistent with the chlorotic appearance of the *L. corniculatus* seedlings grown on this treatment and strongly suggests that the low yields and poor performance of this species were, at least in part, related to the inability of this species to take up sufficient N under the circumneutral pH conditions.

Calcium, magnesium, and potassium

Predictably, the Ca status of the herbage grown on tailings amended with the two highest lime rates were adequate (Table 10.3). The presence of high concentrations of Ca in the tailings appears to have had no major impact on the uptake of Mg, as levels of this micronutrient in the foliage were considered to be satisfactory (Table 10.3). Potassium levels were also generally satisfactory in most treatments. *L. corniculatus* herbage, however, contained low levels of K at both liming rates. The nonmetal-tolerant *F. rubra* cultivar also appeared to exhibit K deficiency at the highest liming rate (Table 10.3). The low K levels in some foliage grown on the heavily limed tailings may reflect Ca-induced K deficiency (Doubleday, 1972; Gemmell, 1981; Costigan et al., 1982).

CONCLUSIONS

The failure of all plant taxa to germinate on the nil lime treatment substantiates chemical characterization of the Tui tailings and confirms that the tailings are an extremely hostile growth medium, even for plants adapted to adverse growing conditions. The pot trial results clearly indicate that without pH modification, attempts to revegetate the tailings via fertilizer application and irrigation alone will be futile.

Germination failure and the observed exacerbation of growing conditions following treatment with lime at a rate of 8.25 Mg ha^{-1} (½ pH-6) indicates that any attempts to rehabilitate the site should ensure that lime is incorporated at rates sufficiently high to ensure sulfide oxidation rates are appreciably reduced. The inability of the ½ pH-6 liming rate to maintain moderately acidic growing conditions for more than a few weeks was disappointing in that it effectively precluded an assessment of the various plant taxa to be made in such conditions.

In the short term, the 1 pH-6 liming rate was identified as being the superior treatment in terms of its ability to maintain the pH of the tailings in a range conducive to plant growth. The ongoing potential of the tailings to reacidify, however, was recognized as a major limitation to the long-term survival of plants established on this treatment. The addition of lime to the tailings at rates $\geq$ 16.5 Mg ha^{-1} was found to facilitate germination and plant growth of all taxa by reducing the pH and bioavailability of metals in the tailings. Plant nutrition, however, was adversely affected by these relatively heavy rates of lime application, particularly at the TAGP liming rate (112 Mg ha^{-1}). Phosphorus deficiency is believed to have been the primary nutrient governing plant growth on the heavily limed tailings.

Both varieties of *A. capillaris* were found to be poorly suited to the moderately alkaline conditions induced by the TAGP lime rate (112 Mg ha^{-1}). Likewise, *L. corniculatus* performed poorly in this treatment, as it did at the lower lime rate (16.5 Mg ha^{-1}). The poor performance of *L. corniculatus* (a legume renowned for its acid tolerance and low fertility requirements; Bennett et al., 1978) was disappointing in that it is recognized that the establishment of a legume on the Tui tailings is fundamentally important to the formation of self-sustaining vegetative cover.

Festuca rubra var. Merlin was the most successful of the plant taxa trialed in terms vigor and biomass production. This taxa formed and maintained what was considered to be a "satisfactory"vegetative cover on both the 1 pH-6 and TAGP lime treatments throughout the two-month growing period. Foliar analysis of the Merlin herbage grown on these treatments indicated that, with the possible exception of Mg, Merlin plants were receiving adequate macronutrients via the application of nutrient solution. The satisfactory performance of *F. rubra* var. Merlin is believed to reflect the fact that this cultivar is a calcicole and is thus relatively well suited to the circumneutral pH conditions induced by the two heavy lime rates. *A. capillaris* var. Parys Mountain, in contrast, is a calcifuge and thus prefers slightly acidic growing conditions (Johnson et al., 1977; Smith and Bradshaw, 1979; Bradshaw and Chadwick, 1980).

Whilst literature pertaining to the reclamation of pyritic mine wastes commonly advocates the use of high rates of lime to facilitate the growth of vegetation and to prevent the generation of acid mine drainage, the results of the present study indicate that there is a trade-off between diminished plant response resulting from the application of high rates of lime and the need to ensure that sufficient lime is added to prevent reacidification of the growing medium and the generation of AMD. Although the addition of lime at the TAGP rate is likely to have effectively prevented reacidification of the tailings, the adverse nutritional effects associated with this lime rate are unlikely to facilitate the establishment of an effective vegetative cover on the Tui mine tailings.

ACKNOWLEDGMENTS

The research presented in this paper formed part of research work conducted by the primary author during his Ph.D. at the Institute of Natural Resources, Massey University, Palmerston North, New Zealand. Support for the preparation of this paper was provided by the Centre for Mined Land Rehabilitation, University of Queensland, Brisbane 4072, Australia.

REFERENCES

Antonovics, J., Bradshaw, A.D., and Turner, R.G. (1971) Heavy Metal Tolerance in Plants. *Adv. Ecol. Res.* **7**: 1–85.

Arand, J.G. (1986) *Mass movement hazard at Te Aroha, North Island, New Zealand*. Unpublished Thesis, University of Waikato, Hamilton, New Zealand.

Backes, C.A., Pulford, I.D., and Duncan, H.J. (1986) Studies on the oxidation of pyrite in colliery spoil. I. The oxidation pathway and inhibition of the ferrous-ferric oxidation. *Reclam. Reveg. Res.* **4**: 279–291.

Baskaran, S., Bolan, N., Rahman, A., and Tillman, R. (1996) Effect of exogenous carbon on the sorption and movement of atrazine and 2,4-D by soils. *Aust. J. Soil Res.* **34**: 609–622.

Bell, L.C., Ward, S.C., Kabav, E.D., and Jones, C.J. (1989) Mine tailings reclamation in Australia - An overview. In *Proceedings of the Conference - Reclamation, a Global Perspective*. Walker, D.G., Powter C.B., and Pole, M.W. (Eds.). 23–31 August, 1989, Calgary, Canada. Alberta Land Conservation and Reclamation Council. Report RRTAC 89–2, Edmonton, Canada, pp. 769–781.

Bennett, F., Tyler, E., Brooks, R. R., Gregg, P., and Stewart, R. (1998) Fertilisation of hyperaccumulators to enhance their potential for phytoremediation and phytomining. In *Plants that Hyperaccumulate Heavy Metals*. Brooks, R.R. (Ed.). Wallingford: CAB International, pp. 249–259.

Bennett, O.L., Mathias, E.L., Armiger, W.H., and Jones, J.N.J. (1978) Plant materials and their requirements for growth in humid regions. In *Reclamation of Drastically Disturbed Lands*. Schaller, F.W., and Sutton, P. (Eds.). American Society of Agronomy, Madison, Wisconsin, pp. 285–306.

Blakemore, L.C., Searle, P.L., and Daly, B.K. (1987) *Methods for Chemical Analysis of Soils*. NZ Soil Bureau Scientific Report 80. Department of Scientific and Industrial Research, Lower Hutt, New Zealand.

Bradshaw, A.D. (1952) Populations of *Agrostis tenuis* resistant to Lead and Zinc. *Nature*. **169**: 1098.

Bradshaw, A.D. and Chadwick, M.J. (1980) The restoration of land: The ecology and reclamation of derelict and degraded land. In *Studies in Ecology*. Vol. 6. Blackwell Scientific, Oxford, 317 pp.

Bradshaw, A.D. and McNeilly, T. (1981) *Evolution and Pollution*. Studies in Biology No. 130. Edward Arnold, London, 76 pp.

Burt, R.A. and Caruccio, F.T. (1986) The effect of limestone treatments on the rate of acid generation from pyritic mine gangue. *Environ. Geochem. Health.* **8**(3): 71–78.

Caruccio, F.T. (1975) Estimating the acid potential of coal mine refuse. In *The Ecology of Resource Degradation and Renewal*. Chadwick, M.J, and Goodman, G.T. (Eds.). Blackwell Scientific Publications, pp. 197–205.

Costigan, P.A., Bradshaw, A.D., and Gemmell, R.P. (1982) The reclamation of acidic colliery spoil. III Problems associated with the use of high rates of limestone. *J. Appl. Ecol.* **19**: 193–201.

Costigan, P.A., Bradshaw, A.D., and Gemmell, R.P. (1984) The reclamation of acidic colliery spoil. IV The effects of limestone particle size and depth of incorporation. *J. Appl. Ecol.*, **21**: 377–385.

Craze, B (1979) Rehabilitation of tailings dumps at Captains Flat, New South Wales. In *Mining Rehabilitation*. Hore-Lacey, I. (Ed.). Australian Mining Industry Council, Canberra, pp. 91–94.

Craze, B. (1977) Restoration of Captains Flat mining area. *J. Soil Conserv. Serv. New South Wales* **33**: 98–105.

Darmer, G. (1973) Grasses and herbs for revegetating phytotoxic material. In *Ecology and Reclamation of Devastated Land*. Hutnik, R.J. (Ed.). Gordon & Breach, London, pp. 91–103.

Doubleday, G.P. (1972) Development and management of soils on pit heaps. *Land. Reclam.* **2**: 25–35.

Edmeades, D.C., Smart, C.E., and Wheeler, D.M. (1983) Aluminium toxicity in New Zealand Soils: preliminary results on the development of diagnostic criteria. *NZ J. Agric. Res.* **26**: 493–501.

Eslick, L.J. (1995) *Heavy metal uptake patterns in riparian vegetation*. Thesis, Boise State University.

Geidel, G. (1979) Alkaline and acid production potentials of overburden material: The rate of release. *Reclam. Rev.* **2**: 101–107.

Gemmell, R.P. (1977) *Colonization of Industrial Wasteland*. Edward Arnold, London, 75 pp.

Gemmell, R.P. (1981) The reclamation of acidic colliery spoil. II The use of lime wastes. *J. Appl. Ecol.* **18**: 879–887.

Harries, J. (1997) Acid mine drainage in Australia: Its extent and potential future liability. Supervising Scientist Report. 125, Office of Supervising Scientist and the Australian Centre for Minesite Rehabilitation Research, Canberra.

Harries, J.R. and Ritchie, A.I.M. (1988) Rehabilitation measures at the Rum Jungle mine site. In *Environmental Management of Solid Waste - Dredged Material and Mine Tailings*. Salomons, W. and Forstner, U. (Eds.). Springer-Verlag, New York, pp. 131–151.

Hoyt, D.B. and Nyborg, M. (1972) Use of dilute calcium chloride for the extraction of plant available aluminium and manganese from acid soil. *Can. J. Soil Sci.* **52**: 163–167.

Hume, L.J., Ofsoski, N.J., and Reynolds, J. (1988) Influence of pH, exchangeable aluminium and 0.02M CaCl$_2$-extractable aluminium on the growth and nitrogen-fixing activity of white clover (*Trifolium repens*) in some New Zealand Soils. *Plant and Soil* **111**: 111–119.

Hunter, G.D. and Whiteman, P.C. (1975) Revegetation of mine wastes at Mount Isa, Queensland. 2. Amendment of nutrient status and physical properties of tailings for plant growth. *Aust. J. Exp. Agri. Animal Husband.* **15**: 803–811.

Johnson, M.S., Cooke, J.A., and Stevenson, J.K.W. (1994) Revegetation of metalliferous wastes and land after metal mining. In *Issues in Environmental Science & Technology*. Hester, R.E. and Harrison, R.M (Eds.). Royal Society of Chemistry, London, pp. 31–48.

Johnson, M.S., McNeilly, T. and Putwain, P.D. (1977) Revegetation of metalliferous mine spoil contaminated by lead and zinc. *Environ. Pollut.* **12**: 261–277.

Koehnken, L. (1997) The Mount Lyell remediation research and demonstration program (MLRRDP) and other Tasmanian acid drainage issues. In *Proceedings of the Third Australian Acid Mine Drainage Workshop*. McLean, R.W. and Bell, L.C.(Eds.). Darwin, 15–18 July 1997, Centre for Minesite Rehabilitation Research, Kenmore, QLD, Australia, pp. 171–181.

MacLean, A.J. and Dekker, A.J. (1976) Lime requirement and availability of nutrients and toxic metals to plants grown in acid mine spoils. *Can. J. Soil Sci.* **56**: 27–36.

McLaren, R.G. and Cameron, K.C. (1996) *Soil Science: Sustainable Production and Environmental Protection.* Oxford University Press, Auckland, 304 pp.

Middleton, K.R. (1973) Nutrient solution. *Plant and Soil* **38**: 219.

Morrell, W.J. (1997) *An assessment of the revegetation potential of base-metal tailings from the Tui Mine, Te Aroha, New Zealand.* PhD. Thesis, Massey University, Palmerston North, New Zealand, 468 pp.

Morrell, W.J., Stewart, R.B., Gregg, P.E.H., Bolan, N.S., and Horne, D. (1996) An assessment of sulfide oxidation in abandoned base-metal tailings, Te Aroha, NZ. *Environ. Pollut.* **94**: 217–225.

Nriagu, J.O. (1984) *Changing Metal Cycles and Human Health.* Springer-Verlag, Berlin.

Pang, L. (1995) Contamination of groundwater in the Te Aroha area by heavy metals from an abandoned mine. *NZ J. Hydrol.* **33**(1): 17–34.

Pulford, I.D. (1991) A review of methods to control acid generation in pyritic coal mine waste. In *Land Reclamation. An End to Dereliction?* Davies, M.C.R. (Ed.). Elsevier Applied Science, London, pp. 269–278.

Searle, P.L. (1979) Measurement of adsorbed sulfate in soils - Effect of varying soil-extractant ratios and method of measurement. *NZ J. Agric. Res.* **22**: 287–290.

Smith, M.R. and Sobek, A.A. (1978) Physical and chemical properties of overburdens, spoils, wastes, and new soils. In *Reclamation of Drastically Disturbed Lands.* Schaller, F.W and. Sutton, P. (Eds.). American Society of Agronomy, Madison, Wisconsin, pp. 149–172.

Smith, R.A.H. and Bradshaw, A..D.(1979) The use of metal tolerant plant populations for the reclamation of metalliferous wastes. *J. Appl. Ecol.* **16**: 595–612.

Thomas, R. (1973) Determination of particle size distribution. New Zealand Soil Bureau Report 10, DSIR, Wellington.

Viets, F. and Boawn, L. (1965) Zinc. In *Methods of Soil Analysis: Part 2 Agronomy Vol. 9.* Black et al. (Ed.), pp. 1090–1101.

Walder, I.F. and Chavez, W.X.J. (1995) Mineralogical and geochemical behaviour of mill tailing material produced from lead-zinc skarn mineralisation, Hanover, Grant County, New Mexico, USA. *Environ. Geol.* **26**: 1–18.

Whiteley, G.M. and Williams, S. (1993) Effects of treatment of metalliferous mine spoil with lignite derived humic substances on the growth responses of metal tolerant and non metal tolerant cultivars of *Agrostis capillaris* L. *Soil Technol.* **6**: 163–171.

Wong, M.H. (1982) Metal cotolerance to copper, lead and zinc in *Festuca rubra. Environ. Res.* **29**: 42–47.

Ziemkiewicz, P.F. and Skousen, J.G. (1995) Prevention of acid mine drainage by alkaline addition. In *Acid Mine Drainage - Control and Treatment.* Skousen, J. and Ziemkiewicz, P. (Eds.). West Virginia University and the National Mine Reclamation Center, Morgantown, West Virginia, pp. 45–56.

11

Groundwater Arsenic Contamination in West Bengal-India and Bangladesh: Case Study on Bioavailability of Geogenic Arsenic

Uttam Kumar Chowdhury[1], Mohammad Mahmudur Rahman[1], Gautam Samanta[1], Bhajan Kumar Biswas[1], Gautam Kumar Basu[1], Chitta Ranjan Chanda[1], Kshitish Chandra Saha[1], Dilip Lodh[1], Shibtosh Roy[2], Quazi Quamruzzaman[2] and Dipankar Chakraborti[1]*

INTRODUCTION

The incidents of arsenic (As) contamination in groundwater and the consequent sufferings of people from chronic As toxicity resulting from drinking the contaminated water from West Bengal, India and Bangladesh are well documented. The clinical manifestations are many, but the most commonly observed symptoms that identify people suffering from chronic arsenic poisoning are arsenical skin lesions. In severe cases, gangrene in limbs and malignant neoplasms are also observed. Major cases of groundwater arsenic contamination have occurred in Taiwan (Lu, 1990); Antofagasta in Chile (Borgono and Greiber, 1971); Mexico (Cebrian et al., 1983), and Argentina (Astolfi et al., 1981). Other incidents involving smaller population groups have been reported from Poland (Geyer, 1898); Minnesota,USA (Feinglass, 1973); Ontario, Canada (Wyllie 1973); Nova Scotia, Canada (Grantham and Jones, 1977); Nakajo, Japan (Terade et al., 1960), Millard County, Utah, USA (Southwick et al., 1983); Lane County, Oregon, USA (Morton et al., 1976); Lassen County, California, USA (Goldsmith et al., 1972), and Fairbanks, Alaska, (Harrington et al., 1978). The principal finding in most of these incidents was the close relationship between the prevalence of cutaneous lesions and the exposure to drinking

[1] School of Environmental Studies, Jadavpur University, Calcutta-700 032, India
[2] Dhaka Community Hospital, Bara Magh Bazar, Dhaka-1217, Bangladesh
* Author to whom all correspondence to be made.
E-mail: dcsoesju@vsnl.com

water containing high levels of As. The comparative results of clinical findings as reported from chronic As poisoning from drinking As-contaminated water show the presence of almost all the stages of arsenical clinical manifestations (Hotta, 1989).

In the last two decades, there has been an alarming increase in the incidence of poisoning resulting from ingestion of As-contaminated groundwater in the Asian countries. As poisoning from drinking water was reported during early 1980s from the Xinjiang province of China (Wang, 1984). In the early 1990s, this type of As poisoning was found in many other places of Inner Mongolia and the Shanxi province. According to various research reports compiled by the end of 1995, As contamination in Inner Mongolia extends to 655 villages of 11 counties and 1774 patients were identified; about 300,000 people are currently estimated to be drinking contaminated water (Luo et al., 1997; Xiao-Juan, 1997). In West Bengal, India, incidents of groundwater As contamination have been reported to be the biggest As calamity in the world (Das et al., 1994; Chatterjee et al., 1995, Das et al., 1995; 1996; Mandal et al., 1996; 1997, 1998, Chowdhury et al., 1997; Guha Mazumder et al., 1997a). The area and population in the As-affected districts of West Bengal are about 38,000 km^2 and 42 million, respectively. However, this does not mean that all people in this area drink contaminated water, but, no doubt, they are at risk. It is estimated that in West Bengal, more than 5 million people drink As-contaminated water and more than 300,000 people suffer from arsenicosis.

The groundwater As problem in Bangladesh was recognized recently (Dhar et al., 1997; Biswas et al., 1998). Of the 64 districts of Bangladesh (total area 148,393 km^2 and total population about 120 million), the concentration of As in groundwater exceeds 0.05 mg/l in 47 districts (the WHO maximum permissible limit). The total area and population of these 47 districts are 112,407 km^2 and 93.4 million, respectively. Thus, in West Bengal, India, and Bangladesh, together more than 100 million people are at risk from groundwater As contamination. In these districts, geogenic As in groundwater is highly bioavailable, as reflected by the clinical symptoms of As poisoning in many people.

Besides providing conclusive evidence of bioavailable geogenic As, this chapter presents data from a case study that was designed to assess As poisoning in West Bengal, India, and Bangladesh. In particular, the case study describes (a) survey studies up to February 2000 of As groundwater contamination in these two countries, (b) As metabolites in urine, (c) As in hair, nail, skin scales, blood, and their correlation with concentration of As in drinking water, (d) dermatological features, (e) detailed study of two As-affected villages, one from West Bengal and another from Bangladesh, and (f) possible sources of As.

MATERIALS AND METHODS

Flow injection hydride generation atomic absorption spectrometry (FI-HG-AAS) was used to analyze water, urine, digested hair, nail, skin-scale, and blood samples. A detailed description of the instrumentation and FI-HG-AAS procedure is described in our earlier communications (Chatterjee et al., 1995; Das et al., 1995; Samanta and Chakraborti 1997). Total As in water and bore-hole sediments was also analyzed spectrometrically using Ag-DDTC in $CHCl_3$ with hexamethylene tetramine (Chakraborti et al., 1982). Sediments were analyzed using X-ray fluorescence (XRF), electron probe microanalysis (EPMA), laser microprobe mass analyser (LAMMA), and X-ray powder diffractometry (XRD). The XRF and EPMA instrumentations have been described previously (Chatterjee et al., 1993; Chakraborti et al., 1992). A LAMMA-500 (Ley Bold, Heraeus Colonge, Germany) was used to measure sediments to get positive and negative mass spectra. A detailed description of LAMMA is given elsewhere (Heines et al., 1979; Das et al., 1996). The sediment samples have also been studied using a Siemens D500 powder diffractometer (Das et al., 1996) with a Cu -target and excitation conditions of 45 KV and 39 mA after the sediment samples have been separated using a permanent magnet to distinguish morphological and magnetic properties.

REAGENTS AND GLASSWARE

All reagents are of Analar quality. A solution of 1.5% $NaBH_4$ (E. Merck, Germany) was prepared in 0.5% NaOH (E. Merck, India) and a 5 M solution of HCl (E. Merck, India) was used for flow injection analysis. Ashing aid suspension was prepared by stirring 10g $Mg(NO_3)_2.6H_2O$ and 1 g MgO in 100 ml of water. Details of the reagents and glassware are given elsewhere (Chatterjee et al., 1995; Das et al., 1995).

SAMPLE COLLECTION AND DIGESTION

Water

Water samples were collected from tubewells (borewells) as given in Chatterjee et al. (1995).

Hair, Nails, Skin-scales, and Urine

Hair, nail, and urine were collected from (a) victims (b) people who drink As-contaminated water, and (c) people who live in As-affected villages and drink safe water. Skin-scales were collected from As victims having hyperkeratosis and keratosis. The procedure for collection, cleaning, and the mode of digestion are described in Chatterjee et al. (1995); Das et al. (1995), and Mandal et al. (1996).

Blood

Blood samples were collected from the arm vein through a stainless-steel needle in a prewashed polyethylene bottle containing heparin as an anticoagulant (Cornelis et al., 1996). The blood samples were kept in a deepfreezer (at minus 20°C) until analysis.

For mineralization of blood samples, 2 ml of blood was taken in a 50-ml borosilicate glass tube and 3 ml ashing aid was added, followed by 3 ml (40% v/v) HNO_3, and evaporation to dryness in a sand bath. When the samples were dry, the beakers were covered with a watch glass and transferred to the muffle furnace where the samples were subjected to ashing as per the following temperature programmes: 150°C for 1 h, 200°C for 30 min, 250°C for 30 min, 250-450°C for 90 min, and 450°C overnight. When a white ash was obtained, it was moistened with water, dissolved in 1 ml 6 M HCl and filtered through millipore filter paper. The volume was made up to 2 ml with 6 M HCl. Triplicate blanks were prepared following the same procedure.

Bore-hole Sediments

Bore-hole sediments were collected from different areas of the affected districts as described by Das et al. (1996).

PROCEDURE FOR DETERMINING As IN WATER, HAIR, NAILS, SKIN-SCALES, BLOOD, AND As METABOLITES IN URINE

A detailed procedure for determining As in water, hair, nails, skin-scales, and sediments has been described earlier (Chatterjee et al., 1995; Das et al., 1995, 1996). In this study, total As in water was determined both by FI-HG-AAS and spectrophotometry. Spectrophotometry (Chakraborti et al., 1982) is a reliable procedure to detect As concentrations greater than 0.04 mg/l whereas by FI-HG-AAS concentrations up to 0.003 mg/l are detected with 95% confidence. Arsenic in hair, nails, blood, and skin-scales was measured by FI-HG-AAS and the procedure described by Das et al. (1995). For the urine samples, only inorganic As and its metabolites were measured by FI-HG-AAS without any chemical treatment, as described by Das et al. (1995) and Mandal et al. (1996).

Arsenic in blood samples was measured by FI-HG-AAS after dry ashing of the blood samples. The accuracy and precision of the procedures was checked by analyzing reference materials. For As, the standard reference materials analysed were bovine liver 1577 (NIST, USA; certified value 0.055 ± 0.005 µg/g), mussel tissue 278 (Community Bureau of Reference,

Belgium, certified value 5.9 ± 0.2 µg/g), and citrus leaves 1572 (NBS, USA; certified value 3.1 ± 0.3 µg/g). The results obtained were as follows 0.057 ± 0.003 µg/g for bovine liver, 5.93 ± 0.06 µg/g for mussel tissue, and 3.12 ± 0.1 µg/g for citrus leaves. These results show that our method was in good agreement with the certified value.

RESULTS AND DISCUSSION

Figure 11.1 shows the 9 districts and 69 blocks of West Bengal that were As-affected. Other than five, all affected blocks are on the right hand side of the river Hooghly. Table 11.1 shows the physical parameters of 9 districts. About 5 million people are drinking As-contaminated water with an As concentration greater than 0.050 mg/l and more than 300,000 people have arsenical skin lesions (Table 11.1). These estimates are based on extrapolation of a three year study in the Deganga block (Table 11.2) of North 24-Parganas district.

Deganga is a medium As-affected block. The total population of Deganga is about 230,000 and there are 15,886 hand tubewells. Almost 100% of the people drink water from tubewells. Analyses of As from 8785 tubewells in Deganga show that 41% of the tubewells (i.e. 3610 tubewells) contain As greater than 0.050 mg/l. Therefore, 95,994 people in Deganga drink As- contaminated water that contains As excess of 0.050 mg/l. Given that 69 blocks are As- affected at present, it may be assumed that 5 million people drink water contaminated with As in excess of 0.050 mg/l. These assumptions have been confirmed by recent studies that demonstrate the risk posed by As to more than 10 million people in these regions.

Chakraborti and Saha (1987) studied the incidences of arsenical dermatosis in 14 villages of West Bengal. Dermatosis was characterized by diffuse and palmoplanter melanosis and keratosis. According to their study, the lowest As concentration in the water capable of producing dermatosis was 0.2 mg/l. In the present study at Deganga block, about 6.3% (i.e. 553 tube wells) of tubewells have an As concentration above 0.2 mg/l. According to the study by Chakraborti and Saha (1987), 15000 people in these regions may show dermatological symptoms. It is reported that symptomatology or arsenical toxicity may develop insidiously after 6 months to 3 years or more depending on the As intake (Lian and Jian, 1994).

The nutritional status may also play an important role in As toxicity. Our study in the affected districts indicates that undernourished people are badly affected (Das et al., 1995; Mandal et al., 1997; Dhar et al., 1997). According to Guha Mazumder (1997a), pigmentation was found in 13 cases and thickening of palms and soles in 4 cases out of 3235 persons drinking water with an As content of 0.010 and 0.050 mg/l. However, the

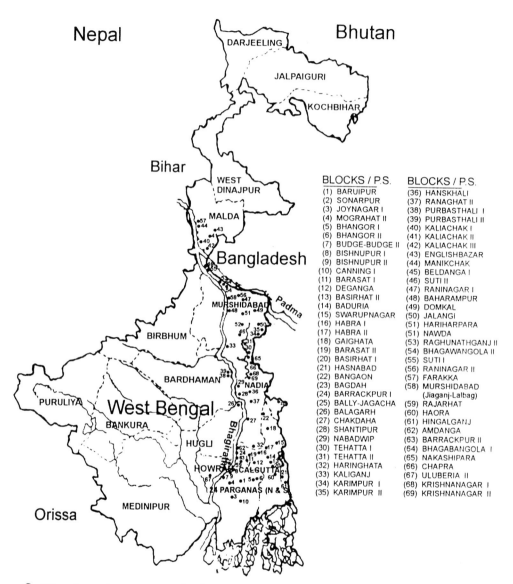

● **Blocks where arsenic in groundwater found above 50 µg/l**

Fig. 11.1: Map showing 69 As-affected blocks in 9 districts of West Bengal having As in groundwater above 0.05 mg/l.

Table 11.1: Physical parameters of nine arsenic affected districts of West Bengal, India

Total area of West Bengal (km²)	Total population of West Bengal	Total area of arsenic affected nine districts of West Bengal (km²)	Total population of arsenic affected 9 districts of West Bengal	Total area of 69 arsenic affected blocks of West Bengal (km²)	Total population of 69 arsenic affected blocks of West Bengal	Total no. of arsenic affected villages/wards	Population drinking arsenic contaminated water above 0.05 mg/l	Population showing arsenic related skin manifestation
89,192.40	67,982,732	38,865	42,700,000	11,495	17,100,000	985	5,000,000 (approx.)	300,000 (approx.)

Table 11.2: Three-year survey report of Deganga block in North 24-Parganas district, West Bengal, India since 1994

Total area (km²)	Total population	Total no. of tubewells present	Total no. of tubewells analyzed	No. of samples having arsenic >0.01 mg/l	No. of samples having arsenic >0.05 mg/l	Distribution of no. of samples in different concentration ranges (mg/l) of arsenic							
						<0.01	0.01-0.049	0.05-0.099	0.10-0.299	0.30-0.499	0.50-0.699	0.70-1.0	>1.0
201.05	234142	15886	8785	5082 (57.85%)	3610 (41.09%)	3703	1472	1713	1577	203	94	19	4

total quantity of As consumed per day and the duration of exposure are important factors (Chakraborti and Saha, 1987; Niu et al., 1997; WHO 1981). Chen and Wu (1962) reported in Taiwan high incidences of hyperkeratosis and skin cancer after people consumed water containing As greater than 0.3 mg/l. Tseng et al. (1961) also observed association of skin cancer with drinking well water containing 0.50 mg/l As. WHO (1981) reported that approximately 1 mg /day As may give rise to skin effects within a few years of exposure. In West Bengal, the affected villagers are farmers who consume a high volume of water (3-6 litre/ day). Considering, on an average, a daily consumption of water of 4 litre, it is likely that people who drink water with 0.25 mg/l of As i.e. 1.0 mg of As/day for a few years may show skin lesions. Several studies worldwide also provide quantitative toxicity data in humans following chronic oral exposure to As. From these studies, it appears that chronic intake of about 0.010 mg As/kg bodyweight/day or higher may result in dermatological and other signs of arsenical toxicity. The study of Chakraborti and Saha identified a lowest observed adverse effects level (LOAEL) of 0.018 mg As/kg/day of dermal and hepatic effects. This value is strongly supported by several other studies that identified similar LOAEL values (Hindmarsh et al., 1977) LOAEL = 0.019 mg As/kg/day; (Cebrain et al., 1983; Abernathy et al., 1989). LOAEL = 0.014 mg/kg/day. If we extrapolate the above data, 15,000 people in Deganga may have dermatological symptoms, and in 69 blocks this may rise to about 1 million. However, assuming conservatively that 30% of the estimate is correct, 300,000 people in 69 blocks may have arsenical skin lesions. Mandal et al. (1996) reported that approximately 20% of the population exposed to As-contaminated water in As-affected villages of West Bengal, have arsenical skin lesions.

The total area and population of Bangladesh are 148,393 km^2 and 120 million, respectively. A preliminary survey of 64 districts in Bangladesh revealed that As concentrations exceeded 0.050 mg/l (the WHO maximum permissible limit) in 47 districts and in 54 districts As exceeded 0.010 mg/l (WHO recommended value for drinking water; WHO, 1993). Figure 11.2 shows that at present groundwater of only 10 districts is safe to drink. Table 11.3 shows the distribution of As in water from 20,987 tubewells from 47 districts of Bangladesh where As exceeded 0.050 mg/l. Table 11.4 shows that 93.4 million people in 47 As-affected districts may be at risk. In those 47 districts, so far we have surveyed 32 districts for As patients, and in 30 districts we have found people suffering from arsenical skin lesions. In the last 5 years, we have analyzed 22,003 water samples from 64 districts of Bangladesh. In 54 districts, we observed arsenic in excess of 0.010 mg/l in groundwater and in 47 districts above

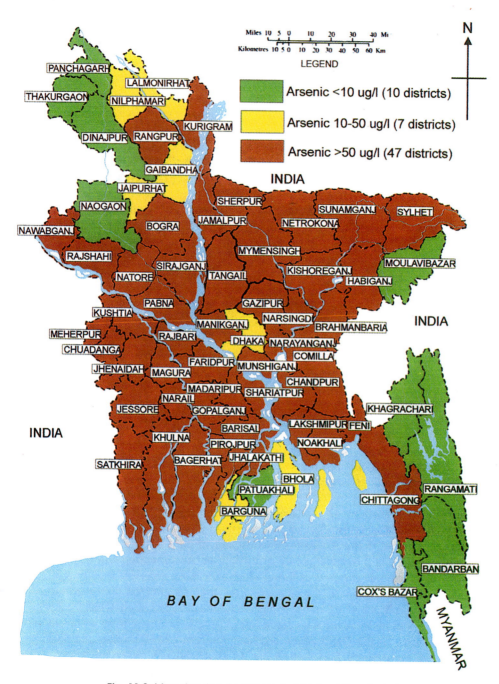

Fig. 11.2: Map showing As-affected districts of Bangladesh.

Table 11.3: Distribution of As in tubewell water of 47 As affected districts of Bangladesh

Total no. of tubewell water samples analyzed	Distribution of no. of tubewell water samples in different concentration ranges (mg/l) of arsenic							
	<0.01	0.01-0.05	0.051-0.099	0.10-0.299	0.30-0.499	0.50-0.699	0.70-0.999	>1.00
20,987	5121 (24.40%)	4332 (20.64%)	2442 (11.65%)	5387 (25.67%)	2098 (10.0%)	882 (4.88%)	465 (2.21%)	258 (1.23%)

0.050 mg/l. However, to get an idea about the number of people who drink As-contaminated water with As concentration above 0.010 mg/l and 0.050 mg/l in Bangladesh, we studied 43 districts in Bangladesh. We got the necessary information on the number of users for each tubewell for 16,410 tubewells out of 22,003 we analyzed from 43 out of 47 districts where groundwater contains arsenic above 0.050 mg/l. Based on the relevant information and arsenic concentration of water from tubewells, the number of users consuming drinking water contaminated with arsenic above 0.010 mg/l and above 0.050 mg/l for each location was calculated. For a district, the calculated value of each location based on actual observations was used to give weightage to the other locations for which sampling was not done. Ultimately, we calculated first the total number of people affected from 10 districts. These results were extrapolated in 33 other districts where we have relevant information/data on number of users and arsenic concentration in tubewells. This shows that the total populations exposed to As above 0.010 mg/l and 0.050 mg/l in 43 districts are 51,590,829 and 25,045,633 respectively, out of 85,397,000 people in 43 districts of Bangladesh.

As IN GROUNDWATER IN WEST BENGAL AND BANGLADESH

Table 11.5 shows the results of the total arsenic analysis of 80,000 water samples collected from West Bengal. Table 11.5 also shows preliminary survey report of tubewells of 64 districts of Bangladesh. The magnitude of As contamination in groundwater is much higher in Bangladesh compared with that in West Bengal, India.

ANALYSIS OF As METABOLITES AND TOTAL As IN HAIR, NAILS AND SKIN-SCALES IN WEST BENGAL AND BANGLADESH

Analysis of urine is a direct indication of As contamination (Cornelis et al., 1996; Buchet and Lauwerys, 1994; Hsu et al., 1997). Total As determination will also account for the unabsorbed nontoxic form of As

Table 11.4: Physical parameters of 47 affected districts of Bangladesh where arsenic was found >0.05 mg/l

Total area (km²)	Total population (million)	Total number of locations	Total number of locations surveyed	Number of locations where arsenic found		Total no. of districts surveyed for patient	Total no. of districts where where patients identified	Total no. of locations surveyed for patients	Total no. of locations where patents identified
				>0.01 mg/l	>0.05 mg/l				
112,407	93.4	357	182	163	149	32	30	74	67

Table 11.5: Distribution of arsenic in tubewells water in nine districts of West Bengal, India and 64 districts of Bangladesh up to February 2000

Country	Total no. of tubewell water samples analyzed	Distribution of no. of tubewell water samples in different concentration ranges (mg/l) of arsenic							
		<0.01	0.01-0.05	0.051-0.099	0.10-0.299	0.30-0.499	0.50-0.699	0.70-1.000	>1.000
West Bengal, India	80,000	38040 (47.55%)	17658 (22.07%)	9980 (12.48%)	11200 (14.0%)	2354 (2.94%)	560 (0.7%)	160 (0.2%)	48 (0.06%)
Bangladesh	22,003	6003 (27.28%)	4400 (20.0%)	2490 (11.32%)	5391 (24.50%)	2102 (9.55%)	885 (4.02%)	470 (2.14%)	262 (1.19%)

coming from seafood (Mohri et al., 1990; Marafante et al., 1987), for example, arsenocholine and arsenobetaine. In the present study, only inorganic As and its metabolites were estimated in urine samples. In our experimental set-up of FI-HG-AAS, arsenocholine and arsenobetaine are non-detectable (Chatterjee et al., 1995).

Tables 11.6 and 11.7 show As metabolites in urine samples from West Bengal and Bangladesh of people from As-affected areas and also of a control population from non affected areas (where As in groundwater is less than 0.01 mg/1). These studies reveal that 85% of the cases investigated in West Bengal, India, and 95% in Bangladesh exceed normal urinary As concentrations (Farmer and Johnson, 1990). A similar trend was reported by Mandal et al. (1998), who found that although some people were drinking safe water in the affected villages, their urine As level was somewhat elevated. The study shows that the contamination is unavoidable and that it may come from surfaces of edible herbs, vegetables, and contaminated water used for washing, food preparation, etc.

Tables 11.6 and 11.7 show elevated levels of As in hair and nails measured for a few thousand people of West Bengal, India, and Bangladesh. There is no recommended value of As in skin-scales. Higher skin-scales As indicates obvious body burden of As. In the current field study, it has been observed that in affected villages, many people drink contaminated water but show no arsenical skin lesions, although their hair and nails contain elevated levels of As. From a medical point of view, they may be classified as subclinically affected.

BLOOD As CONCENTRATIONS: A CASE STUDY AT ARSENIC-AFFECTED VILLAGES OF WEST BENGAL

Accumulation of As in the blood does not occur in humans. A major part of both inorganic and organic As in blood is cleared rapidly in humans (WHO, 1981). For this reason, blood As reflects exposure for only a short period following absorption and is time-dependent. Only if exposure is continued, a steady state of As in blood is reached and there is a possibility of finding a relationship between As in blood and exposure. Only in case of continued As exposure, blood As levels may be a useful indicator of its presence in the body (Morton and Dunnette, 1997).

Arsenic in blood tends to be associated with globulins from which it is cleared into tissues within 24 h (Ellenhorn and Barceloux, 1988). The rate of decline of As in the erythrocytes is comparable with that in plasma but the red cells contain about three times more As than the plasma does (WHO, 1981). But in rats, As is associated with erythrocytes (WHO, 1981).

Arsenic in whole blood was measured from some people of few villages of West Bengal who drank As-contaminated water and showed arsenical skin lesions on their body, along with As in blood, As in urine, hair and

Table 11.6: Parametric presentation of As in urine, hair, nails, and skin-scales samples from the As affected victims and control population of West Bengal

Parameter	As victims with skin-lesions				People without skin-lesions drinking As contaminated water			People drinking safe water with respect to As			Controlled population		
	Urine[a] (µg/l)	Hair[b] (mg/kg)	Nail[c] (mg/kg)	Skin-scale[d] (mg/kg)	Urine[a] (µg/l)	Hair[b] (mg/kg)	Nail[c] (mg/kg)	Urine[a] (µg/l)	Hair[b] (mg/kg)	Nail[c] (mg/kg)	Urine[a] (µg/l)	Hair[b] (mg/kg)	Nail[c] (mg/kg)
No. of samples	855	814	802	165	6688	4765	4786	2252	1559	1793	75	75	75
Mean	462	1.48	4.56	6.82	286	1.33	2.78	86.59	0.939	1.603	16	0.341	0.748
Maximum	3893	20.34	44.89	15.51	3947	6.39	21.36	149.99	1.99	2.94	41	0.499	1.066
Minimum	55	0.96	0.88	1.28	10	0.18	0.38	17.10	0.21	0.40	10	0.217	0.540
Median	293	1.32	3.87	4.46	184	1.16	2.31	93.75	0.83	1.46	15	0.338	0.743
Standard deviation	579	2.20	3.98	4.75	342	0.79	2.00	39.30	0.421	0.614	10.5	0.103	0.107
% of samples having As above normal/ toxic level	89	68	79		81	46	62	48	26	34			

[a] Normal excretion level of As in urine range from 5 to 40 µg/day (1.5 l)
[b] Normal amount of As in hair is about 0.08 to 0.25 mg/kg with 1.0 mg/kg being the indication of toxicity.
[c] Normal As content in nail is 0.43 to 1.08 mg/kg
[d] There is no available data of normal level of As for skin-scales in literature.

Table 11.7: Parametric presentation of As in urine, hair, nail, and skin-scales samples from the As affected victims of 30 affected districts and control population of Bangladesh

Parameters	Subjects from As affected areas				Controlled population		
	Urine[a] (µg/l)	Hair[b] (mg/kg)	Nail[c] (mg/kg)	Skin-scale[d] (mg/kg)	Urine[a] (µg/l)	Hair[b] (mg/kg)	Nail[c] (mg/kg)
No. of sample	1084	4386	4321	705	62	62	62
Mean	280	3.39	8.57	5.73	31	0.41	0.83
Maximum	3086	28.06	79.49	53.39	94	0.85	1.58
Minimum	24	0.28	0.26	0.60	6	0.21	0.09
Median	115.78	2.34	6.40	4.80	29	0.395	0.812
Standard deviation	410	3.33	8.73	9.79	20	0.18	0.68
% of samples having arsenic above normal/ toxic level	95.11	83.15	93.77	97.44			

[a] Normal excretion level of arsenic in urine range from 5 to 40 µg/day (1.5 l)
[b] Normal amount of arsenic in hair is about 0.08 to 0.25 mg/kg with 1.0 mg/kg being the indication of toxicity.
[c] Normal arsenic content in nail is 0.43 to 1.08 mg/kg
[d] There is no available data of normal level of arsenic for skin-scale in literature.

nails of these people was also measured. Data from these studies show elevated level of As was found in most of the biological samples. Table 11.8 shows the results.

Table 11.8: Arsenic concentrations in whole blood, urine, hair, and nails of some arsenic-affected people of West Bengal

Parameters	Blood (µg/l)	Urine (µg/l)	Hair (mg/kg)	Nail (mg/kg)
No. of observations	99	83	83	83
Mean	30.1	545.15	3.08	9.76
Maximum	145.5	235.2	9.85	22.31
Minimum	5.29	7.81	0.69	1.69
Median	21.18	405.72	2.31	9.78
Standard deviation	26.86	525.87	2.14	5.24
% of samples having arsenic above normal/ toxic level		98	98	100

The mean blood As concentration was 30.1 µg/l (n = 99, range 5.29-45.5 µg/l). The mean As concentration of water is 0.52 µg/l. The whole blood As concentration (60 µg/l) of black foot disease patients and members of their families living in endemic areas of Taiwan (Heyborn, 1969) was higher than that of As-affected patients in West Bengal. Vahter et al. (1995) collected blood samples from women from Andean villages of Argentina. The mean blood As at San Antonio de Las Cabres was 8 µg/l and mean As in water was 0.19 mg/l; at Santa Rosa de los Pastos Grands, mean blood As was 1.5 µg/l and mean As in water was 0.037 µg/l. Unexposed persons have low levels of As in their blood. The blood As of women from the small communities of Santa Rosa de los pastos Grandes, Olacapato, and Tolar Grande was considerably lower (1.0-2.4 µg/l; Vahter et. al., 1995) and similar to the reported average blood As concentration in nonsmoking women in USA (1.5 µg/l; Kagey et al., 1977) and children in prior Czechoslovakia (1.5 µg/l; Bencko and Symon, 1977). Mean blood As have been reported from Denmark (mean 2.5 µg/l; Heyborn, 1969). In West Bengal, the mean blood As concentration in control people is 2.77 µg/l (n = 10, range 1.07-16 µg/l), which is comparable to levels reported in other countries.

DERMATOLOGICAL FEATURES FROM 151 VILLAGES IN 7 As-AFFECTED DISTRICTS OF WEST BENGAL AND 214 VILLAGES IN 30 ARSENIC-AFFECTED DISTRICTS OF BANGLADESH

In 200 villages of 7 districts of West Bengal, 29,035 people from As-affected villages were examined at random. Of these, 4361 (15%) people from 151 villages were found to have arsenical skin lesions (Fig. 11.3). In Bangladesh, 17,896 people from 241 villages were examined at random

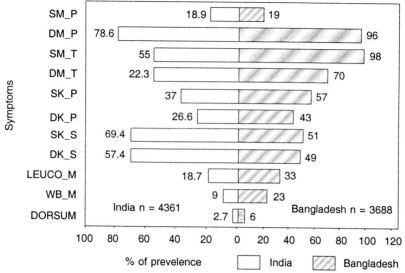

Fig. 11.3: Comparison of dermatological symptoms among the arsenic affected people of West Bengal, India and Bangladesh. SM_P: spotted melanosis on palm; DM_P: diffuse melanosis on palm; SM T: spotted melanosis on trunk; DM_T: diffuse melanosis on trunk; SK_P: spotted keratosis on palm; DK_P: diffuse keratosis on palm; SK_S: spotted keratosis on sole; DK S: diffuse keratosis on sole; LEUCO M: leuco melanosis; WB_M: whole body melanosis.

and 3688 (20.6%) people from 214 villages were found to have arsenical skin lesions. The high percentage of people affected may be attributed to the bias in our survey focusing on areas that had people suffering from arsenical skin lesions.

Survey and assessment of people suffering from As poisoning is often challenging, given the social implications of such studies, including examination of women. The slow progress in these studies may be attributed to the following : (a) social problems because women do not want to be examined, especially young women (b) field visits to affected villages are during the day when men go to their workplace, and (c) transportation problems for people living away from the examination site. These are only few of many reasons for affected people not coming for investigation. However, there are also exceptions where local people take the initiative in making these investigations successful.

In As-affected villages in West Bengal and Bangladesh, all possible symptoms of As toxicity (Mandal et al., 1996; Dhar et al., 1997) such as diffuse melanosis; spotted melanosis; leucomelanosis; mucus membrane melanosis on tongue, and lips; diffuse and nodular keratosis on palm and sole; nonpetting edema; asceitis; squamous cell carcinoma; basal cell carcinoma, and bowens disease were recorded. Further evidence is accumulating on people having arsenical skin lesions also suffering from gangrene and carcinoma affecting lung, bladder, genitourinary tract, etc. Figures 11.4-11.10 show the photographs of seven patients from West Bengal and Bangladesh suffering from arsenicosis; Table 11.9 shows the dermatological symptoms of these seven patients.

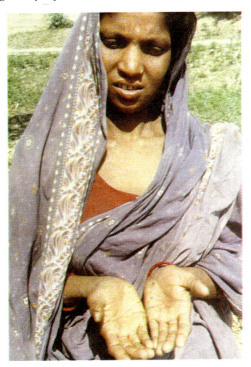

Fig. 11.4: Village: Madanpur, Dist: Murshidabad of West Bengal. Due to her skin lesions her husband after marriage sent her back to her parents.

DETAILED STUDY OF As-AFFECTED TWO VILLAGES: FAKIRPARA FROM WEST BENGAL AND SAMTA FROM BANGLADESH

Both in West Bengal and Bangladesh, the As-affected areas are vast, and it will take a long time to know the actual magnitude of the problem. To get an indication of the magnitude of this problem, a detailed study was conducted in some As-affected villages in West Bengal and Bangladesh. Data from the detailed study of Fakirpara of Deganga block in North 24-Parganas district of West Bengal and Samta village of Sarsa in Jessore district of Bangladesh are presented here. The detailed physical parameters of these two villages are given in Table 11.10.

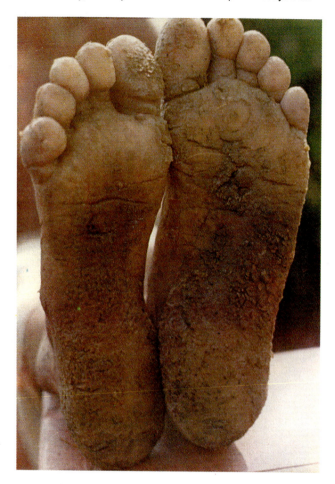

Fig. 11.5: Village: Uttar Kolsur, District: North 24-Parganas of West Bengal. She has become almost invalid.

Figures 11.11 and 11.12 show the distribution of contaminated tubewells in these two villages. Figure 11.13 shows the comparative study of the As concentration range (mg/l) against percentage of the samples. In Fakirpara village, all 46 tubewells that people were using for drinking were analyzed and all 265 tubewells of Samta village were analyzed.

Table 11.11 shows the parametric presentation of As in urine (n = 331), hair (n = 260), and nails (n = 285) of out of a total population of 764 (collected at random) of Fakirpara village, and urine (n = 301), hair (n = 293), and nails (n = 242) out of a total population of 4841 in the Samta village. Table 11.11 also shows that majority of the people, As in hair, nails, urine is above the normal/toxic level. Although many of them have

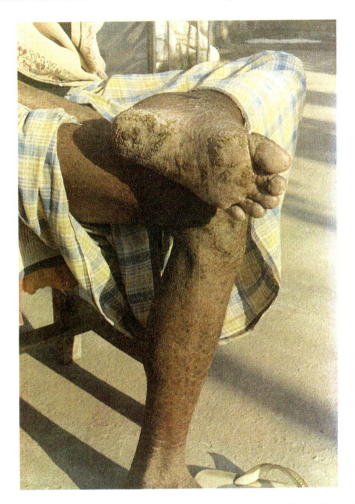

Fig. 11.6: Village: Ambikanagar, District: North 24-Parganas of West Bengal. In his family all seven members are affected and his elder brother died of skin cancer (SCC). The patient was also operated for skin cancer.

no arsenical skin manifestation, they may be subclinically affected. Arsenic in urine reflects recent exposure. Therefore, a correlation can be found between As in urine and As in water. Regression analysis has been carried out to find the correlation between As concentration in urine and water of the Samta village. Figure 11.14 shows a strong linear correlation ($r^2 = 0.89$ and $P < 0.0001$) between the average As concentration in urine ($n = 301$) and water ($n = 265$). A similar strong linear relationship was recorded in Fakirpara, West Bengal ($r^2 = 0.87$ and $P < 0.0001$) between As in urine and water. Deposition and release of As in hair and nails are a

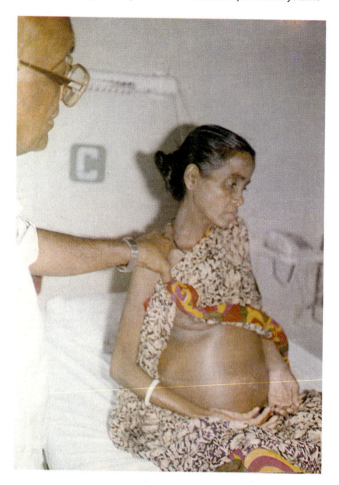

Fig. 11.7: Village: Palpara, District: Narayanganj of Bangladesh. In her family all five members are affected. She was suffering from asceitis.

slow process; therefore, change of water for a short period does not reflect immediately on the As concentration in hair and nails (Mandal et al., 1998) but reflects on urinary arsenic concentration (Mandal et al., 1998). Regression analysis has been carried out for the arsenic concentrations in hair, nails and water. The linear regression show a good correlation between arsenic in water and hair ($r^2 = 0.91$, $P < 0.0001$) and nails ($r^2 = 0.94$, $P < 0.0001$) (Figs. 11.15 and 11.16). In Fakirpara, West Bengal, good correlations among arsenic in hair, nails and water have been found.

Examination of people in Fakirpara village showed high incidences of arsenical skin lesions (23%). Similar studies in Samta village showed that over 50% of the people examined had arsenical skin lesions. Figure 11.17

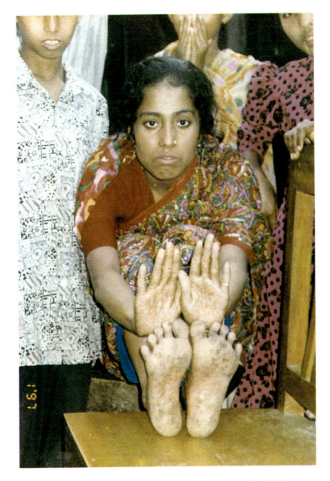

Fig. 11.8: Village: Sripur, District: Lakshmipur of Bangladesh. At the age of 21 years she was divorced because of her skin lesions. In her father-in-laws house no one used to take food from her hand.

shows the dermatological symptoms of the affected people in the two villages.

SOURCE OF As AND REASON OF DECOMPOSITION

Das et al. (1996) and SOES (1991) reported that a single rural water supply scheme (RWSS) in one of the nine As-affected districts (Malda) withdraws from groundwater 147 kg As/year, indicating a geological origin. Later studies by Chakroborti and coresearchers in West Bengal and the British Geological Survey in collaboration with the Bangladesh Geological Survey have confirmed this hypothesis. Calculations show that from a single pump at Paranpur village, Raninagar Block-II, Murshidabad, 200 g of As

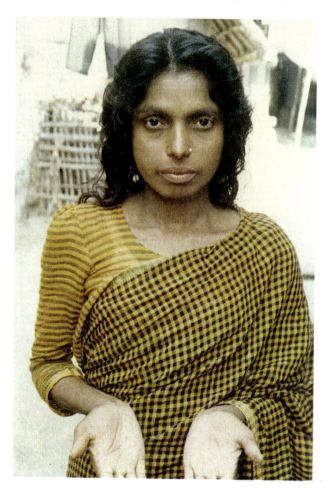

Fig. 11.9: Village: Khazadanga, District: Khulna of Bangladesh. Villagers used to think she was a leprosy patient and did not allow her to take water from common tubewells.

is emerging out with groundwater withdrawal for a continuous period of 8 h.

So far, in the nine affected districts of West Bengal, 112 boreholes have been made and 2235 sediment samples collected at 3.3 m intervals from six out of the nine affected districts.

Figure 11.18 shows the locations of these bore-holes. Of these, 86 samples show As levels ranging from 10 to 373 mg/kg and the number decreased to 48 after samples were homogenized. The homogenized samples show arsenic levels 10 to 196 mg/kg (Table 11.12). Some opaque particles collected from these parent samples show much higher concentrations of As (Table 11.12).

Fig. 11.10: Village: Krishnakati, District: Bagerhat of Bangladesh. Once in his village he was known best sportsman ; but now he has become almost crippled.

Electron probe micro analysis (EPMA) of the enriched particle shows particles that are rich in iron, sulfur, and As (Das et al., 1996). The laser microprobe mass analysis (LAMMA) showed that As-rich particles of bore-hole sediments come from iron pyrite (Das et al., 1996). Some enriched As samples were analyzed by an X-ray diffractometer and showed the presence of hematite, magnetite, pyrite, quartz, and $FeSO_4$.

Although the origin of As in groundwater is not clear, many scientists suggest that heavy groundwater withdrawal may be one of the reasons for As in groundwater. The United States Geological Survey (USGS) in one of their publications (Welch et al., 1988) comments that mobilisation

Table 11.9: Dermatological features of some patients (shown in Figs. 10.4 to 10.10) in West Bengal, India and Bangladesh.

SL. No.	Sex and Age	Melanosis						Keratosis					Others
		Palm		Trunk		Leuco	Whole Body	Palm		Sole		Dorsum	
		Spotted	Diffuse	Spotted	Diffuse			Spotted	Diffuse	Spotted	Diffuse		
Fig. 10.4	F/28	–	++	+++	+	–	+	++	+	++	+	–	–
Fig. 10.5	F/32	+	+	+++	–	+	+	+++	+	+++	++	+	–
Fig. 10.6	M/32	–	–	+++	+	+	+	+++	++	+++	++	–	SCC
Fig. 10.7	F/47	–	++	++	+	+	+	++	+	++	+	–	Asceitis
Fig. 10.8	F/24	+	+	+++	+	+	+	+++	+	+++	+	–	–
Fig. 10.9	F/25	+	+	+++	+	–	++	++	+	++	+	–	–
Fig. 10.10	M/26	+	+	+++	+	–	++	+++	+	+++	+	–	–

Note. + Mild; ++ Moderate; +++ Severe; SCC : Squamous Cell Carcinoma

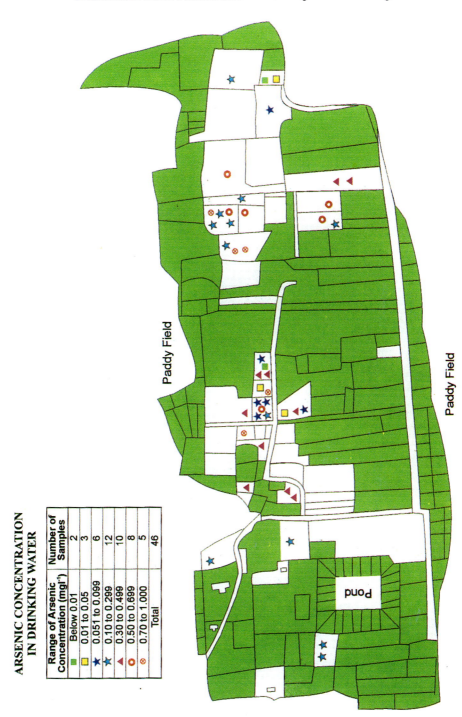

ARSENIC CONCENTRATION IN DRINKING WATER

Range of Arsenic Concentration (mgl⁻¹)		Number of Samples
Below 0.01	🟩	2
0.01 to 0.05	🟨	3
0.051 to 0.099	⭐	6
0.10 to 0.299	⭐	12
0.30 to 0.499	◀	10
0.50 to 0.699	⊙	8
0.70 to 1.000	⊗	5
Total		46

Fig. 11.11: Map of village Fakirpara, Deganga Block, North 24-Parganas, West Bengal, showing the range of arsenic concentration in water.

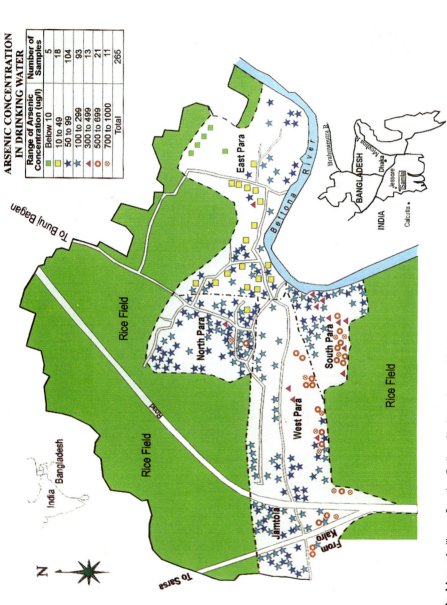

Fig. 11.12: Map of village Samta, Police Station: Sarsa, District: Jessore of Bangladesh showing the range of arsenic concentration in water.

Table 11.10: Physical parameters of two villages, Fakirpara, Deganga Block / Police Station, North 24-Parganas, West Bengal, India and Samta, Sarsa Police Station, Jessore district, Bangladesh

Village	Area (km²)	Population	Male	Female	Total no. of tubewells	% of tubewell having arsenic above 0.05 mg/l	No. of people examined	No. of arsenical patients	% of prevalence
Fakirpara (India)	0.5	764	371	393	46	89	764	175	22.91
Samta (Bangladesh)	3.2	4841	2517	2324	265	91	600	330	55

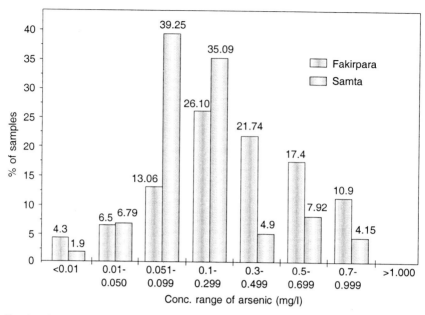

Fig. 11.13: Comparative study of arsenic concentration in groundwater in Samta village of Bangladesh and Village Fakirpara of West Bengal.

of As in sedimentary aquifers may be, in part a result of changes in the geochemical environment due to agricultural irrigation. Another view is that due to high use of phosphate fertilizer, arsenate in soil might have leached to underground aquifers. Scientists believe this may be true for highly porous soils, but for heavier soils little leaching is likely. Moreover, iron and aluminium present in soil will absorb arsenate, thereby minimizing leaching. According to Kinniburgh et al. (1994), abstraction from chalk aquifers over many decades has resulted in the dewatering of the overlying Basal Sand Aquifers allowing air entry and has led to the localised oxidation of pyrite and the acid released during oxidation of pyrite is responsible for the poor quality pore water in the Basal Sands. The rate of oxidation is probably greatest near to wells and bore-holes which allowed ready access of air. On the basis of available literature and our bore-hole sediment data, one reason of the presence of As in groundwater of West Bengal may be heavy groundwater withdrawal that aerates the aquifer leading to the decomposition of the pyrite (FeS_2) rich in As and the acid released leached out the As in soluble form in groundwater. The decomposition of pyrite (FeS_2) rich in As may be by the following mechanism:

$$2\ FeS_2 + 2H_2O + 7O_2 \Rightarrow 2Fe^{2+} + 4HSO_4^-$$
$$4\ Fe^{2+} + O_2 + 4H^+ \Rightarrow 4Fe^{3+} + 2H_2O$$
$$FeS_2 + 14Fe^{3+} + 8H_2O \Rightarrow 15Fe^{2+} + 2SO_4^{2-} + 16H^+$$

Table 11.11: Parametric presentation of arsenic in urine, hair, and nail of the population of Fakirpara, Deganga block, North 24-Parganas district, West Bengal, India and Samta village, Jessore district, Bangladesh

Parameter	Fakirpara, West Bengal, India			Samta, Bangladesh		
	Urine (µg/l)	Hair (mg/kg)	Nail (mg/kg)	Urine (µg/l)	Hair (mg/kg)	Nail (mg/kg)
No. of Samples	331	260	285	301	293	242
Mean	528.08	3.34	8.19	484.7	2.38	7.19
Maximum	4411.76	18.53	44.89	3085.73	9.48	29.6
Minimum	7.81	0.32	0.85	23.68	0.46	0.26
Median	318.75	2.56	6.58	295.24	2.37	6.06
Standard deviation	629.09	2.85	6.45	497.25	1.58	4.88
% of samples having arsenic above normal/ toxic level	98	91	99	98	95	100

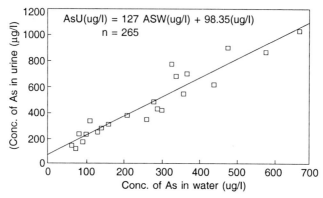

Fig. 11.14: Correlation between inorganic arsenic and its metabolites in urine and arsenic in water.

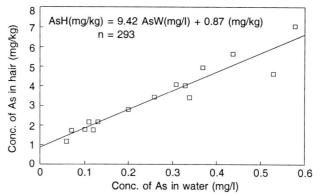

Fig. 11.15: Correlation between arsenic in hair and arsenic in water of Samta village in Jessore district of Bangladesh.

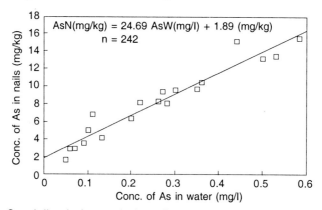

Fig. 11.16: Correlation between As in nail and As in water of Samta village in Jessore district of Bangladesh.

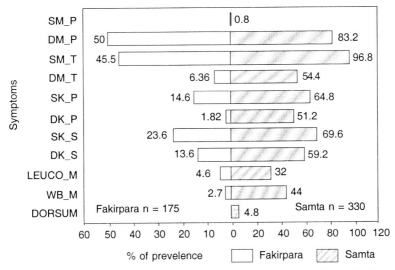

Fig. 11.17: Comparative study of dermatological symptoms of the affected people of Samta village of Bangladesh and Fakirpara, West Bengal, India.

The Fe^{3+} ions formed act as a catalyst for the further decomposition of pyrite.

The mechanism is similar to that reported by Kinniburgh et al., (1994) for presence of arsenic in ground water in London. The only difference being that the London sediment pyrite was not rich in As. Another possible reason could be biotic degradation of pyrite. When conditions become sufficiently acidic (pH < 4), the acidophilic pyrite-oxidizing bacterium *Thiobacilus ferrooxidans* can increase the rate of pyrite oxidation by several orders of magnitude. At higher pH, these organisms including *Metallogenium* and other *Thiobacillus* species may be come important. As huge amounts of nitrate fertilizer are used, its role for pyrite breakdown needs to be investigated also. The possibility of bacteria-mediated denitrification in the presence of pyrite has also been reported. Since the above mechanisms were postulated, numerous groups of scientists have postulated other mechanisms that negate the pyrite-oxidation hypothesis. For example, desorption from or reductive dissolution of Fe-oxyhydroxides in a reducing aquifer environment (Bhattacharya et al., 1997). Although both the mechanisms continue to receive much attention the enhanced bioavailability of these geologic materials following the installation of tube wells is certain, and to be studied in detail.

CONCLUSIONS

From the study of As groundwater contamination and suffering of people in West Bengal, India, and Bangladesh, it appears that the overall situation

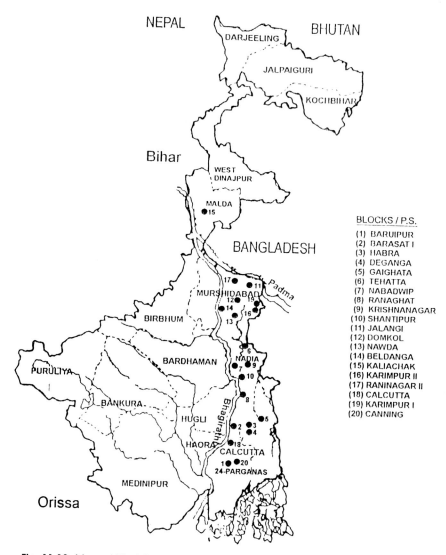

Fig. 11.18: Map of West Bengal showing blocks where bore-holes were made.

is serious. More As-affected areas and patients are becoming evident with the increasing number of studies. Although in Bangladesh the study started 5 years ago, the available data indicates that the situation is more serious than in West Bengal.

The solution to the current problem is to probably to seek alternative sources of water to underground water. West Bengal has 2000 mm of average rainfall, about 4000 km² of wetland, hundreds of Ox-Bow Lakes, and vast river basins flooded almost every year. For Bangladesh, the

Table 11.12: Arsenic in bore-hole sediments at different arsenic affected Blocks of West Bengal, India

Districts	South 24-Parganas	North 24-Parganas	Nadia	Murshidabad	Malda	Calcutta	Total
No.of blocks	2	4	7	5	1	1	20
No. of bore-hole examined	5	14	75	13	4	1	112
No. of sample analysed	108	620	1000	314	130	63	2235
No. of bore-hole found arsenic concentration > 0.01 µg/g	2	8	20	9	2	Nil	41
No. of sample found arsenic concentration > 10 µg/g	2	24	41	15	4	Nil	86
Maximum concentration (µg/g) in original sample (without sorting)	16.88 (P) 12.81 (H)	60.50 (P) 69.47 (H)	373.80 (P) 196.75 (H)	77.60 (P) 31.19 (H)	122.47 (P) 208.80 (H)	4.44	
Depth at which maximum concentration found in original sample (m)	41.81-97.27	208.02-211.02	136.36-139.39	136.36-151.51	78.78-81.81	44.24-47.27	
High concentration of some selected particle from parent sample i.e. original sample without sorting		<10 (P) 115.67 (S)	(1) <10 (P), 130.98 (S), (2) 19.2 (P), 78.8 (S), (3) 16.45 (P), 68.51 (S), (4) <10 (P), 148.61 (S), (5) 85.72 (P), 1763.57(S), (6) 373.80(P), 2777.78(S)	(1) 15.22 (P), 2250.98 (S), (2) 77.7 (P), 2116.53 (S)			
Sample depth at which the selected particles sorting out (m)		172.72-175.75	(1) 60-70, (2) 70-80, (3) 90-100, (4) 90-100, (5) 218.18-221.21, (6) 136.36-139.39	(1) 154.54-193.93, (2) 136.36-151.51	—	—	—

Note. P: parent sample; H: homogenized sample; S: selected particle from parent sample.

natural water resources available are one of the highest in the world. In the As-affected area of Ronpiboon, Thailand, people preserve rainwater in cement jars and use it instead of As-contaminated groundwater. They use rainwater for drinking and cooking for almost 6 to 8 months of the year. We should conserve our water resources. A proper watershed management is required. Due to our negligence, most water bodies are becoming dry. If these water bodies are used properly, not only will many villages get As-free water but they will also use pisciculture, and duck-breeding, grow vegetables at the banks of these water bodies, and improve their economic condition. Ground water may be tapped only after proper utilization of these available surface water resources. This As issue of West Bengal and Bangladesh, the biggest in the world, should not be neglected. The world should learn a lesson from this problem, that any country where water extraction from underground goes similarly unchecked could be leaving itself open to similar problems.

REFERENCES

Abernathy, C. O., Marcus, W., Chen, C., Gibb, H. and White, P. (1989) *Report on arsenic workgroup meeting*. Office of Drinking Water, Office of Research and Development, USEPA, Memorandum to Cook P and Preuss P, Office of Regulatory Support and Scientific Management, USEPA.

Astolfi, E., Maccagno, A., Fernandez, J. C. G., Vaccara, R. and Stimola, R. (1981) Relation between arsenic in drinking water and skin cancer. *Biol. Trace Elem. Res.* 3: 133 to 143.

Bencko, V. and Symon, K. (1977) Health aspects of burning coal with a high arsneic content. *Environ. Res.* 13: 378–385.

Bhattacharya P, Chatterjee D and Jacks G. (1997) Occurrence of arsenic contaminated groundwater in alluvial aquifers from Delta Plains, Eastern India: Options for safe drinking water supply. Int. Jour. Water Resources Management 13(1): 79–92.

Biswas, B. K., Dhar, R. K., Samanta, G., Mandal, B. K., Chakraborti, D., Farak, I., Islam, K. S., Chowdhury, M. M., Islam, A. and Roy, S. (1998) Detailed study report of Samta, one of the arsenic-affected villages of Jessore District, Bangladesh. *Cur. Scie.* 74(2): 134–145.

Borgono, J. M. and Greiber, R. (1971) Epidemiological study of arsenicism in the city of Antofagasta. In *Trace Substances in Environmental Health-V. A symposium*, Hemphill D D (Eds.). University of Missouri Press, Columbia, pp. 13–24.

Buchet, J. P. and Lauwerys, R. (1994) Inorganic arsenic metabolism in humans. In *Arsenic, Exposure and Health*, Chappell, W. R., Abernathy, C. O. and Cothern, C. R. (Eds.). Science and Technology Letters, Northwood, pp. 181–189.

Cebrain, M. E., Albores, A., Aguilar, M. and Blakely, E. (1983) Chronic arsenic poisoning in the north of Mexico. *Human Toxico.* 2: 121–133.

Chakraborti, D., Valentova, M. and Sucha, L. (1982) *Decomposition of materials containing traces of arsenic and its spectrophotometric determination*. Prague Institute of Chemical Technology of Czechoslovakia Analytical Chemistry, H-17, 31–41.

Chakraborti, A. K. and Saha, K. C. (1987) Arsenical dermatosis from tubewell water in West Bengal. *Indian J. Med. Res.* 85: 326–334.

Chakraborti, D., Das, D., Chatterjee, A., Jin, Z. and Jiang, S. G. (1992) Direct determination of some heavy metals in urban air particulates by electrothermal atomic absorption spectrometry using Zeeman background correction after simple acid decomposition. Part IV: Applications to Calcutta air particulates. *Environ. Technol.* 13: 95–100.

Chatterjee, A., Das, D. and Chakraborti, D. (1993) A study of ground water contamination by arsenic in the residential area of Behala, Calcutta due to industrial pollution. *Environ. Pollut.* **80(1):** 57–65.

Chatterjee, A., Das, D., Mandal, B. K., Chowdhury, T. R., Samanta, G. and Chakraborti, D. (1995) Arsenic in groundwater in six districts of West Bengal, India: the biggest arsenic calamity in the world. Part I: Arsenic species in drinking water and urine of the affected people. *The Analyst* **120:** 643–650.

Chen, K. P. and Wu, H. Y. (1962) Epidemiological studies on black foot diseases: II. A study of source of drinking water in relation to the disease. *J. Formosan Med. Assoc.* **61:** 611–617.

Chowdhury, T. R., Mandal, B. K., Samanta, G., Basu, G. K., Chowdbury, P. P., Chanda, C. R., Karan, N. K., Dhar, R. K., Lodh, D., Das, D., Saha, K. C. and Chakraborti, D. (1997) Arsenic in groundwater in seven districts of West Bengal, India: the biggest arsenic calamity in the world. The status report up to August, 1995. In *Arsenic: Exposure and Health Effects,* Abernathy, C. O., Calderon, R. L. and Chappell, W. R. (Eds.). Chapman and Hall, New York, pp. 91–111.

Cornelis, R., Heintow, B., Herber, R. F. M., Christensen, J. M., Poulsen, O. M., Sabbioni, E., Templeton, D. M., Thomassen, Y., Vather, M. and Vesterberg, O. (1996) Sample collection guideline for trace elements in blood and urine. *J. Trace Elem. Med. Biol.* **10:** 103–127.

Das, D., Chatterjee, A., Samanta, G., Mandal, B., Roy Chowdhury, T., Samanta, G., Chowdhury, P. P., Chanda, C., Basu, G., Lodh, D., Nandi, S., Chakraborti, T., Mandal, S., Bhattacharya, S. M. and Chakraborti, D. (1994) Arsenic contamination in groundwater in six districts of West Bengal, India: the biggest arsenic calamity in the world. *The Analyst* **119:** 168N–175N.

Das, D., Chatterjee, A., Mandal, B. K., Samanta, G., Chakraborti, D. and Chanda, B. (1995) Arsenic in groundwater in six districts of West Bengal, India: the biggest arsenic calamity in the world. Part II: Arsenic concentration in drinking water, hair, nail, urine, skinscale and liver tissue (Biopsy) of the affected people. *The Analyst,* **120:** 917–924.

Das, D., Samanta, G., Mandal, B. K., Roy Chowdhury, T., Chanda, C. R., Chowdhury, P. P., Basu, B. K. and Chakraborti, D. (1996) Arsenic in groundwater in six districts of West Bengal, India. *Environ. Geochem. Health* **18:** 5–15.

Dhar, R. K., Biswas, B. K., Sarnanta, G., Mandal, B. K., Chakraborti, D., Roy, S., Jafar, A., Islam, A., Ara, G., Kabir, S., Khan, A. W., Ahmed, S. A. and Hadi, A. (1997) Groundwater arsenic calamity in Bangladesh. *Curr. Sci.* **73(1):** 48–59.

Ellenhorn, M. J. and Barceloux, D. G. (1988) *Medical Toxicology: Diagnosis and Treatment of Human Poisoning.* Elsevier, New York, pp. 1012–1016.

Farmer, J. G. and Johnson, L. R. (1990) Assessment of occupational exposure to inorganic arsenic based on urinary concentrations and speciation of arsenic: *Brit. J. Ind. Med.* **47:** 342–348.

Feinglass, E. J. (1973) Arsenic intoxication from well water in the United States. *New Eng. J. Med.* **288(16):** 828–830.

Geyer, L. (1898) Uber die chronischen Hautueranderungen becin Arsenicismus and Betrachtungen iiber die Massenerkrankungen in Reichenstein in Schlesien. *Arch. Derm. Syphilol.* **43:** 221–280.

Goldsmith, J. R., Deane, M., Thom, J. and Gentry, G. (1972) Evaluation of health implications of elevated arsenic in well water. *Water Res.* **6:** 1133–1136.

Grantham, D. A. and Jones, J. F. (1977) Arsenic contamination of water wells in Nova Scotia. J. Am. Water Works Assoc. **69:** 653–657.

Guha Mazumder, D. N. (1997) Chronic arsenic toxicity in West Bengal. *Curr. Sci.* **72(2):** 114.

Guha Mazumder, D. N., Das Gupta, J., Santra, A., Pal, A., Ghosh, A., Sarkar, S., Chattopadhaya, N. and Chakraborti, D. (1997) Non-cancer effects of chronic arsenicosis with special reference to liver damage. In *Arsenic Exposure and Health Effects*, Abernathy, C. O., Calderon, R. L. and Chappell, W. R. (Eds.). Chapman and Hall, New York, pp. 112–123.

Harrington, J. M., Middaugh, J. P., Morse, D. L. and Housworth, J. (1978) A survey of a population exposed to high concentrations of arsenic in well water in Faribanks, Alaska. *Am. J. Epidemiol.* **108(5)**: 377–385.

Heines, H. J., Meier, S., Vogt, H. and Wechsung, R. (1979) Laser induced mass spectrometry of organic and inorganic compound with LAMMA. *Conference proceeding of 8th Triannual International Mass Spectrometry*, Oslo, Norway, p. 19.

Heyborn, K. (1969) Environmental variation of arsenic levels in human blood determined by neutron activation analysis. *Clin. Chim. Acta* **28**: 349–357.

Hindmarsh, J. T., McLetchie, O. R., Heffernan, L. P. M., Hayne, O. A., Ellenberger, H. A., McCurdy, R. F. and Thiebaux, H. J. (1977) Electromyographic abnormalities in chronic environmental arsenicalism. *J. Anal. Toxicol.* **1**: 270–276.

Hotta, N. (1989) clinical aspects of chronic arsenic poisoning due to environmental and occupational pollution in and around a small refining spot. *Jap. J. Constit. Med.* **53(1-2)**: 49–70.

Hsu, K. H., Froines, J. R. and Chen, C. J. (1997) Studies of arsenic ingestion from drinking water in Northern Taiwan: chemical speciation and urinary metabolites. In *Arsenic: Exposure and Health Effects*, Abernathy, C. O., Calderon, R. L. and Chappell, W. R. (Eds.) Chapman and Hall, New York, pp. 190–209.

Kagey, B. T., Bumgarner, J. E. and Creason, J. P. (1977) Arsenic levels in maternal - fetal tissue sets. In *Trace Substances in Environmental Health XI, A Symposium*, Hemphill, D. D. (Eds.). Columbia University of Missouri Press, Missouri, pp. 252–256.

Kinniburgh, D. C., Gale, I. N., Smedley, P. L. and Darling, W. G. (1994) The effects of historic obstruction of groundwater from the London Basin aquifers on groundwater quality. *App. Geochem.* **9**: 175–195.

Lian, F. W. and Jian, Z. H. (1994) Chronic arsenism from drinking water in some areas of Xinjiang, China. In *Arsenic in the Environment. Part II: Human Health and Ecosystem Effects*, Nriagu, J. O. (Eds.). John Wiley and Sons, Inc., New York, pp. 159–172.

Luo, Z. D., Zhang, Y. M., Ma, L., Zhang, G. Y., He, X., Wilson, R., Byrd, D. M., Griffths, J. G., Lai, S., He, L., Grumske, K. and Lamm, S. M. (1997) Chronic arsenicism and cancer in Inner Mongolia - consequences of well-water arsenic levels greater than 50 μg/l. In *Arsenic: Exposure and Health Effects*, Abernathy, C. O., Calderon, R. L. and Chappell, W. R. (Eds.). Chapman and Hall, New York, pp. 55–68.

Lu, F. J. (1990) Blackfoot disease: arsenic or humic acid ? *The Lancet* **336**: 115–116.

Mandal, B. K., Chowdhury, T. R., Sa.manta, G., Basu, G. K., Chowdhury, P. P., Chanda, C. R., Lodh, D., Karan, N. K., Dhar, R. K., Tamili, D. K., Das, D., Saha, K. C. and Chakraborti, D. (1996) Arsenic in groundwater in seven districts of West Bengal,India - the biggest arsenic calamity in the world. *Curr. Sci.* **70(2)**: 976–986.

Mandal, B. K., Chowdhury, T. R., Samanta, G., Basu, G. K., Chowdhury, P. P., Chanda, C. R., Lodh, D., Karan, N. K., Dhar, R. K., Tamili, D. K., Das, D., Saha, K. C. and Chakraborti, D, (1997) Chronic arsenic toxicity in West Bengal. *Curr. Sci.* **72(2)**: 114–117.

Mandal, B. K., Roy Chowdury, T., Samanta, G., Mukherjee, D. P., Chanda, C. R., Saha, K. C. and Chakraborti, D. (1998) Impact of safe water for drinking and cooking on five arsenic-affected families for 2 years in West Bengal, India. *Sci.Total Environ.* **218**: 185–201.

Marafantle, E., Vahter, M., Norin, H., Envall, J., Sandstrom, M., Christakopoulos, A. and Ryhage, R. (1987) Biotransformation of dimethylarsenic acid in mouse, hamster and man. *J. App. Toxicol.* **7:** 111–117.

Mohri, T., Hisanga, A. and Ishinishi, N. (1990) Arsenic intake and excretion by Japanese adults: A 7-day duplicate diet study. *Food Chem. Toxicol.* **7:** 521–529.

Morton, W. E., Starr, G., Pohl, D., Stoner, J., Wagner, S. and Weswig, P. (1976) Skin cancer and water arsenic in Lane county Oregon. *Cancer* **37:** 2523–2532.

Morton, W. E. and Dunnette, D. A. (1997) Health effects of environmental arsenic. In *Arsenic in the Environment, Part II: Human Health and Ecosystem Effects*, Nriagu, J. O. (Eds.). John Wiley and Sons, Inc., New York, pp. 17–34.

Niu, S., Cao, S. and Shen, E. (1997) The status of arsenic poisoning in China. In *Arsenic: Exposure and Health Effects, Abernathy*, C. O., Calderon, R. L. and Chappell, W. R. (Eds.). Chapman and Hall, New York, pp. 78–83.

Samanta, G. and Chakraborti, D. (1997) Flow-injection atomic absorption spectrometry for the standardization of arsenic, lead and mercury in environmental and biological standard reference materials. *Fresenius J. Anal. Chem.* **357:** 827–832

SOES (School of Environmental Studies) (1991) *Groundwater arsenic contamination episode in five districts of West Bengal - A preliminary study*. School of Environmental Studies, Jadavpur University, Calcutta, India, 32 pp.

Southwick, J. W., Western, A. E., Beck, M. M., Whitley, T., Isaacs, R., Petajan, J. O. and Hansen, C. D. (1983) An epidemiological study of arsenic in drinking water in Millard County Utah. In *Arsenic: Industrial, Biomedical, Environmental perspectives*, Lederer, W. H. and Fensterheim, R. L. (Eds.). Van Nostrand Reinhold Company, New York, pp. 210–225.

Steering Committee (1991) *Arsenic pollution in groundwater in West Bengal. Arsenic Investigation roject*, PHE Dept., Government of West Bengal, India,157 pp.

Terade, H., Kalsuta, K., Sasagawa, T., Saito, H., Shirata, H., Fukuchi, K., Sekiya, T., Yokoyama, Y., Hirokawa, S., Watanabe, Y., Hasegawa, K., Oshina, T. and Sekiguchi, T. (1960) A Clinical observation of chronic toxicity by arsenic. *Nihon Rinsho* **118:** 2394–2403.

Tseng, W. P., Chen, W. Y., Sung, J. L. and Chen, J. S. (1961) A clinical study of blackfoot disease in Taiwan: An endemic peripheral vascular disease. *Mem. Coll. Med. Natl. Taiwan Univ.* **7:** 1–18.

Vahter, M., Concha, G., Nermell, B., Nilsson, R., Dulout, F. and Natarajan, A. T. (1995) A unique metabolism of inorganic arsenic in native Andean women. *Eur. J. Pharmacol. Environ. Toxicol. Pharmacol. Sec.* **293:** 455–3426.

Wang, G. (1984) Arsenic poisoning from drinking water in Xinjiang. *Chinese J. Preven. Med.* **18(2):** 105–107.

Welch, A. H., Lico, M. S. and Hughes, J. L. (1988) Arsenic in groundwater of the Western R United States. *Groundwater* **26(3):** 334–347.

WHO (1981) *Arsenic: Environmental Health criteria 18*. World Health Organisation, Geneva, Switzerland.

WHO (1993) *Guideline for drinking water quality : recommendations* (Vol. 1, 2nd Ed.), World Health Organisation, Geneva, p. 41.

Wyllie, J. (1973) An investigation of the source of arsenic in a well water. *Can. Pub. Health J.* **28:** 128–135.

Xiao, J. G. (1997) Report from Inner Mongolia, China. *Asia Arsenic Network Newslett. Japan* **2:** 7–9.

Bioavailability, Toxicity and Risk Relationships in Ecosystems: The Path Ahead

R. Naidu[1], N.S. Bolan[2] and D.C. Adriano[3]

Remediation of contaminated sites is often established on the basis of risk assessments that rely on both toxicological impact on ecosystem and human health. The success of bioremediation, contaminated site rehabilitation and production of contaminant free food is largely dependent on the availability of the contaminants for plants, animals and microbial biota in the terrestrial environment. Consequently 'bioavailability' is used as the key indicator of potential risk that contaminants pose to both environmental and human health. However, the definition of the term 'bioavailability' and the concept on which it is based are unclear, the methods adopted vary throughout the world and therefore there is no single standard technique for the assessment of either plant availability of contaminants or their ecotoxicological impacts on soil biota. In this book, while attempting to define the bioavailability, we have taken into consideration that bioavailability is a function of both soil, nature of contaminant, species/receptor organisms and environmental perturbations including the duration of contamination (i.e., contaminant ageing).

DEFINITION

Bioavailability of a chemical in the soil environment has been defined as the fraction of the total contaminant in the interstitial porewater (i.e. soil solution) and surfaces of soil particles that is available to the receptor organism. This suggests that depending on the solubility of the contaminant, there is a continuum between zero and 100% bioavailability and within this spectrum the pool of contaminant available to receptor

[1] CSIRO Land and Water, Private Mail Bag. No. 2, Glen Osmond, Adelaide, South Australia 5064, Australia
[2] Department of Soil Science, Massey University, Palmerston North, New Zealand
[3] Savannah River Ecology Laboratory, Drawer E Aitken, South Carolina 29802, USA

organisms may vary with time, the nature of soil types and organisms, and the perturbations imposed by the environment. Considerable controversy exists in the literature relating to "what constitutes the bioavailable fraction" including the definition itself and the methods used for its measurements. For instance, microbiologists often regard the concentration that can induce a change either in morphology or physiology of the organism as the bioavailable fraction, soil plant scientists regard the plant available pool as bioavailable fraction. Consequently, terms such as 'bioavailable', 'phytoavailable' and 'available' are in use in the literature. Thus there is no single adequate definition of bioavailability. Moreover soil chemists and plant scientists have often used a single chemical extraction as an index of bioavailability assuming that bioavailability is a static phenomenon. Recent studies have, however, indicated that the transformation of contaminants in soils is a dynamic process which means bioavailability may change with time.

A more generic definition of *bioavailability* is the potential for living organisms to take up chemicals from food (i.e., oral) or from the abiotic environment (i.e., external) to the extent that the chemicals may become involved in the metabolism of the organism. More specifically, it refers to biologically available chemical fraction (or pool) that can be taken up by an organism and can react with its metabolic machinery (Campbell et al. 1995); or it refers to the fraction of the total chemical that can interact with a biological target (Vangronsveld and Cunningham, 1998). In order to be bioavailable, the contaminant (e.g., metals) have to come in contact with the plant (i.e., physical accessibility). Moreover, metals need to be in a particular form (i.e., chemical accessibility) to be able to enter a plant root. In essence, for a metal to be bioavailable, it will have to be mobile and transported and be in an accessible form to the plant.

By synthesizing and refining several definitions given in the literature, it is possible to propose two complementary definitions of bioavailability of metals, one focusing on the level of bioavailable metal species at the soil-root/soil-microbe interface, the second relating to the effect on the biota. A specific case is offered for the soil-plant system.

(1) Definition based on the dynamics of replenishment/maintenance of metal activities at the soil-root interface: *The bioavailable fraction of a trace metal is defined as the sum of the activities of all bioeffective species that can be maintained at the soil-root interface during a given time against the depletion induced by plant uptake or other plant-induced-changes* (e.g., speciation) *rendering the metal less mobile.*

It is assumed that a higher activity of a relevant metal species is more effectively supporting passive or active uptake and/or is more phytotoxic. Moreover, it implies that when comparing two soils for a given plant at a given time, the soil with the larger rate of metal delivery as a function of soil buffer power, microbial turnover and mobility (transport) will be able to maintain the higher concentration of the metal at the soil-root interface. In this sense, the definition given above is dynamic. This concept is operational, because in principle, it is possible to assess the activities of relevant metal species at the soil-root interface as well as the resupply using predictive modeling (rhizosphere models) if the underlying processes are known, or to design experimental systems that allow their measurement.

(2) Definition based on effects: *The bioavailable fraction of a trace metal is the sum of all species that, when their concentration or pool size is changed, have a significant/measurable effect on one or more of the following plant characteristics: (i) metal uptake into the plant (concentration); (ii) plant growth and biomass (yield); and (iii) root characteristics (morphology), leaf characteristics (necrosis, chlorosis) and photosynthesis and protoplast assays.*

Hence, the bioavailability of a metal in soil customarily is defined by the amount of metal absorbed by growing plants or by concentration in the harvested plant tissue. Plants typically absorb << 1% of the metals present in soil (Adriano, 2001). This expression does not indicate the extent on which metals present in soils are bioavailable nor the duration of bioavailability. Therefore, bioavailability of metals must account for not only the plant uptake in one growing season, but also the total amounts available over time i.e., the total labile pool from the various soil components (solid phases). The metal uptake by plants is determined by the kinetics of metal mobilized in the soil solution in rhizosphere. In this manner, bioavailability of metals in soils may be defined in terms of a capacity factor (dissolved metal in soil solution), which describes how much is available, and a rate factor (the rate of metal dissolution), which relates to the amounts of metal that is available for plant uptake.

Indicators of Bioavailability

Bioavailability of metals in soils can be examined using chemical extraction and bioassay tests. Chemical extraction tests include single extraction and sequential fractionation (Ruby et al., 1996; Basta and Gradwohl, 2000). Bioassay involves plants, animals and microorganisms (Yang et al., 1991).

Chemical

Several chemical techniques have been used to assess the bioavailable fraction (to plants) of metals in soils, but these have serious limitations in providing a good estimate of the pool of metal ions that are accessible to plants and microbes. A range of chemical extractants including mineral acids (e.g., 1 N HCl), salt solutions (e.g., 0.1 M $CaCl_2$), buffer solutions (e.g., 1 M NH_4OAc) and chelating agents (e.g., DTPA) have been used to predict the bioavailability of metals in soils (Sutton et al., 1984; Payne et al., 1988; van der Watt et al., 1994). Chelating agents, such as EDTA and DTPA have often been found to be more reliable in predicting the plant availability of metals (Sims and Johnson, 1991), since they are more effective in removing soluble metal-organic complexes that are potentially bioavailable. However, it should not be readily assumed that these chelating agents actually measure availability. For example, although the DTPA-extractable Cu was linearly related to the amount of Cu applied to three soils it was unrelated to Cu level in grain or leaf tissue of corn (Payne et al., 1988; Zhu et al., 1991). Similarly, while DTPA extractability of Pb and Cu may increase with increasing pH, plant uptake often decreases with pH (Merry et al., 1986). Also, diminution in plant uptake of Cd has been noted in successive years of cropping on sludge-amended soils despite steady levels of DTPA-extractable Cd (Bidwell and Dowdy, 1987). Since there is a tendency for DTPA-extractable metals to increase with metal level in soil (Martinez and Peu, 2000), and for uptake in crops to increase in response to increased metal level in the soil, the correlation between DTPA-extractable metals and plant uptake may in some cases be fortuitous.

Sequential fractionation schemes are often used to examine the redistribution or partitioning of metals in various chemical forms that typically include soluble, adsorbed (exchangeable), precipitated, organic and occluded. Although the extraction procedures vary between chemical fractionation schemes, generally the solubility and bioavailability of metals in soils decrease with each successive step of the scheme (Basta and Gradwohl, 2000). Specific chemical pools measured by chemical fractionation schemes have been correlated with plant uptake of metals and have been successful in predicting the plant availability of metals in soils (Shuman, 1991). However, the ability of chemical extractants to dissolve metals is matrix dependent and is sensitive to several factors such as equilibration time, vigor of shaking, temperature etc. Therefore chemical extraction should be validated for different metals sources, such as inorganic fertilizer, and organic sewage sludge and manure byproducts.

The diversity of reagents used to extract specific metal forms from contaminated soils makes comparison of results cumbersome. Even when the same reagent is employed, the efficiency of extraction depends on the nature of sample, its particle size distribution, duration of extraction, pH, temperature, strength of extractant, and solid : solution ratio (Miller et al., 1986). Chemical reagents used for the extraction may themselves alter the indigenous speciation of metals, and in general less vigorous extractants will probably be more selective for specific fractions than more severe reagents, which may extract other forms, although the overall efficiency of extraction may be lower (Lake et al., 1984; Ross, 1994). It is important to stress that often, the failure of chemical extraction techniques is related to their improper application rather than the efficacy of the test *per se.*

Plants take up most of the metals including the micro nutrients from soluble and exchangeable fractions (Adriano, 2001). Furthermore, metal phytotoxicity in soils is determined by the fraction(s) of the metal that is bioavailable. For example, addition of manure to soils almost invariably results in shifting the most dominant metal form to the organic-bound, metal bioavailability can be expected to decline. This has implications to our current regulatory guidelines that are generally based on total metal content. It is important to emphasize that there is a dynamic equilibrium between these fractions, and any depletion of the available labile pool (soluble and exchangeable fractions) due to plant uptake or leaching losses will result in the continuous release from other fractions to replenish the available 'pool' (Adriano, 2001).

The bioavailability of metals in soils has recently been examined using the physiologically-based in-vitro chemical fractionation schemes that include Physiologically Based Extraction Test (PBET), Potentially Bioavailable Sequential Extraction (PBASE) and Gastrointestinal (GI) Test. These innovative tests predict the bioavailability of metals in soils when ingested by animals and humans. As in the case of traditional sequential extraction scheme, the ability of the PBET extractant to solubilize metals increases with each successive extraction step. Metals extracted earlier in the PBET sequence are more soluble and, therefore, more potentially bioavailable than metals extracted later by the more aggressive extractants (Basta and Gradwohl, 2000). Despite the recognized nonspecific (i.e., operational) nature of chemical extraction methods, their analytical simplicity and rapidity renders them most suitable for routine metal form identification and estimation of the bioavailability of metals under field conditions. For example, such scheme could be useful in assessing the shift in the chemical forms (not species) of metals in response to addition of soil ameliorants such as biosolids, lime, etc. However, the distribution of a given metal among the various fractions can only be considered as estimates at best due to the subjectivity of the steps involved.

Biological

In situ techniques involving growing of the biota of interest in the contaminated medium and quantifying the uptake of metal into the biota or assessing the toxicological response are being used by many researchers as bioindicators of contaminants. Measurements of metal bioavailability and toxicity in soils using soil microorganisms are receiving increasing attention, as microorganisms are more sensitive to metal stress than plants or soil macrofauna. The methods using microflora and protozoa have the potential to provide a measure of bioavailability of metals in the short-term and even facilitate the measurement of temporal changes. In contrast responses by mesofauna (microarthropods) and macrofauna (enchytridae, invertebrates and earthworms) are cumulative effects. These methods, however, are time consuming and can only provide an overall effect of metal bioavailability to the species tested. Bioavailability, as determined by plant studies should be more correctly termed "phytoavailability" as these pools are not necessarily available to the same degree to other soil organisms. There is no single microbiological property that is ideal for monitoring soil pollution caused by different types of metals.

- Problems in interpretation of environmental measurements are common because of lack of suitable control, or baseline, measurements.
- There are advantages in using measurements that have some form of internal control e.g. microbial biomass as a percentage of total soil organic matter, as it helps side-step the lack of environmental control data.

The rapid development of molecular techniques and their continued successful application to the study of soil microbial ecology and function, provides significant future potential for the monitoring of soil pollution impacts. However, molecular tools are comparatively expensive, the level of information these techniques provide, over and above more traditional microbial indicator techniques, needs to be clearly demonstrated.

Role of Speciation in Bioavailability

As discussed in Chapter 2, bioavailability is also dependent on the nature of chemical species. The role of chemical speciation is critical in mediating the bioavailability of most metal contaminants in soils. Toxicological studies should measure the chemical speciation and all the relevant physicochemical parameters. Chemical speciation is often explored using a chemical equilibrium approach, but the dynamic aspects of many kinetically limited reactions need further study. The research aimed at improving our understanding of chemical speciation also needs to be coupled with toxicological assays so that we may better understand the links and feedback mechanisms between the chemical speciation dynamics of contaminants and their toxicological effects.

Manipulating Bioavailability to Manage Remediation

With growing public awareness of the implications of contaminated soil environment on human and animal health there has been increasing interest in the science community in developing technologies for remediating contaminated sites. Unlike organic contaminants, most metals do not undergo biological or chemical degradation and therefore persist in soils for a long time after their introduction. The mobilization of metals in soils for plant uptake and leaching to groundwater can, however, be minimized by reducing the 'bioavailability' of metals through chemical and biological immobilization. Recently there has been increasing interest in the immobilization of metals using a range of inorganic compounds, such as lime, phosphate fertilizers (e.g., apatite rocks) and alkaline waste materials and organic by-products, such as 'exceptional quality' alkaline biosolids.

Reducing metal availability and optimizing plant growth through inactivation may prove to be an effective method of in situ soil-metal remediation on industrial, urban, smelting, and mining sites. In addition, these stabilization techniques can occur as part of a treatment train with other remediation methods, including those now under development, the most intriguing of which is 'biomining' or 'phytomining' the available fraction of metals with plants.

Since bioavailability is a key factor for remediation technologies in situ immobilization using inorganic and organic ameliorants may offer a promising option. For example, lime stabilized biosolids are increasingly being used to immobilize metals in soils, thereby reducing their bioavailability for plant uptake. But the use of alkaline biosolids may result in the generation of dissolved organic carbon to form soluble complex with the metals, thereby facilitating their transport. Also the alkaline nature of some biosolids to limit the uptake of certain metals, such as Cd and Zn, may result in the solubilization of humic compounds. Another major inherent problem associated with immobilization techniques is that although the metals are less bioavailable, the contaminant concentration in soils remains unchanged. The immobilized metal may become more mobile and bioavailable with time through natural weathering processes. With biosolid, one of the main concerns about its long term efficiency is the potential for metal to mobilize if and when the organic matter undergoes significant oxidation.

CONCLUSION

There is considerable inconsistency in the definition and concept of bioavailability related to what constitutes the bioavailable fraction depending on the view of the individuals. Thus the literature clearly shows the inadequacy of the current definitions for bioavailability and recognises the need for a more coherent and universally acceptable definition atleast in the field of ecology and environmental science. Considering this and keeping in view of the bioavailability as a dynamic process and inconsistencies and vagueness of the existing definitions a new definition of bioavailability is proposed. Although more tailored for the soil-plant system, our version could be adaptible to other biota such as microbes and soil invertebrats. Clearly there is a need for biological assays in conjunction with chemical assays in order to assess the toxicity and bioavailability. Currently there are no appropriate methods for assessing the dynamic nature of the processes controlling in situ bioavailability and thus research needs to be focussed in this direction. Further study is required to examine the mechanisms involved in the sequestration and immobilization of heavy metals by inorganic and organic compounds.

REFERENCES

Adriano, D. C. 2001. *Trace Elements in Terrestrial Environments: Biogeochemistry, Bioavailability and Risks of Metals*, 2nd Ed., Springer—New York, NY.

Basta, N.T. and Gradwohl, R. 2000. Estimation of Cd, Pb, and Zn bioavailability in smelter-contaminated soils by a sequential extraction procedure. *J. Soil. Cont.* **9**, 149–164.

Bidwell, A. M. and Dowdy, R. H. 1987. Cadmium and zinc availability to corn following termination of sewage-sludge applications. *J. Environ. Qual.* **16**, 438–442.

Campbell, P.G.C. 1995. Interactions between trace metals and aquatic organisms: A critique of the free-ion activity model. p. 45–102. *In* A. Tessier and D. R. Turner (ed.) Metal speciation and bioavailability in aquatic systems. John Wiley & Sons, New York, NY.

Lake, D. L., Kirk, P. W. W., and Lester, J. N. 1984. Fractionation, characterization, and speciation of heavy metals in sewage sludge and sludge-amended soils. *J. Environ. Qual.* **13**, 175–183.

Martinez, J. and Peu, P. 2000. Nutrient fluxes from a soil treatment process for pig slurry. *Soils Use Manage.* **16**, 100–107.

Merry, R. H., Tiller, K. G., and Alston, A. M. 1986. The effects of soil contamination with copper, lead and arsenic on the growth and composition of plants. 2. Effects of source of contamination, varying soil-pH, and prior waterlogging. *Plant Soil* **95**, 255–269.

Miller, W. P., Martens, D. C., Zelazny, L. W., and Kornegay, E. T. 1986. Forms of solid-phase copper in copper-enriched swine manure. *J. Environ. Qual.* **15**, 69–72.

Payne, G. G., Martens, D. C., Kornegay, E. T., and Lindemann, M. D. 1988. Availability and form of copper in three soils following eight annual applications of copper-enriched swine manure. *J. Environ. Qual.* **17**, 740–746.

Ross, S. M. 1994. Retention, transformation and mobility of toxic metals in soils. In: *Toxic Metals in Soil-Plant System*, pp. 63–152. (Ross, S. M., Ed.) John Wiley & Sons, New York.

Ruby, M. V., Davis, A., Schoof, R., Eberle, S., and Sellstone, C. M. 1996. Estimation of biovailability using a physiologically based extraction test. *Environ. Sci. Technol.* **30**, 420–430.

Shuman, L. 1991. Chemical forms of micronutrients in soils. In: *Micronutrients in Agriculture*, 2nd Ed. pp. 113–144. (Mortvedt, J. J., Ed.) Soil Sci. Soc. Am. Inc., Madison, WI.

Sims, J. T. and Johnson, G. V. 1991. Micronutrient soil test. In: *Micronutrients in Agriculture,* 2nd p. 427–476. (Mortvedt, J. J., Ed.) Soil Sci. Soc. Am. Inc., Madison, WI.

Sutton, A. L., Nelson, D. W., Mayrose, V.B., Kelly, D. T., and Nye, J. C. 1984. Effect of copper levels in swine manure on corn and soil. *J. Environ. Qual.* **13**, 198–203.

US-EPA. 1999. Use of monitored natural attenuation at superfund, RCRA corrective action, and underground storage tank sites. Directive number 9200. 4-17P. Washington, D.C. EPA, Office of Solid Waste and Emergency Response.

van der Watt, H. V. H., Summer, M. E., and Cabrera, M. L. 1994. Bioavailability of copper, manganese and zinc in poultry manure. *J. Environ. Qual.* **23**, 43–49.

Vangronsveld, J., and Cunningham, S. D. 1998. Metal-contaminated soils: In-situ inactivation and phytoremediation. Springer-Verlag, Berlin.

Yang, J. E., Skogley, E. O., Georgitis, S. J., and Ferguson, A. H. 1991. Phytoavailability soil test: Development and verification of theory. *Soil Sci. Soc. Am. J.* **55**, 1358–1365.

Zhu, Y. M., Berry, D. F., and Martens, D. C. 1991. Copper availability in two soils amended with 11 annual applications of copper-enriched hog manure. *Commun. Soil Sci. Plant Anal.* **22**, 769–783.

Index

Absorption 146
  Cr(III) and Cr(VI) from soils 164
  plant species 165
  tag-along 146
AEC measurements 99
Acid mine drainage 271
  pyrite 272, 273, 275, 276
Adsorption
  non-specific 146
  soil factors influencing 151
  specific 146
Algae 35, 109
  bioaccumulation 122
  functional attribute 109
  indicators 125
  metal binding mechanisms 110,
    118, 124
    surface binding and precipitation
    119-121
  native 118
  resistance to heavy metals 118
  uptake of metals 121
ANZECC 220
Antagonistic 91
Arsenic 291
  blood 291, 301, 302, 304, 305, 306,
    307, 321, 323
  drinking water 294, 317
  groundwater 92, 292, 299, 300, 301,
    303, 313, 314
  hair 294, 301, 302, 304, 305, 306,
    307, 321
  metabolites 294, 301, 302, 304, 305,
    306, 307, 321
  mobilisation 315

  nails 294, 301, 302, 304, 305, 306,
    307, 321
  poisoning 291, 292, 299
    countries 291, 292
    social implications 308
    symptoms 291, 308
  skin-scales 294, 301, 302, 304, 305,
    306, 307, 321
  source 313, 314, 315, 320
    pyrite 320
  urine 294, 301, 302, 304, 305, 306,
    307, 321
ARMCANZ 220
ATP 97, 98
Availability 11

Bacteria 35, 43, 44
  inorganic nutrients 44
  metabolites 44
  microbial activity 44
  rhizosphere 44
Bangladesh 291-299, 301, 302, 307, 309
Bioaccessibility 257
  IEUBK model 259
  physiologically-based extraction
    (PBET) 257
  rat model 257, 258
  swine model 257, 258, 263
Bioaccumulation 9, 10
Bioassay 92
Bioavailable fraction 4, 13
Bioavailability 4, 6, 7, 21, 22, 39, 40,
  41, 42, 58, 61, 87, 88, 89, 90, 91, 117,
  121, 122, 123, 126, 131, 134, 257, 271,
  278, 291, 292, 323, 331-338

binding strength   42
  Cd, Zn   42
  microbial processes   44
  plant physiological   46
  rhizosphere   43
    Cu, Mn, Zn   43
    pH   41, 44
    soil solution   43
blood Pb   259
  pregnant women   260
definition   21, 22, 26, 331, 332
environmental factors   126-132
indicators   333
  chemical   333
low molecular weight metabolite
  effect   123
manipulation   336
methods   21
measurement   27
  biosensors   39
    lux   39
  chemical techniques   27, 28
    chelating agents   28, 31
    dilute salt   28, 32
    limitations   31
  microbial measures   35
    soil microflora   36
    soil microfauna   37
    soil mesofauna   37
    soil macrofauna   37
  molecular techniques   40
  plant measurements   34
  regulatory guidelines   47
  soil ingestion   47
soil Pb   258, 259, 263
solubility/bioavailability research
  consortium   258
Biochemical markers   97
  bacteria biomass   99
  fungal biomass   99, 101
Biolog   104
Biomagnification   9
Biomarker   4, 39
Biomethylation   118
Bioremediation   134-136
  biosorbent   135

Biosensors   39
Biosolid   24
Bioavailable   39
Bivalent cations   134

C^{14} 93
C ratios   93
Capacity   77
Carcinoma   302, 308
Chelating agents   28, 30, 31, 33
Chelators   130
Chromium   145, 196, 197, 198
  accumulation   166
  compartmentation in plants   149
  Cr(III)   198
  Cr(VI)   198
  plant yield   198, 199, 200
  redox transformations   146
  role in plants and animals   146
  speciation   146
  solubility and adsorption   146
  toxicity   152
Clover   96
Complexation   73
Contaminant   62, 213
  annual loads   215
    metals   219
      temporal patterns   224
    sulphate   215
  anthropogenic   62, 63
  baseline studies   213
  ecological impacts   214
  frequency distribution   220
  load in river water   213
  natural   62
  speciation   62, 63
Copper-uranium mine   208
Cr(III)   146, 149, 152, 157, 158, 159, 165
  localization in plant   162
Cr(VI)   146, 149, 150, 152, 163, 165
  localization in plant   162
  reduction   160
Cyanobacteria   94, 95, 109

Dermatosis   295, 307, 316
Dehydrogenase activity   92, 102
Detoxification   150

Direct microscopy   100
Dissolved metals   150
Dissolved organic matter   243
DNA   102
Dolly the sheep   102
DTPA   146, 334

Earthworms   38
Ecotoxicological   35
EDTA   122, 146, 158, 159, 166, 167, 189,
    190, 197
EDTA-extractable   95
Environmental
    definition
        ANZECC   16
        NEPC   15
        NSW   15
        queensland   15
    healthy   15
    toxicology   3, 8
Enzyme functions   91
Eukaryotic   109
European Union (EU)   85, 91, 92
Exposure   4, 258, 299
    nonresidential   260

Fertilizers   24, 25
Finnis river   209, 220, 223, 224, 225
Free Ion model   122
Functional groups   93
Fungi   35

Hard acid   150
Hazard   3
    evaluation   3
Health
    animal   24
    human   24
Heavy metals   23, 25, 85, 86, 87, 112-
    119, 207, 254
    acetic acid extractable   29
    anthropogenic   23
    background concentrations   68
    bioaccumulation   229
    bioavailability   254, 256
        leaching   256
        mamallian   256

reduction   257
Cd   24, 29, 32, 34, 37, 46, 334, 337
Cd, Cu, Pb   63, 68, 69, 71
Cd, Cu, Cr, Ni, Pb, Zn   179, 180,
    186, 188, 189, 194, 196, 198
CH$_3$COONH$_4$ 29
Complexed   28
complexes   239
Cu   24, 29, 32, 34, 37, 67, 334, 337
Cu, Cd, Zn, Ni, Pb, Hg   115-118, 12
Cu, Ni, Cd, Cr, Zn, Pb   85, 86, 88,
    89, 90, 92, 102, 103
Cu, Zn, Mn, U, Al, Co, Pb, Fe,
    Ca   209, 212, 223, 224
DTPA   29, 30, 33
Ecological effects   208, 222, 224
EDTA   29, 30, 33, 122, 146, 158, 159,
    166, 167
effects on microbial
    communities   87, 88
Essential trace metals   207, 208
exchangeable   28
fate   25
hazard   254
Input   207, 210, 212, 213
    mining   207
        toxic effects   207
        waste dumps   212
metal-ligand complex   123
mixtures   116
mobility   196
Ni   149, 150
Operationally defined fractions
    179
    carbonate bound   179, 183
    exchangeable   179, 182, 183, 184
    organic   179, 186
    oxide bound   179, 183
    residual   179, 183
pedogenic   23
plant uptake   30
plant uptake   181, 190, 191
        Zn   181, 191
    chemical availability   189, 190
    concentration factor   192
    leaching   187
    micronutrient   197

phytotoxicity   197
sediment concentrations   221
sequential chemical extraction   256
spatial and seasonal changes   223
speciation and bioavailability   239, 240, 241
sources   23
threshold concentrations   68
toxicity   91, 112, 113, 122, 125, 131
water-soluble   28
Human health   17
Hydroponic studies   152, 153, 167
Hyperaccumulation   150
Hyperaccumulator   149, 150, 151

IINERT   267
Immobilisation   91
Indian mustard   150
Indicator
heavy metal   97
metabolic activity   98
soil pollution   96
Inhibition   95
Inhibitors   91
Inplace inactivation   253, 261
cost benefit analyses   261, 264
field studies   263
limitations   261
plants as a tool   261
soil amendments   261, 262
Intensity   77
Iron   149

Keratosis   302, 307-310

Legume-rhizobium symbiosis   96
Lime   273, 274, 277, 278, 284
Lux genes   103
Lysosomes   39

Macrocosms   193
field experiments   193
Management   13
Manure   88
Melanosis   316
Mesocosms
large scale columns   184

advantages and limitations   184
lysimeters   185
soil monoliths   185, 188, 189
Metabolic energy   99
Metabolic reactions   207
Metabolites   44
Metal gradient   95
Metal-xenobiotic interactions   132, 133
Methylation   45
Microbial
activities   90, 91
biomass   86, 87, 88, 89, 90, 91, 97
communities   87
ecosystem   86
specific activity measurements   92
Microbial battery test system   103
Microbial biomass   336
Microcosms   176
Advantages and limitations   176
Micronutrients   136
Micro-organisms   85, 86
Microtox   39
Mineralisation   87, 91, 93, 103
Mine Wastewaters   229-232
biological assessment   229
effect on aquatic ecosystems   230
toxicity testing   230, 231, 232
Minewastes   271
metal-contaminanted   273
Mixtures   11, 17
Mobility   254
Modeling   14, 63
Models   70
aquatic bioassays   70
free ion activity model   70
semi-mechanistic   75
soils   71
Moisture holding capacity   92
Monovalent cations   134
Mycorrhizae   96
Mycorrhizal fungi   45

N^{15} 96
N_2-fixation   94
autotrophic   94
heterotrophic   94
symbiotic   96

Nitrogenase activity   94
Non-bioavailable   59
Non-hyperaccumulator   149
Novel approaches   102
Nutrition   291, 295
   role in arsenic toxicity   295

Organics   65
   K_{oc} 65
   K_{ow} 65
Organic complexes   151, 157

Partitioning   66
   metals   66, 67
      columns   66
      soil profiles   66
Partition coefficient   65
PBET   335
PCR   103
pH   151
Phospholipid fatty acid   100
Phosphatase activity   101
Physiologically based extraction
   test   335
Phytoavailability   76
Phytochelatins   122, 123, 124, 150
Phytoplankton   116, 121, 122
Phytoremediation   150
Phytotoxic   332
Phytotoxicity   96
Phytotrophic   109
Plant nutrition   283
   potential problems   283
      calcium deficiency   285
      magnesium deficiency   285
      nitrogen deficiency   285
      phosphorus deficiency   283, 284
      Potassium deficiency   285
Plant uptake   147
   Cr source and concentration   152,
   153, 154
      Translocation in plants   155
   Effect of organic acid   157
   Nutrient deficiency effect   156
Pollutant indicators in river waters
   217, 229-230
   algae   223

bivalves   223
fish   217
insects   217
local organisms   244
molluscs   217
Pollutant loading   86
Pore water   69
Protozoa   37, 38
Pyrite   320
   decomposition   323
   oxidizing bacteria   323

Ranger   210, 226
   Community structure   236
      monitoring   236
   Magella creek   226-229, 236, 237-
   239, 242, 243, 244
      water quality   229, 244
         seasonal variations   244
      toxicity testing   232, 233, 234, 235,
      236
Redox   45
Regulatory guidelines   47
   NHMRC   47
   USEPA   47
Rehabilitation   272
   metal-tolerant grass   273, 278, 279,
      283, 284
   seed germination   275, 277, 278,
      279, 282
   strategies for metalliferous mining
      272
Remedial studies   213-224
   post-remedial   219
      mean annual loads metals   219
      pH   219
   pre-remedial   213
      mean annual loads metals   213
      pH   213
Remediation   253, 334
   asphalt capping   266
   costs   266
   inplace inactivation   253, 254, 255,
      266
   metal   337
   reducing metal availability   337

soil washing 254
    costs 265, 266
solidification 265, 266
    costs 265, 266
stabilization 265, 266
    costs 265, 266
Respiration 91, 92
Rhizobium 94, 96, 103, 281, 282
Rhizosphere 33, 46
Risk Assessment 3, 4, 200
Root exudates 33, 46
Root-soil interface 32
Rum jungle 208, 220

Salinity 130
Sediments 66
Sequential fractionation 334
Sewage sludge 91, 98, 186, 188, 193
Sludge 24
Siderphores 123
Soft base 150
    metals 120
Soil 66
    agricultural 86
    ameliorants
    amendments 255
    columns 176
    enzymes 101
    ergosterol content 101
    fertility 87
    flu dust 180
    iron oxyhydroxide 255
    lime 178, 180, 183
    microbial populations 102
    organic carbon 90
    organic matter 87, 88,255
    plant yield 180, 194, 195
    phosphate fertilizer 255
    solution 28, 64, 190, 192
Solution culture 175

Speciation 58, 59, 131, 338
    bioavailability 60
    chemical equilibrium model 73
    solution phase 59
Spiked 77, 78
Substrate utilization 103
Symbiosis 94
Synchrotron X-ray fluorescence 163

Tailings 271, 273, 274, 275, 285, 296
    characterization and monitoring
        275, 276
TCA 98
TCLP 254, 257, 264
Threshold 24
Toxicity 45, 111, 117, 277
    Cu, Fe, Zn 278
    heavy metals 112, 113, 122
    tests 111, 112
Toxicology 3, 6
    evaluation 3
Toxin 12
Translocation 147

Uranium 210, 226, 229, 230
    biological response of freshwater
        237
        sediment 238
        surface water 237
    ecotoxicology 230
    geochemistry 229, 230
    speciation and bioavailability 239,
        240, 241
        modeling 242
USEPA 85

Water holding capacity 92
West Bengal 307

XANES 162

Zn 24, 29, 32, 34, 37, 149, 150